Nutritional Anemia in Preschool Children

Anil Gupta

Nutritional Anemia in Preschool Children

Springer

Anil Gupta
Department of Physiology and Biochemistry
Eklavya Dental College and Hospital
Kotputli, Rajasthan, India

ISBN 978-981-10-5177-7 ISBN 978-981-10-5178-4 (eBook)
DOI 10.1007/978-981-10-5178-4

Library of Congress Control Number: 2017945829

Printed on acid-free paper

This Springer imprint is published by Springer Nature
The registered company is Springer Nature Singapore Pte Ltd.
The registered company address is: 152 Beach Road, #21-01/04 Gateway East, Singapore 189721, Singapore

With humility and sincerity, I dedicate this book to the memory of my younger brother, the late Shri Sudhir Gupta, who struggled for life and left for heavenly abode for eternal peace.

Preface

Nutritional anemia is a major public health hazard with global prevalence affecting children in preschool age and women during pregnancy and lactation. Population in rural areas, slums and remote settings have higher predisposition to deficiency of iron, folic acid and macronutrients and infection owing to illiteracy, poverty, limited access to medical facility, poor hygiene, improper sanitation and lack of motivation. In a report by WHO in 1980, it was estimated that about 700 million population, globally, suffered from anemia. Moreover, around 50 % of the anemia prevalence was attributed to iron deficiency. However, the prevalence of nutritional anemia is variable all over the world due to wide variations in dietary pattern, literacy rate, socioeconomic status, medical facility and environmental conditions. Developing countries have higher vulnerability towards anemia. Millennium Development Goals were laid against worldwide poverty and hunger, HIV infection, malaria and child mortality and in favour to achieve gender equality, universal primary education and environmental sustainability.

Despite the large accomplishments pertaining to worldwide targets set forth, however, a big proportion of the population in distant regions and the proportion who suffered discrimination owing to gender disparity, geographical location and ethnicity have still to be benefitted. Achievements have been skewed throughout the world. According to the Millennium Development Goals report of 2015, about 800 million population today survive in utmost poverty and severe inadequacy of food. Around 160 million children below the age of 5 years suffer from stunted growth due to insufficient food intake. Diarrhoea and malaria in children under 5 years inflict a high disease burden responsible for high child mortality and morbidity. Still, another 2.4 billion population do not have access to proper sanitation, and a large segment of population still defecates in the open. The developing world harbours around 880 million population living in highly pitiable state in slum areas.

A huge population is struggling for survival on the thinnest and unpalatable sources of food per day in the deserted and isolated areas, where it must face extreme hot and cold weather. This section of human community is living in highly deplorable conditions without even knowing the Millennium Development Goals set forth for its betterment.

I feel immense pleasure and exhilaration on the completion of the book *Nutritional Anemia in Preschool Children*. Submissively, I would admit that extensive and intensive approaches have been practised for writing the book. It is designed to furnish valuable minute details relevant to the topic. It incorporates a concise text on the introduction and epidemiology of nutritional anemia in preschool children. Dietary iron deficiency is inflicting the poverty-stricken population in developing countries. An exhaustive information has been provided over the role of iron in the body, iron metabolism and the aetiology of iron deficiency anemia in children. Implications of geophagy, *Plasmodium falciparum* infection and HIV infection have been broadly discussed pertaining to their involvement in the pathogenesis of anemia.

Chapters focusing on laboratory methods for the screening of haemoglobin; role of vitamin B_{12} and folic acid, copper, zinc and selenium; and impact of vitamin A illustrate fine details of the subject. Remote predisposing factors such as diarrhoea, malnutrition and the effect of cow milk on the occurrence of anemia in children have been incorporated in the book.

The present book is the result of my knowledge gained through postdoctoral research on the nutritional status of children between 2 years and under 5 years of age. I hope the book will be helpful in the dissemination of facts, ideas, concepts and knowledge among academicians, research scholars and students. Efforts are taken to provide an error-free book; however, I welcome any criticism, comment or suggestion for the improvement of the forthcoming edition of the book.

Kotputli, Rajasthan, India
15 August 2016

Anil Gupta

Acknowledgements

Almighty Shri Shirdi Sai Baba bestowed upon me knowledge and perseverance for writing the book. My father, Shri Ved Parkash Gupta, always inspires me for high endeavour. I am highly indebted to my father for instilling the habit of learning in me, since my school days.

Throughout the period of writing, I had encountered a number of obstacles. However, my wife has been a source of motivation to me to come out successfully through thick and thin.

Further, I owe my gratitude to Dr. Abhinav Shreshtha, associate editor of biomedicine at Springer (India) Pvt. Ltd., for his support, cooperation and gentle attitude which helped me a lot in the completion of the book.

Contents

About the Author

Dr. Anil Gupta is employed as associate professor in the Department of Biochemistry at Eklavya Dental College and Hospital, Kotputli, under the Rajasthan University of Health Sciences, Jaipur, India. He graduated in biosciences from Punjab University in 1989. He obtained his bachelor's degree in dental surgery from the University of Poona in 1984. Later on, he did his master's degree in biochemistry in 2009 and completed his Ph.D. in biochemistry in SJJT University in 2012. Persistently, for the past 5 years, he has been pursuing postdoctoral research, independently, related to the nutritional status of children. His 27 publications have been accepted and published. He presented papers in Thapar University, the Birla Institute of Technology and Science, Punjabi University, M.D. University and Arya P.G. College under Kurukshetra University. He has been accorded with merit certificates, scholarships and medals during study. He has teaching experience of more than 11 years as a mentor to dental students and a teacher to premedical examination aspirants and students in nursing and polytechnic colleges. He is a Ph.D. guide to research scholars working in distinctive fields like heavy metal contamination of water, drinking water quality analysis and diabetes mellitus and dyslipidaemia. He serves as reviewer and member of the editorial board in national and international journals.

Chapter 1
Introduction

Abstract Anemia is characterized by a decline in the concentration of haemoglobin below a reference. Nutritional anemia is a type of anemia which becomes apparent owing to inadequate intake of minerals in diet. Deficiency of iron, vitamin B_{12} and folic acid is rampant in the daily diet leading to manifestation of nutritional anemia, especially in children. Iron deficiency and iron deficiency anemia are the major concerns worldwide, pertaining to the health of children. It may either stem from poor intestinal absorption of iron necessary to supplement increased iron demand in the body during periods of active growth, and/or chronic loss of iron from the body may eventually result into iron deficiency. Iron deficiency anemia is integrated with high burden of diseases which are responsible for high morbidity and mortality in growing children especially in developing countries.

Non-nutritional factors like helminthic infestation, malaria, tuberculosis and diarrhoea affecting the children are equally responsible for nutritional anemia. Other than nutritional factors and infection, malnutrition influences the body metabolism and is involved in the high mortality among children under 5 years of age in developing countries. Malnutrition is responsible for anemia and recurrent infections in growing children. The number of population inflicted with iron deficiency is so large that it can be declared as global health hazard.

1.1 Definition

Nutritional anemia is a global health hazard affecting children in infancy, early childhood and adolescence period along with its dreaded impact on the health of women during pregnancy and lactation. Generally, nutritional anemia is the consequence of inadequate micronutrients in diet. Deficiency of iron, vitamin B_{12} and folic acid in food is mainly responsible for nutritional anemia in children. Nevertheless, the role of trace minerals like selenium, zinc and copper in the cause of nutritional anemia has been studied extensively in the past by researchers. Nowadays, it is well established that trace minerals influence the erythropoiesis and have been instituted as additional factors in the aetiology of nutritional anemia in children.

Anemia is a clinical condition that signifies a decline in the concentration of haemoglobin below a predetermined value. Its value swerves with age and gender;

A. Gupta, *Nutritional Anemia in Preschool Children*,
DOI 10.1007/978-981-10-5178-4_1

hence, a cut-off value is established pertaining to age group and gender of the individual (WHO 2001).

According to a WHO (1968), Rodak (2007) and Beutler and Waalen (2006), cut-off value for the haemoglobin was determined which is used to compare the haemoglobin level of a patient and coin a diagnosis for anemia and is profoundly important for screening of anemia in children. Hb is <11 g/100 ml in 6 months to 6 years age group, <12 g/100 ml in 6 years to 14 years age group, <13 g/100 ml in adult male, <12 g/100 ml in nonpregnant adult female and <11 g/100 ml in pregnant women.

Later on DeMeyer et al. (1989) forwarded another classification of anemia which was based on a definite value of haemoglobin. Its level in percentage determines the severity of anemia.

Accordingly, Hb concentration above 80% of the definite value is considered as mild anemia. Whereas, ChB concentrations in a range between 80 and 60% and less than 60% of the definite values are established as moderate and severe anemia.

Conclusively, mild anemia is associated with haemoglobin level above 10 g/100 ml but essentially below a definite value. Moreover, moderate and severe anemia are related to haemoglobin concentrations in the range 7–10 g/100 ml and below 7 g/100 ml, respectively.

Afterward, the US Agency for International Development (United States Agency for International Development 1997) described another category of anemia as very severe. It was characterized by haemoglobin concentration below 4 g/100 ml. According to Murray and Lopez (1966), Hb concentration below 12 g/100 ml and 11 g/100 ml in developed countries and developing countries, respectively, is considered as anemia in children below 5 year of age.

Iron deficiency and iron deficiency anemia are two main concerns worldwide regarding the health of children. Iron deficiency is the state of inadequate iron in the body which impedes the normal physiological functioning of body organs. It may either stem from poor intestinal absorption of iron necessary to supplement increased iron demand in the body during periods of active growth, and/or chronic loss of iron from the body may eventually result into iron deficiency. Contrarily, iron depletion refers to the state in which body iron store is either completely exhausted or near exhaustion; nevertheless, the physiological functioning of body organs is maintained. Functional iron deficiency is a state, wherein, despite the adequate iron store, the iron transportation to body tissues is impaired resulting into deranged physiological functioning of body organs (Bootha and Aukettb 1997).

Iron deficiency, if persists for a longer period, finally is responsible for a fall in Hb concentration in the body and hence is associated with iron deficiency anemia. However, ID can also exist in the absence of anemia (WHO and UNICEF 2001).

1.2 Overview

Iron deficiency anemia is integrated with high disease burden which is responsible for high rate of morbidity and mortality in growing children, especially, in developing countries.

Iron deficiency is a global public health hazard (Christofides et al. 2005) distressing (Domellof and Hemell 2002) around 51% and 12% of children below age of 4 years in developing and developed countries, ultimately, crippling the physical, mental and intellectual health of children in the active growth period. According to Dallman (1990), iron deficiency is a clinical condition associated with decreased or absence of mobilization of iron from body stores coupled with its diminished bioavailability to tissues and the bone marrow for erythropoiesis. Author hypothesized three stages of iron deficiency in the body, wherein each stage has its own characteristics. In the first stage, iron store of the body is exhausted. However, iron deficiency is not associated with any significant abnormality in structure and or functioning of iron-dependent compounds. It is followed by a stage of iron deficiency without anemia. Biosynthesis of haemoglobin is adversely affected due to iron deficiency; nevertheless, the normal concentration of Hb, in this stage, remains unaffected, and sign of anemia is unseen. Zinc protoporphyrin concentration in RBCs increases (Paszkowska et al. 1997). The third stage is marked by further deterioration of Hb synthesis and is manifested by a fall in Hb concentration below a set mark characterized as anemia. Morphology of RBCs is affected; thereby, mean cell volume decreases.

Non-nutritional factors like helminthic infestation, malaria, tuberculosis and diarrhoea among children are equally responsible for nutritional anemia.

Infection triggers immune response (Schaible and Kanufmann 2007) in the body whose sustenance needs a lot of energy consumption. Any inconsistency in the supply of nutrients to growing children, additionally, puts a high burden on the energy balance of the body so infection and energy deficiency derange the erythropoiesis together with functioning of vital organs of the body which manifests into anemia, poor psychomotor skills, wasting, stunting and compromised immunity.

Malarial fever is the main factor for the high morbidity and mortality according to WHO report on malaria. It inflicts around 216 million population (WHO 2011) of world and has a high mortality rate (655,000) among African children.

Anemia is a common consequence of malaria in children. It is noteworthy and pitiable that severe anemia, as a secondary cause, is responsible for high death rate of malaria-inflicted children. Malaria disturbs the homeostasis of the body of children through its effects on the functioning of body systems including the bone marrow and brain. Later on, authors (Murphy and Bregman 2001) extensively studied the impact of malarial burden on children. According to authors, cerebral malaria exerts devastating health effects over children under 5 years of age, and the effects were studied by authors in African population. Accordingly, around 110,000 children with malaria have high fatality rate per year and die due to the complications secondary to cerebral malaria. It was reported by authors that severe anemia as a consequence of malaria affects large population of children under 5 years of age.

Apart from nutritional factors and infection, malnutrition influences the body metabolism and is involved in the high mortality among children under 5 years of age in developing countries. Malnutrition is responsible for (WHO 2005) anemia and recurrent infections in growing children.

Anemia is the associated comorbidity with malnutrition and collectively impairs the body growth and development, especially, in infancy and early childhood period. Anemia and malnutrition exert serious effects on the children during the formative years as well as in the adolescent period which may be labelled as short-term and long-term harmful effects. Any impairment in the growth of children in earlier life continues to have its ill effects in later period of life even after the termination of the cause (Pelletier 1994).

Iron deficiency is the most prevalent form of nutritional disorder affecting population of children in growing phase and women in pregnancy and lactation in developing as well as developed countries.

Number of population inflicted with iron deficiency is so large that it can be declared as global health hazard. According to WHO report, iron deficiency exerts a high disease burden in the form of malnutrition, impaired growth and development and loss of intellectual, psychological, emotional, social and economic life of children and individuals (WHO 2016).

References

Beutler E, Waalen J (2006) The definition of anemia: what is the lower limit of normal of the blood hemoglobin concentration? Blood 107(5):1747–1750

Bootha W, Aukettb MA (1997) Iron deficiency anemia in infancy and early childhood. Arch Dis Child 76:549–554

Christofides A, Schauer C, Zlotkin SH (2005) Iron deficiency anemia among children: addressing a global public health problem within a Canadian context. Pediatr Child Health 10(10):597–601

Dallman PR (1990) Iron. In: Brown ML (ed) Present knowledge in nutrition. ILSI Press, Washington, DC

DeMeyer EM, Dallman P, Gurney JM, Hallberg L, Sood SK, Srikantia SG (1989) Prevention and controlling iron deficiency anemia through primary health care: a guide for health administrators and programme managers. WHO, Geneva

Domellof M, Hemell O (2002) Iron deficiency anemia during the first two years of life. Scand J Nutr 46(1):20–30

Murphy SC, Bregman JG (2001) Gaps in the childhood malaria burden in Africa: cerebral malaria, neurological sequelae, anemia, respiratory distress, hypoglycemia, and complications of pregnancy. Am J Trop Med Hyg 64(1–2 suppl):57–67

Murray CJL, Lopez AD (1966) The global burden of disease: a comprehensive assessment of mortality and disability from diseases, injuries, and risk factors in 1990 and projected to 2020. Harvard University Press, Cambridge, MA, p 43

Paszkowska M, Jezynska A, Krol M, Rokicka-Piotrowicz M, Kuratowska Z (1997) Zinc protoporphyrin (ZPP) in diagnosis of anemia. Pol Arch Med Wewn 98(12):520–526

Pelletier DL (1994) The relationship between child anthropometry and mortality in developing countries: implications for policy, programs and future research. J Nutr 124(10 Suppl):2047S–2081S

Rodak BF (2007) Hematology: clinical principles and applications. Saunders, Philadelphia, p 220

Schaible UE, Kanufmann SHE (2007) Malnutrition and infection: complex mechanisms and global impacts. PLoS Med 4(5):e115

United States Agency for International Development (1997) Anemia detection methods in low-resource settings: a manual for health workers. PATH, Seattle

WHO (Retrieved 2016) Micronutrient deficiencies: iron deficiency anemia. Available at: http://www.who.int/nutrition/topics/ida/en/

WHO, UNICEF (2001) Iron deficiency anemia: assessment, prevention, and control. A guide for programme managers. WHO, Geneva

World Health Organization (1968) Nutritional anemias: report of a WHO scientific group. WHO, Geneva, technical report series, no. 405

World Health Organisation (2001) Iron deficiency anemia: assessment, prevention and control: a guide for programme managers. WHO, Geneva

World Health Organization (2005) Nutrition: challenges. Available at: http://www.who.int/nutrition/challenges

World Health Organization (2011) World Malaria report. WHO, Geneva. Available at: http://www.who.int/malaria/world_malaria_report_2011/en/

Chapter 2
Epidemiology of Nutritional Anemia

Abstract Iron deficiency is a pandemic clinical condition affecting large proportion of population around the world. In India, iron deficiency is the main cause of nutritional anemia. The prevalence of anemia in rural children was higher in comparison to urban children.

Iron deficiency anemia is highly prevalent in children in age group from 6 to 24 months. However, it affects preschool children, pregnant women and old-aged persons. Nutritional anemia is a major public health hazard in the developing countries.

2.1 Global Prevalence

Iron deficiency is a worldwide clinical condition affecting (Allen et al. 2006) more than two billion population around the world.

According to Stoltzfus and Dreyfuss (1998), iron deficiency anemia is highly prevalent in children in age group from 6 to 24 months. However, it affects preschool children, pregnant women and old-aged persons.

According to WHO (2008) global database, the highest prevalence of anemia was observed in preschool children (47.4%), and the least prevalence of anemia (12.7%) was seen in adult males during a survey conducted during 1993–2005. As shown in a flow chart, schoolchildren suffered from 25.4% prevalence of anemia (Fig. 2.1).

According to WHO (2011) report, prevalence of anemia in children was 42.6%. Further, higher than 50% of children in Southeast Asian and African regions suffered from anemia. Moreover, severe anemia affected around 3.6% of children in African regions. The WHO report affirmed that the predisposing factors for anemia were extremely prevalent in African regions. The commonest were malaria, thalassaemia and sickle cell anemia (Modella and Darlison 2008).

A. Gupta, *Nutritional Anemia in Preschool Children*,
DOI 10.1007/978-981-10-5178-4_2

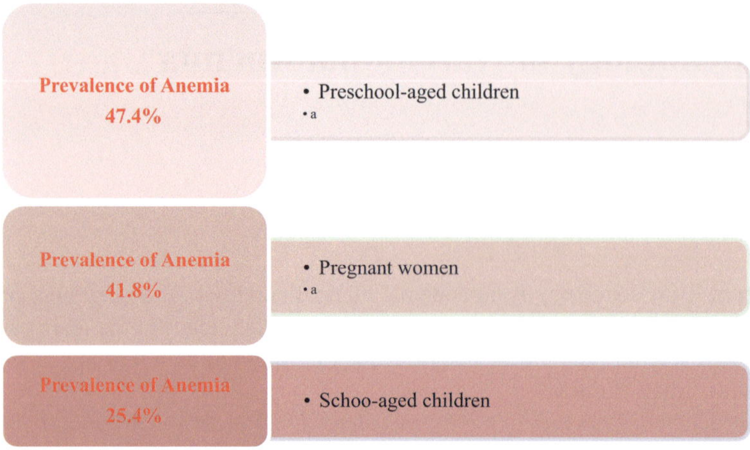

Fig. 2.1 Worldwide prevalence of anemia in different groups (Source: WHO 2008)

2.2 Prevalence of Nutritional Anemia in India

The International Institute for Population Sciences and ORC Macro (2000) conducted a National Family Health Survey during 1998–1999 in India. The survey covered the married women in age group between 15 and 49 years and the children below 3 years of age. It was a first effort (IIPS and ORC Macro 2000) to assess the anemia in married women and children through the estimation of haemoglobin. According to IIPS and ORC Macro (2000), around 74.2% of the children suffered from anemia, while 52% of women in pregnancy were affected with anemia.

IIPS and ORC Macro (2000) reported a minor difference in the prevalence of anemia in rural and urban children. It was 75% in rural children, whereas urban children had 71% of iron deficiency anemia.

Parental education is necessary for the prevention of anemia in children as well as for seeking medical guidance and treatment for the management of iron deficiency anemia. It was found in the survey (IIPS and ORC Macro 2000) that the prevalence of anemia was high (78%) in children of illiterate mothers in comparison to the prevalence of anemia (62%) in children belonging to mothers who were educated up to high school level. Another research work by authors (Rahman and Chowdhury 2006; Choi et al. 2011) also confirms the determining role of mother education on the health of children. Income and standard of living modulate the dietary habits, hygiene and sanitation. These factors, collectively, establish the health of children. In the NFHS-2 conducted by IIPS and ORC Macro (2000), it was scrutinized that children of low-income families had high prevalence of anemia (79%) in contrast to the children of high-income families who had 67% of anemia.

Nutritional anemia is a major public health hazard in the developing countries. In India, iron deficiency is the main cause of nutritional anemia. A report (NFHS-3, 2005-06) provides valuable data regarding anemia in children under the age of

5 years. Accordingly, around 78.9% of the children under 5 years of age during 2005–2006 were observed to be anaemic. The report further described that the prevalence of anemia in rural children was higher in comparison to urban children during 2005–2006. Anemia prevalence was 72.2% in urban children, while it was 80.9% in rural children. Comparison of the reports of NFHS-2 (1998–1999) and NFHS-3 (2005–2006) confirmed that prevalence of nutritional anemia in under-five aged children has increased during 2005–2006 in India.

References

Allen L, de Benoist B, Dary O, Hurrell R (2006) Guidelines on food fortification with micronutrients. World Health Organization and Food and Agricultural Organization of the United Nations, Geneva

Choi HJ, Lee HJ, Jang HB, Park JY, Kang JH, Park KH, Song J (2011) Effects of maternal education on diet, anemia, and iron deficiency in Korean school-aged children. BMC Public Health 11:870

International Institute for Population Sciences (2005) National Family Health Survey (NFHS-3) 2005-06. IIPS, India

International Institute for Population Sciences, ORC Macro (2000) National Family Health Survey (NFHS-2)1998–99. IIPS, India

Modella B, Darlison M (2008) Global epidemiology of haemoglobin disorders and derived service indicators. Bull WHO 86(6):417–496

Rahman A, Chowdhury S (2006) Determinants of chronic malnutrition among preschool children in Bangladesh. J Biosoc Sci 39:161–173

Stoltzfus RJ, Dreyfuss ML (1998) Guidelines for the use of iron supplements to prevent and treat iron deficiency anemia. INACG/WHO/UNICEF, Washington, DC

World Health Organization (2008) In: de Benoist B, McLean E, Egli I, Cogswell M (eds) Worldwide prevalence of anemia 1993–2005. World Health Organization, Geneva

World Health Organization (2011) The global prevalence of anemia in 2011. WHO Press, Geneva

Chapter 3
Physiological Contemplation of Iron in the Body

Abstract Iron has high biological significance owing to its role in wider physio-biochemical activities in the human body. Iron is an essential mineral for human and other organisms except for a few microbes. Iron is the structural and functional part of haemoglobin, myoglobin, cytochromes, catalase, peroxidase, FAD, NAD, ferritin, transferrin and porphyrin macromolecules. Term and healthy infants have adequate body iron store that may be necessary for growth as early as 4–6 months. Moreover, exclusive breast-feeding in this period can help to supplement 0.3 mg/L of the iron in the body of infants. Iron deficiency in tender age group between 1 and 3 years is highly injurious to overall health and development of children. It may permanently impair psychomotor development in children. Exclusively and prolonged breast-fed infants are more liable to iron deficiency owing to least iron content of breast milk and rapid growth of the body and increased iron demand.

3.1 Overview

Iron element has an atomic number of 26 and atomic weight of 55.847. It is a transition metal. It is one of the prime elements of the earth's crust and is the most copious in its distribution in nature (Bothwell et al. 1979). Iron has high biological significance due to its involvement in diverse physio-biochemical activities in the human body. It is the primitive element that has a role in the origin of molecules that were essential for the appearance of life and further evolution of higher life through its unique redox property (Bothwell 1979).

Iron is an essential mineral for the human body and many other organisms with the exception of a few microbes. This mineral has been extensively investigated in relation to its impact on health and diseases (Donovan et al. 2006; Beard 2001; Brock et al. 1994).

Iron is the structural and functional part of haemoglobin, myoglobin, cytochromes, catalase, peroxidase, FAD, NAD, ferritin, transferrin and porphyrins. Iron is the gold of life (Mascoti et al. 1995; FAO/WHO 1988).

A. Gupta, *Nutritional Anemia in Preschool Children*,
DOI 10.1007/978-981-10-5178-4_3

3.2 Average Iron Content in the Human Body

Total body iron has been estimated to be approximately 3.8 g in males and 2.3 g in females (Bothwell and Charlton 1981; Kohogo et al. 2008).

Difference in body iron store in males and females is associated with much lesser iron reserve in females and a corresponding low Hb and RBC count in female in comparison to male. Iron of the body exists in functional form and in storage form. Approximately 75% of the total iron of the human body (about 30–40 mg/kg body weight of the iron) is present in functional form, and the remaining 25% (0–20 mg/kg) body weight of iron exists in storage form as ferritin and haemosiderin, primarily, in the liver, muscles, intestinal mucosa, reticuloendothelial cells and bone marrow (Bothwell 1995).

3.2.1 *Infancy*

At birth, a baby emerges into an oxygen-rich environment from a comparatively hypoxic environment that prevailed in the uterus. Erythropoietin level undergoes decline that is the regulator of the RBC synthesis (Burman 1971a, b, 1972; Beal et al. 1962).

Haemoglobin of the infant goes down from a concentration of about 17 g/100 ml to a concentration of 12 g/100 ml. Decline of Hb is noticeable during initial 1.5 months of life, and this physiological condition stimulates the release of erythropoietin. Hence, in the next 2 months, between the second and fourth month, the fall in haemoglobin level starts increasing (Dallman 1993; Kotecha 2011; Domellof and Hernell 2002).

The infant continues to grow and the body volume increases. Iron circulates from body store to blood circulation. Therefore, full-term, normal weight and healthy infant has adequate body iron store that can meet the iron need of baby up to the first 4–6 months. Moreover, exclusive breast-feeding during this period can help to supplement 0.3 mg/L of the iron in the body of the baby (Rao and Georgieffa 2006). Human breast milk, though poor in iron (0.2–0.4 mg/L, appx 0.35 mg/L), is efficiently absorbed. It helps to supplement body iron store of the breast-fed infant (Prentice 1996; Nagin 2009; Saarinen et al. 1977).

Other authors (Siimes and Salmenperal Perheenpupa 1984; Meinzen-Derr et al. 2006; CDC 1998; Lind et al. 2004; Marques et al. 2014) described the low content of iron in human breast milk. According to them, it was in the range 0.0 mg/100 ml–0.09 mg/l00 ml; however, authors posited that the probable reason for this characteristic of human breast milk is attributed to its components like lactoferrin, low calcium and phosphate.

Iron absorption from breast milk is variable and is dependent on the age of infant. According to authors (Domellof et al. 2002), regulatory method for haemoglobin synthesis is not properly developed in infants.

Authors (Saarinem et al. 1977) concluded through a radioisotopic study in infants that iron absorption from breast milk is around 49% by breast-fed infants. This value is much higher in comparison to iron absorption from cow milk and asserted by authors that iron abortion from cow milk is only 10%. Therefore, authors suggested that full-term infants, who are breast-fed, may not require iron supplementation in routine life.

Human breast milk is a panacea for infant and unanimously regarded as safe and essential exclusive breast-feed up to the age of 6 months (AAP 1997; WHA 2001), although it is low in iron concentration, which is around 0.2–0.4 mg/L (AAP 1997; WHA 2001). However, high iron absorption in infants from breast milk is possible due to high bioavailability of iron in breast milk. Probably, it is a compensatory mechanism of the body for low iron content in human breast milk (Saarinem et al. 1977).

According to the author (Oski 1993), it has been assumed that an infant in age group between 4 and 6 months can absorb around 0.12 mg of iron from about three-fourths of a litre of breast milk, provided the infant is exclusively breast-fed. In the infant, significant amount of iron during iron turnover is deposited in growing tissue of the body (Oski 1993). According to APA (1997), in the age of 6 months, around 0.5 mg per day of iron is required for body tissue growth, and another 0.1 mg per day of iron is needed for growth of muscles. According to authors, a 6-month-aged infant has iron loss around 0.15 mg per day, and it is compensated by daily intake of 0.12 mg of iron through iron absorption from breast milk (Domellof and Hernell 2002; WHA 2001). According to authors, iron need of infant rises to around 0.75 mg per day after the age of 6 months. Therefore, additional feeding is necessary to account for a rise in the demand of iron in infancy. Hence, it is advisable to start supplementary feeding in addition to breast-feeding after age of 6 months to fulfil extra demand for iron (Michaelsen et al. 2000).

Authors hypothesized that a sustained diet high in bread and low in meat and fish is responsible for decreased serum ferritin levels. This pattern of diet in infants predisposes to higher incidence of iron store depletion in infancy (Michaelsen et al. 2000).

The time at which IDA will develop in exclusively breast-fed infants depends on the size of iron stores at birth. According to authors (Domellof et al. 2001), IDA in most developing countries develops soon after the age of 4–6 months in exclusively breast-fed infants.

Further, between 6 and 12 months, there is rapid growth of the body of the infant, and the iron requirement rises. It is around 0.9–1.3 mg per kg per day which is estimated to be in the range of 7.8–11 mg per day. Owing to rapid body growth and common prevalence of dietary deficiency of iron, the requirement of iron rises in the body of children below age of 2 years. Children in this age group are highly predisposed to iron deficiency and associated anemia (Yip 1989; Hallberg 1992; Soh et al. 2004; Michaelsen et al. 1995).

Iron deficiency in tender age group between 1 and 3 years is highly injurious to overall health and development of children. It may permanently impair psychomotor development in children (Stolfus 2001; Grantham-Mcgregor and Ani 2001).

According to the Committee on Nutrition of the American Academy of Pediatrics, term infants should be given 1 mg/kg/day of the iron to the maximum of 15 mg per day, and iron supplement should be started at the age of 4 months and be continued up to 3 years of life. However for the preterm low-weight infant, according to authors (Domellof and Hernell 2002), iron supplement should be 2 mg/kg/day to 15 mg/day and should be started from the second month onwards and be continued up to 3 years of life.

Exclusively and prolonged breast-fed infants are more liable to iron deficiency owing to least iron content of breast milk and rapid growth of the body and increased iron demand. Infants should be given proper solid foods in addition to breast-feeding after the age of 6 months. Nutrients like zinc, iron and vitamin D are in minimum concentration in breast milk. Infants below 6 months of age can adequately be nurtured on breast milk. However, beyond this age, the infant's body iron store is depleted. The body grows rapidly. Iron demand increases. To keep pace with the iron demand and supply, food supplements should be started besides breast-feeding after the age of 6 months (Callahan 2011; Dewey and Chaparro 2007).

After the age of 2 years till the start of puberty, the body grows steadily, and iron requirement of the body also decreases slightly during this period. Depending on the age and weight, daily iron intake exhibits variability.

3.2.2 *Childhood*

Physical growth of the body occurs between the 2nd year and 12th year in boys and 10th year in girls. Growth was estimated by Bothwell et al. (1979) and was found in the range 2.50–2.75 kg/year. Iron need of the body rises in the range between 0.5 and 0.8 mg per day. In schoolchildren, according to authors (Bothwell et al. 1979), haemoglobin level increases by about 1 g/100 ml.

3.2.3 *Adolescence*

Child age after 10–11th year in females and 11–12th year in males is characterized by rapid growth of the body, called as growth spurts. It results in increased body weight (Bothwell et al. 1979), which continues throughout the puberty, and it is around 4.5 kg per year in males and 4 kg per year in females. Additionally, haemoglobin level also rises to a value of about 2 g/100 ml in males and 1 g/100 ml in females. These are age-related changes in weight and haemoglobin concentration (Kail and Cavanaugh 2010) (Table 3.1).

Coincidence of growth spurt and an ascending Hb level, there is a raised daily iron requirement for male and females, which is 1.2 mg and 1.68 mg per day, respectively. Girls have comparatively higher iron need per day owing to menarche (FAO/WHO 1988).

Table 3.1 Daily iron requirement in different age groups (mg/day)

Age	Male	Female
Birth to 6 months	0.27	0.27
7–12 months	11	11
1–3 years	7	7
4–8 years	10	10
9–13 years	8	8

Source: Institute of Medicine, Food/Nutrition Board (2001) and WHO/FAO (2004)

3.2.4 Females in Reproductive Age Group

Menstrual cycle starts in females at the adolescence period. It is the major reason for loss of iron. Females lose iron around 0.5–1 mg per day up to the menopause. Further, pregnancy demands a high iron requirement in females. Every pregnancy costs around 1,000 mg of iron in the life of women. Hence daily iron demand in females is about 1.5 mg (Table 3.1).

3.2.5 Adult Males and Postmenopausal Females

Daily iron demand in adult males and females in the menopausal stage continues to remain steady provided the daily intake of food is balanced, and they should not suffer from any disease. Chronic diseases like uncontrolled diabetes mellitus, chronic infections and peptic ulcer result in loss of iron, and demand for it rises (Table 3.1).

3.3 Daily Iron Requirement

Iron requirement per day is dependent on age, gender, weight, health and physiological status of individuals. About 3–4 mg of iron is passed from maternal circulation into the foetus during the last trimester of pregnancy. This physiological process helps in the attainment of a body iron store of about 75 mg/kg of body weight of the infant (250–300 mg) (Fig. 3.1).

Therefore, full-term (41 weeks), normal weight (3.4 kg average with a range of 2.7–4.7 kg) infants have adequate body iron store and do not suffer from iron deficiency at birth. On the contrary, preterm infants (37 weeks) are deprived of time to build up body iron store and become susceptible to iron depletion and iron-deficiency anemia. Full-term but low-birth-weight infants have reduced body iron store in comparison to normal weight infants. Further, as the body weight of the infant rises

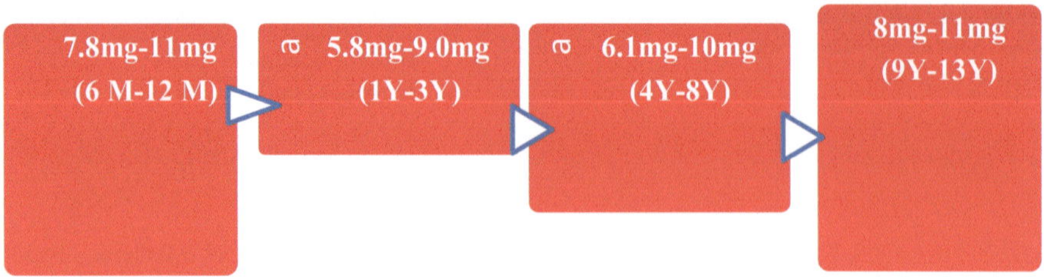

Fig. 3.1 Daily iron requirement in different age groups (mg/day) (Source: Institute of Medicine, Food/Nutrition Board 2001; WHO/FAO 2004)

and the body volume grows, there is a surge in the requirement of daily iron for the synthesis of haemoglobin. It has been found that about 40 mg of iron is needed for every 1 kg of weight gain during early period of childhood (Dallman 1990).

According to the Institute of Medicine/Food and Nutrition Board (2001), daily iron intake is expected to be 7.8–11 mg at 6–12 months, 5.8–9.0 mg at 1–3 years, 6.1–10 mg at 4–8 years and 8–11 mg/day at 9–13 years, corresponding to 0.9–1.3, 0.5–0.8, 0.3–0.5 and 0.2–0.3 mg/kg/day. Further, daily iron intake is maximum (7.8–11 mg) at the age between 6 and 12 months, whereas its intake becomes minimum (5.8–9 mg) between 1 and 3 years of age. Moreover, iron intake assumes almost equal value in the age groups 6–12 months and 9–13 years (Fig. 3.1).

3.4 Essential and Functional Iron Compounds

Essential or functional iron compounds are helpful in carrying out normal metabolic functions of the body. These compounds are protein with nonprotein moiety as prosthetic group.

3.4.1 Haemoglobin and Myoglobin

Globin proteins in these compounds are conjugated with porphyrin moieties. Haemoglobin contains four polypeptide chains linked to four haem moieties and can transport four oxygen molecules from the surface of the respiratory membrane to tissues. About 70% of the total body iron is focused within haemoglobin (Marengo-Rowe 2006).

Haem contains iron in ferrous state that is held coordinatively within a heterocyclic nucleus called as porphyrin. It is made of four pyrrole rings linked together through methene bridges. Porphyrin molecules harbour Fe^{2+} ion in its centre (Perutz et al. 1960) (Table 3.3).

Four nitrogen atoms of pyrrole rings bind with Fe^{2+} ion by forming coordinative bonds.

Ferrous ion is covalently linked to the nitrogen atom of the imidazole ring of histidine moiety within the polypeptide chain. Sixth oxidation state of iron is utilized to link outwardly with water molecule in the deoxygenated Hb, whereas in oxygenated Hb, the oxygen molecule replaces water molecule and bonds itself reversibly and coordinatively with Fe^{6+} (Allison 1954).

Myoglobin is another chromoprotein containing iron in ferrous state. It is found in the muscles of all mammals. Mb concentration of skeletal muscles, smooth muscles and cardiac muscles is about 2.5 g/100 ml, 0.3 g/100 ml and 1.4 g/100 ml, respectively (Nelson and Cox 2000) (Table 3.2).

It is made of single polypeptide chain of 152 amino acids (Ordway and Garry 2004).

According to Lodish et al. (2000), it possesses a haem moiety around which polypeptide chain (globin) undergoes folding and refolding forming eight alpha helices and a hydrophobic core. Human globin is composed of 154 amino acids (Lippard and Berg 1994; Kendrew et al. 1960; Kendrew 1961).

Mb is the oxygen carrier protein of muscles and has higher affinity for oxygen than Hb. It is seen in terms of difference in saturation of Hb (only 50%) and Mb (90%) at pO2 of 30 mm of Hg within tissues. Mb lacks coordinative bonding with oxygen molecule since it is a monomeric entity (Millikan 1939; Flogel et al. 2001).

Myoglobin is a reservoir and carrier of oxygen within skeletal and cardiac muscle tissues in humans and mammals. It is necessary to maintain adequate cytoplasmic oxygen supply. It carries oxygen from sarcolemma to mitochondria (Sidell and O'Brian 2006; Millikan 1939; Wittenberg and Wittenberg 1985, 1987, 1989; Williams and Neufer 1996; Yan et al. 2001).

Myoglobin maintains inflow of oxygen into muscle cells and the consumption of oxygen by the mitochondrial cytochrome oxidase enzyme (Wittenberg and Wittenberg 2003; Mammen et al. 2003; Antunes et al. 2004).

3.4.2 Catalases

Catalase is an important antioxidant enzyme which is present in all aerobic organism. It converts hydrogen peroxide which has toxic potential into water and oxygen. Catalase is protective in nature against the oxidative stress induced by reactive oxygen species (Chelikani et al. 2004).

In humans, it is present in the blood, kidneys, marrow, mucosa and liver. It serves to catalyse reduction of hydrogen peroxide, produced in consequence of metabolic processes in tissues, into oxygen and water (Johansson and Borg 1988; Lardinosis and Rouxhet 1996). Catalase is a tetramer consisting of four polypeptide chains, where each one is made up of 500 amino acids. Each chain is linked to haem moiety (Boon et al. 2007) (Table 3.3).

Peroxidative action of catalase was first described by Keilin and Hartree (Andreoli et al. 1990).

$$2H_2O_2 \xrightarrow{\text{catalase}} 2H_2O + O_2$$

Peroxidatic activity of catalase is discernible at comparatively low concentration of hydrogen peroxide in tissues (Anderson et al. 1997).

3.4.3 *Peroxidases*

Peroxidases constitute a large family of conjugated enzymes that contain haem as prosthetic group loosely attached to apoprotein or cysteine residue. These enzymes can catalyse hydrogen peroxide into water and oxygen, and a typical reaction that can be catalysed by peroxidases is represented as follows:

$$\text{ROOR'} + \text{electron donor } (2\ e-) + 2\text{H}+ \rightarrow \text{ROH} + \text{R'OH}$$

$$\underset{\text{glutathione reduced}}{2GSH} + H_2O_2 \xrightarrow{\text{Glutathione peroxidase}} \underset{\text{glutathione disulphide}}{GS\text{–}SG} + 2H_2O$$

Glutathione peroxidase is an example of peroxidases and it works in combination with the glutathione reduced, G-SH. Glutathione peroxidase family has eight isoforms in humans.

Glutathione peroxidase acts as electron donor and can reduce the hydrogen peroxide and organic hydroperoxides. Glutathione peroxidases (Gpx), Gpx1, Gpx2, Gpx3 and Gpx4, have selenium as prosthetic group (Epp et al. 1983).

3.4.4 *Cytochromes*

These are iron-containing pigmented compounds which are focused primarily in the mitochondria responsible for operation of electron transport system generating ATP molecules in the cells. Cytochromes contain iron metal in two states as Fe^{2+} (reduced state) and Fe^{3+} (oxidized state), and cytochrome is called as ferrocytochrome and ferricytochrome, respectively, and hence can catalyse oxidation and reduction reactions in the body (Reedy and Gibney 2004) (Table 3.2).

Cytochromes exhibit polymorphism based on the type of haem group, the type of apoprotein and the nature of binding between them (Tables 3.2 and 3.3).

Table 3.2 Types of cytochromes

Cytochrome b	**Haem b (protoporphyrin IX)**	Representative structure
Cytochrome c	Haem c	Two vinyl side chains of haem b are replaced by thioether linkage
Cytochrome a	Haem a	Methyl side chain at position 8 oxidized to formyl group and isoprenoid unit attached to vinyl group at position 2 of haem b
Cytochrome a_3	Haem a	Same structure

Source: Wikipedia (2015)

Table 3.3 Iron distribution in the body

Type of compound	Iron content	% of total body iron
Haemoglobin	**2500 mg**	**60–70%**
Myoglobin	**400 mg**	**5–10%**
Catalase, peroxidase	**2–3 mg**	**1%**
Cytochromes	**4–5 mg**	**1%**
Ferritin	**300–700 mg**	**10–15%**
Transferrin	**6–8 mg**	**1%**

Source: Chatterjea and Shinde (2002)

3.5 Iron Storage Compounds

3.5.1 Ferritin

Ferritin is a universal iron storage protein. It is present in the archaebacteria, plants and animals including humans. It is an intracellular protein and found predominantly in the liver, spleen, bone marrow and intestinal mucosa. It releases iron in a controlled way into the circulation. Ferritin is a spherical molecule and made up of 24 peptide monomeric units (Langlosis et al. 2004). Ferritin is an intracellular protein meant for iron storage. Many authors that worked in the laboratory suggested that ferritin might inhibit the differentiation of B lymphocytes into plasma cells. This process interferes with antibody formation (Morikawa et al. 1995; Aisen 1995; Casiday and Frey 2000) (Table 3.3).

These are arranged symmetrically in a way to form a hollow spherical shell. The outer diameter of this apoprotein spherical shell is 12 nm (Theil 1987; Andrews et al. 1992).

Internally, it encloses a chemically inert iron core having a diameter of 6 nm. These peptide subunits are folded into four parallel helices, and another fifth helix is placed at an angle of 60° (Andrews et al. 1992). Small channels are present at the intersection of these peptide subunits which allow the transportation of ions or molecules. Ferritin has a molecular weight of 450 kDa, and it can encapsule 2,000–4,300 iron atoms in ferric form that are finely arranged in crystalline lattice network, called ferrihydrite phosphate $(Fe(O)(OH)_8\ (FeOPO_3\ H_2).n(H_2\ PO_4)$ (Wikibooks 2014; Frey et al. 1995; Aisen and Listowsky 1980; Recalcati et al. 2008; Fisher et al. 2007; Theil 1990; Harrison and Arosio 1996).

Iron in tissues has the potential to induce oxidative damage. It may disturb ionic transport across the cell membrane and hence manifest into increased membrane permeability. It can damage protein and enzyme molecules and hence interfere with biochemical reactions. Also, it can induce chromosomal aberrations and negatively affect cell functions (Gardiner 1989; Halliwell 1992; Lesnefsky 1994; Loeb et al. 1988; Stadtman 1990). According to Witte et al. (1996), haemochromatosis in the body is associated with higher occurrence of malignancy in individuals.

The body protects itself through antioxidant enzymes and by ferritin protein against oxidation by free radicals. Superoxide dismutase, peroxidases, catalases and glutathione are the molecules in the body that change superoxide free radicals and peroxides into least harmful molecules that are removed from the body (Basaga 1990; Floyd 1990; Sies 1993).

Ferritin is another protective conjugated protein in the body. It can sequester free iron from tissues. Hence, it minimizes the oxidative potential of free iron and limits the formation of other free radicals. Ferritin is a cryoprotectant (Crichton 1990).

3.5.2 Haemosiderin

Haemosiderin is another storage form of iron which is strictly confined within cells. It is a complex of ferritin and denatured ferritin. Iron stored inside haemosiderin is least available to tissues, circulation and the bone marrow for Hb synthesis. It appears during iron overload in the body when the apoprotein content of the body is exhausted, and the condition is called haemosiderosis, wherein the tissues are without any injury. However, haemosiderosis associated with cell injury is called haemochromatosis. This condition can damage organs.

Quantity of ferritin in the body is larger than the haemosiderin under the condition of normalcy. Otherwise in iron overload, amount of haemosiderin iron becomes larger than ferritin and the ratio of haemosiderin iron to ferritin iron builds up in the body during periods of iron overload.

According to Cook (1929), haemosiderin is unique in itself and highly different in its chemical nature. Its granules are analogous to RBCs without any characteristic size or shape. These granules possess iron containing reddish to brown pigment. Haemosiderin, according to Cook (1929), is devoid of carbon or nitrogen and cannot be described as an organic compound as per the author. According to an assumption by the author, haemosiderin may be an inorganic compound of iron, wherein it neither shows properties of ferrous nor ferric form of iron. The author concluded that the iron must be present in haemosiderin in oxide or hydroxide state.

The pigment is of peculiar nature. It contains no carbon or nitrogen and therefore cannot be an organic compound. Its iron content is very high and aside from iron consists only of hydrogen and oxygen (or water). It must be an inorganic compound of iron, yet its reactions are not those characteristic of ferric iron. It gives none of the reactions of whatever ferrous iron.

3.6 Labile Iron Pool

Labile iron pool (LIP) is a chelatable and redox-active iron, which serves as an essential source of iron in cells for cellular iron metabolism (Kakhlon and Cabantchik 2002). Labile iron pool in the body is marked by anatomic boundary. Iron is imported and exported across the body tissues with the help of proteins that enable its transportation and accessibility, called as labile iron pool (Kruszewski 2003).

Labile iron pool is confined within the cytoplasm of cells. It is constituted of apoferritin, a protein having affinity for iron and metabolically active form of iron that remains in ferric state. Exogenous iron is absorbed by GIT and moves into the LIP, and its concentration within cells is maintained by the iron requirement in the body. Also intracellular extra iron in the body is deposited in LIP (Epsztejn et al. 1997).

Hence, LIP maintains a steady state. It prevents free state of iron that has high oxidative potential and can regenerate hydroxyl free radicals in the body via Fenton reaction. Content of iron in LIP is maintained partly by negative feedback that curtails the absorption of dietary iron at the intestinal mucosa and partly by transcriptional regulation of apoferritin and is called iron haemostasis of the body (Kruszewski 2003).

Iron has high potential for generation of free radicals. A small amount of free iron in the body can catalyse synthesis of hydroxyl radical via Fenton/Haber–Weiss reaction cycle. The free radical is highly injurious to tissues of the body (Fenton 1894). Availability of free iron in the body constitutes labile iron pool. It is necessary for metabolic reactions involving the iron-dependent enzymes. LIP is a source of free iron that might participate in free radical formation. Hence, the amount of LIP is minimized through the regulation of transcription and translation of protein necessary for maintaining iron level in the body (Kruszewski 2003; Crichton et al. 2002: Lipinski et al. 2010).

3.7 Physiological Iron Turnover

3.7.1 *From Senile RBC*

Daily, a small amount of Fe is released through the phagocytosis of senile RBCs by the macrophages of the reticuloendothelial system especially the spleen, liver and lymph nodes. Haemoglobin becomes free in circulation which could be easily excreted in urine owing to its small (64,000 Da) molecular weight. It is barred by another biochemical reaction in circulation. Haptoglobin is an alpha-2 globulin having two light chains and two heavy chains in its structure with a molecular weight of 90 kD (Dobryszycka 1997).

Haptoglobin is an acute phase protein, and its level increases during inflammation and infection in the body. Haemoglobin is immediately captured by haptoglobin in blood circulation preventing its excretion and results in Hb–Hp complex which is further phagocytosed by Kupffer cells in the liver (Wassell 2000).

Haem is detached from globin and released in circulation. Haem-bound iron has high toxicity and can cause lipid peroxidation thus resulting in oxidative stress and releasing hydroxyl radicals. Haemopexin is a beta-glycoprotein with a molecular weight of 57,000 Da. Its serum level is around 50–100 mg/dl in normal and healthy person and has a half-life period of 8 days. Haemopexin molecule has the highest affinity towards haem and forms a haem–haemopexin complex, hence preventing the haem-bound Fe-induced oxidative stress and loss of haem from the body. It delivers haem to hepatocytes and acts as scavenger (Tolosano and Altruda 2002).

Porphyrin is removed from haem in the liver, and Fe is released in circulation and carried by transferring protein to the bone marrow for biosynthesis of Hb. Also part of Fe is stored in ferritin form within cells. Daily, around 15–20 mg of Fe is released by catabolism of RBC (Fig. 3.2).

3.7.2 In the Body

Iron in the body moves from plasma and returns to plasma through the bone marrow and spleen. Body iron turns over from plasma to plasma via the bone marrow and spleen in harmony with RBC iron recycling. Iron daily need depends on the age, sex and physiological status of individual. Dietary sources vary in their iron content. Leafy vegetables contain 20 mg/100 g of the iron and are good source of non-haem iron. About 1 mg of iron is absorbed through GIT daily. Excretion of iron from the body does not follow the conventional pattern removal of any substance. Faeces contain around 80–90% of the dietary iron that goes unabsorbed in the small intestine. Excretion of iron via urine is almost (0.08 mg) insignificant. Around 30% (50–80 g) of the epithelial cells of GIT are exfoliated and replenished. So the loss of iron through desquamation of these cells is high. Also the iron is lost through the shedding of the cells of the skin that contain ferritin.

Finch (1959) estimated daily iron loss from the body of an adult person with 70 kg of weight. It was through blood 0.35 mg, mucosa 0.10 mg, bile 0.20 mg, urine 0.08 mg and skin 0.20 mg (Finch 1959).

Conclusively, about 1 mg of iron is lost from the body of an adult person daily, and this loss is miniscule which is replenished by absorbing nearly equal amount (1 mg) of dietary iron by GIT.

Further, erythropoiesis in the bone marrow demands about 20 mg of iron daily, wherein the catabolized RBC delivers around 20–25 mg of iron in the circulation that is channelized to the bone marrow; hence, an equilibrium is maintained between the demand and supply of iron in the body (Fig. 3.2).

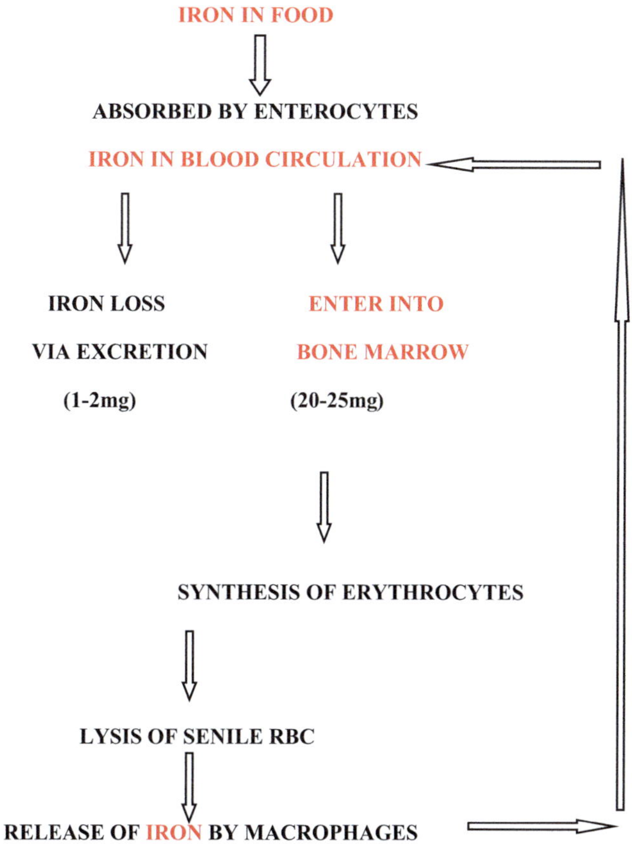

Fig. 3.2 Diagram showing iron distribution in human body

In the absence of this turn of iron, it becomes impossible to meet the high demand of iron for erythropoiesis and redox reactions by dietary intake. In females during the period of menarche and premenopause, loss of blood through each menstrual cycle is high and demands supplementation by appropriate vehicle. Each menstrual cycle retrocedes about 20–40 mg of iron from the body. Women in their reproductive years need to absorb approximately 2 mg of iron per day. However, when dietary iron is more abundant, absorption is appropriately attenuated (Fig. 3.2).

References

Aisen P (1995) Ferritin receptors and the role of ferritin in iron transport. Target Diagn Ther 4:339–354

Aisen P, Listowsky I (1980) Iron transport and storage proteins. Annu Rev Biochem 49:357–393

Allison AC (1954) Protection afforded by sickle-cell trait against subtertian malarial infection. Br Med J 1:290–294

American Academy of Pediatrics (1997) Work group on breastfeeding. Breastfeeding and the use of human milk. Pediatrics 100:1035–1039
Anderson HR, Nielsen JB, Nielsen F, Grandiean P (1997) Antioxidative enzymes in human erythrocytes. Clin Chem 43:562–568
Andreoli TE, Bennett JC, Carpenter CJ, Plum F (1990) Hematology. Cecil Essent Med 3:817–1002
Andrews SC, Arosio P, Bottke W, Briat JF, von Darl M, Harrison PM, Laulhere JP, Levi S, Lobreaux S, Yewdall SJ (1992) Structure, function, and evolution of ferritins. J Inorg Biochem 47(3–4):161–174
Antunes F, Boveris A, Cadenas E (2004) On the mechanism and biology of cytochrome oxidase inhibition by nitric oxide. Proc Natl Acad Sci U S A 101:16774–16779
Basaga HS (1990) Biochemical aspects of free radicals. Biochem Cell Biol 68:989–998
Beal VA, Meyers AJ, McCammon RW (1962) Iron intake, hemoglobin, and physical growth during the first two years of life. Pediatrics 30:518–539
Beard JL (2001) Iron biology in immune function, muscle metabolism and neuronal functioning. J Nutr 131:568S–579S
Boon EM, Downs A, Marcey D (2007) Catalase: H2O2 oxidoreductase. Catalase Structural Tutorial Text, pp 02–11
Bothwell TH (1979) Iron metabolism in man. Blackwell Scientific Publications, London
Bothwell TH (1995) Overview and mechanisms of iron regulation. Nutr Rev 3(9):237–245
Bothwell TH, Charlton RW (1981) Iron deficiency in women. The Nutrition Foundation, Washington, DC
Bothwell TH, Charlton RW, Cook JD, Finch CA (1979) Iron metabolism in man. Blackwell Scientific, Oxford
Brock JH, Halliday JW, Powell LW (1994) Iron metabolism in health and disease. Saunders WB, London
Burman D (1971a) Anemia in infancy – is iron important? Proc R Soc Med 64(5):579–580
Burman D (1971b) Iron requirements in infancy. Br J Haematol 20(3):243–247
Burman D (1972) Haemoglobin levels in normal infants aged 3 to 24 months, and the effect of iron. Arch Dis Child 47(252):261–271
Callahan A (2011) Why is breast milk so low in iron? Available at: http://scienceofmom.com/2011/10/12/why-is-breast-milk-so-low-in-iron/
Casiday R, Frey R (2000) Iron in biology: study of the iron content in ferritin, the iron-storage protein. Available at: http://www.chemistry.wustl.edu/~edudev/LabTutorials/Ferritin/Ferritin.html
Centers for Disease Control and Prevention (1998) Recommendations to prevent and control iron deficiency in the United States. MMWR Recomm Rep 47:1–29
Chatterjea MN, Shinde R (2002) Text book of medical biochemistry. Jaypee Brothers Medical Publishers, New Delhi, p 536
Chelikani P, Fita I, Loewen PC (2004) Diversity of structures and properties among catalases. Cell Mol Life Sci 61(2):192–208
Cook SF (1929) The structure of and composition of hemosiderin. Available at: http://www.http://www.jbc.org/content/82/3/595.full.pdf
Crichton RR (1990) Proteins of iron storage and transport. Adv Protein Chem 40:281–363
Crichton RR, Wilmet S, Legssyer R, Ward RJ (2002) Molecular and cellular mechanisms of iron homeostasis and toxicity in mammalian cells. J Inorg Biochem 91(1):9–18
Dallman PR (1990) Iron. In: Brown ML (ed) Present knowledge in nutrition. ILSI Press, Washington, DC
Dallman PR (1993) Nutritional anemias in childhood. In: Suskind RM, Lewinter-Suskind L (eds) Textbook of pediatric nutrition. Raven Press, New York, pp 91–105
Dewey KG, Chaparro CM (2007) Mineral metabolism and body composition iron status of breastfed infants. Proc Nutr Soc 66(3):412–422
Dobryszycka W (1997) Biological functions of haptoglobin-new pieces to an old puzzle. Euro J Clin Chem Biochem 35(9):647–654
Domellof M, Hernell O (2002) Iron deficiency anemia during first two years of life. Scand J Nutr 46(1):20–30

Domellof M, Cohen RJ, Dewey KG (2001) Iron supplementation of breast-fed Honduran and Swedish infants from 4 to 9 months of age. J Pediatr 138:679–687
Domellof M, Lonnerdal B, Abrams SA, Hernell O (2002) Iron absorption in breast-fed infants: effect of age, iron status, iron supplements and complementary foods. Am J Clin Nutr 76(1):198–204
Donovan A, Roy CN, Andrews NC (2006) The ins and outs of iron homeostasis. Physiology 21:115–123
Epp O, Ladenstein R, Wendel A (1983) The refined structure of the selenoenzyme glutathione peroxidase at 0.2-nm resolution. Eur J Biochem 133(1):51–69
Epsztejn S, Kakhlon O, Glickstein H, Breuer W, Cabantchik ZI (1997) Fluorescence analysis of the labile iron pool of mammalian cells. Anal Biochem 248:31–40
FAO/WHO (1988) Requirements of vitamin A, iron, folate and vitamin B12. FAO, Rome
Fenton HJH (1894) Oxidation of tartaric acid in presence of iron. J Chem Soc Trans 65(65):899–911
Finch CA (1959) Body iron exchange in man. J Clin Invest 38(2):392–396
Fisher J, Devraj K, Ingram J, Slagle-Webb B, Madhankumar AB, Liu X, Klinger M, Simpson IA, Connor JR (2007) Ferritin: a novel mechanism for delivery of iron to the brain and other organs. Am J Phys Cell Phys 293(2):C641–C649
Flogel U, Merx MW, Godecke A, Decking UKM, Schrader J (2001) Myoglobin: a scavenger of bioactive NO. Proc Natl Acad Sci U S A 98:735–740
Floyd RA (1990) Role of oxygen free radicals in carcinogenesis and brain ischemia. FASEB J 4:2587–2597
Frey R, Donlin M, Bashkin J (1995) Ferritin molecular-graphics tutorial. Available at: http://www: LabTutorials/Ferritin/FerritinTutorial.html
Gardiner HW (1989) Oxygen radical chemistry of polyunsaturated fatty acids. Free Radic Biol Med 7:65–86
Grantham-McGregor S, Ani C (2001) A review of studies on the effect of iron deficiency on cognitive development in children. J Nutr 131:649s–668S
Hallberg L (1992) Iron requirements: comments on methods and some crucial concepts in iron nutrition. Biol Trace Elem Res 35:25–45
Halliwell B (1992) Reactive oxygen species and the central nervous system. J Neurochem 59:1609–1623
Harrison PM, Arosio P (1996) The ferritins: molecular properties, iron storage function and cellular regulation. Biochim Biophys Acta 1275(3):161–203
Institute of Medicine/Food and Nutrition Board (2001) Dietary reference intakes for vitamin A, vitamin K, arsenic, boron, chromium, copper, iodine, iron, manganese, molybdenum, nickel, silicon, vanadium, and zinc. National Academy Press, Washington, DC
Johansson LH, Borg LAH (1988) A spectrophotometric method for determination of catalase activity in small tissue samples. Anal Biochem 174:331–336
Kail RV, Cavanaugh JC (2010) Human development: a life span view. Wadsworth, Cengage Learning, Australia, pp 296
Kakhlon O, Cabantchik ZI (2002) The labile iron pool: characterization, measurement, and participation in cellular processes(1). Free Radic Biol Med 33(8):1037–1046
Kendrew JC (1961) The three-dimensional structure of a protein molecule. Sci Am 205(6):96–110
Kendrew JC, Dickerson RE, Strandberg BE, Hart RG, Davies DR, Phillips DC, Shore VC (1960) Structure of myoglobin. Nature 185:422–427
Kohogo Y, Ikuta K, Ohtake T, Torimoto Y, Kato J (2008) Body iron metabolism and pathophysiology of iron overload. Int J Hematol 88(1):7–15
Kotecha PV (2011) Nutritional anemia in young children with focus on Asia and India. Indian J Community Med 36(1):8–16
Kruszewski M (2003) Labile iron pool: the main determinant of cellular response to oxidative stress. Mutat Res 531:81–92

Langlois de Estaintot B, Santambrogio P, Granier T, Gallois B, Chevalier JM, Precigoux G, Levi S, Arosio P (2004) Crystal structure and biochemical properties of the human mitochondrial ferritin and its mutant Ser144Ala. J Mol Biol 340(2):277–293

Lardinois OM, Rouxhet PG (1996) Peroxidatic degradation of azide by catalase and irreversible enzyme inactivation. Biochem Biophys Acta 1298:180–190

Lesnefsky EJ (1994) Tissue iron overload and mechanisms of iron-catalyzed oxidative injury. Adv Exp Med Biol 366:129–146

Lind T, Hernell O, Lonnerdal B, Stenlund H, Domellof M, Persson LA (2004) Dietary iron intake is positively associated with hemoglobin concentration during infancy but not during the second year of life. J Nutr 134:1064–1070

Lipinski P, Starzynski RR, Stys A, Straciło M (2010) Iron homeostasis, a defense mechanism in oxidative stress. Postepy Biochem 56(3):305–316

Lippard SJ, Berg JM (1994) Principles of bioinorganic chemistry. University science book, Mill Valley

Lodish H, Berk A, Zipursky LS, Matsudaira P, Baltimore D, Darnell J (2000) Evolutionary tree showing the globin protein family members myoglobin and hemoglobin. Mol Cell Biol Freeman WH, pp 3136–3

Loeb LA, James EA, Waltersdorph AM, Klebanoff SJ (1988) Mutagenesis by the autoxidation of iron with isolated DNA. Proc Natl Acad Sci U S A 851:3918–3922

Mammen P, Kanatous SB, Yuhanna IS, Shaul PW, Garry MG, Balaban RS, Garry DJ (2003) Hypoxia-induced left ventricular dysfunction in myoglobin-deficient mice. Am J Phys 285:H2132–H2141

Marengo-Rowe AJ (2006) Structure-function relations of human hemoglobins. Proceeding 19(3):239–245

Marques RFSV, Taddei JAAC, Lopez FA, Braga JAP (2014) Breastfeeding exclusively and iron deficiency anemia during the first 6 months of age. Rev Assoc Med Bras 60(1)

Mascotti DP, Rup D, Thach RE (1995) Regulation of iron metabolism: translational effects medicated by iron, heme and cytokines. Annu Rev Nutr 15:239–261

Meinzen-Derr JK, Guerrero ML, Altaye M, Ortega-Gallegos H, Ruiz-Palacios GM, Morrow AL (2006) Risk of infant anemia is associated with exclusive breast-feeding and maternal anemia in a Mexican cohort. J Nutr 136(2):452–458

Michaelsen KF, Milman N, Samuelson G (1995) A longitudinal study of iron status in healthy Danish infants: effects of early iron status, growth velocity and dietary factors. Acta Paediatr:1035–1044

Michaelsen KF, Weaver L, Branca F, Robertson A (2000) Feeding and nutrition of infants and young children. Guidelines for the WHO European Region. WHO, Copenhagen

Millikan GA (1939) Muscle hemoglobin. Physiol Rev 19:503–523

Morikawa K, Oseko F, Morikawa S (1995) A role for ferritin in hematopoiesis and the immune system. Leuk Lymphoma 18(5–6):429–433

Nagin MK (2009) Minerals in breast milk. Available at: http://www.about.com/

Nelson DL, Cox MM (2000). In: Lehninger principles of biochemistry. Worth publisher, New York, pp 206

Ordway GA, Garry DJ (2004) Myoglobin: an essential hemoprotein in striated muscle. J Exp Biol 207:3441–3446

Oski FA (1993) Iron deficiency in infancy and childhood. N Engl J Med 329:190–193

Perutz MF, Rossmann MG, Cullis MG, Muirhead H, Will G, North ACT (1960) Structure of haemoglobin. A three-dimensional Fourier synthesis at 5.5Å resolution, obtained by X-ray analysis. Nature 185:416–422

Prentice A (1996) Constituents of human milk. Food Nutr Bull 17(4):1–14

Rao R, Georgieffa MK (2006) Iron in fetal and neonatal nutrition. Semin Fetal Neonatal Med 12(1):54–63

Recalcati S, Invernizzi P, Arosio P, Cairo G (2008) New functions for an iron storage protein: the role of ferritin in immunity and autoimmunity. J Autoimmun 30(1–2):84–89

Reedy CJ, Gibney BR (2004) Heme protein assemblies. Chem Rev 104(2):617–649
Saarinen UM, Siimes MA, Dallman PR (1977) Iron absorption in infants: high bioavailability of breast milk iron as indicated by the extrinsic tag method of iron absorption and by the concentration of serum ferritin. J Pediatr 91:36–39
Sidell B, O'Brian KM (2006) When bad things happen to good fish: the loss of hemoglobin and myoglobin expression in Antarctic icefishes. J Exp Biol 209:1791–1802
Sies H (1993) Strategies of antioxidant defense. Eur J Biochem 215:213–219
Siimes MA, Salmenperal Perheenpupa J (1984) Exclusive breast feeding for 9 months: risk of iron deficiency. J Pediatr 104:196–199
Soh P, Ferguson EL, McKenzie JE, Homs MYV, Gibson RS (2004) Iron deficiency and risk factors for lower iron stores in 6–24-month-old New Zealanders. Eur J Clin Nutr 58:71–79
Stadtman ER (1990) Metal ion-catalyzed oxidation of proteins: biochemical mechanism and biological consequences. Free Radic Biol Med 9:315–325
Stoltzfus R (2001) Summary: implications for research and programs. J Nutr 131:697S–701S
Theil EC (1987) Ferritin: structure, gene regulation, and cellular function in animals, plants, and microorganisms. Annu Rev Biochem 56(1):289–315
Theil EC (1990) The ferritin family of iron storage proteins. Adv Enzymol Relat Area Mol Biol 63:421–449
Tolosano E, Altruda F (2002) Hemopexin: structure, function, and regulation. DNA Cell Biol 21(4):297–306
Wassell J (2000) Haptoglobin: function and polymorphism. Clin Lab 46(11–12):547–552
Wikibooks (2014) Structural biochemistry-ferritin. Available at: https://en.wikibooks.org/wiki/Structural_Biochemistry/Ferritin_and_transferrin
Wikipedia (2015) Heme A. Available at: https://en.wikipedia.org/wiki/Heme_A
Williams RS, Neufer PD (1996) Regulation of gene expression in skeletal muscle by contractile activity. In: Rowell LB, Shepherd JT (eds) Handbook of physiology. Exercise: regulation and integration of multiple systems. Oxford University Press, New York, pp 1124–1150
Witte DL, Crosby WH, Edwards CQ, Fairbanks VF, Mitros FA (1996) Practice guideline development task force of the College of American Pathologists. Hereditary hemochromatosis. Clin Chim Acta 245:139–200
Wittenberg BA, Wittenberg JB (1985) Oxygen pressure gradients in isolated cardiac myocytes. J Biol Chem 260:6548–6554
Wittenberg BA, Wittenberg JB (1987) Myoglobin-mediated oxygen delivery to mitochondria of isolated cardiac myocytes. Proc Natl Acad Sci U S A 84:7503–7507
Wittenberg BA, Wittenberg JB (1989) Transport of oxygen in muscle. Annu Rev Physiol 51:857–878
Wittenberg JB, Wittenberg B (2003) Myoglobin function reassessed. J Exp Biol 206:2011–2020
World Health Assembly (2001) Infant and young child nutrition. Resolution WHA 54:2
World Health Organization/Food and Agriculture Organization of the United Nations (2004) Vitamin and mineral requirements in human nutrition. WHO, Geneva
Yan Z, Serrano AL, Schiaffino S, Bassel-Duby R, Williams SR (2001) Regulatory elements governing transcription in specialized myofiber subtypes. J Biol Chem 276:17361–17366
Yip R (1989) The changing characteristics of childhood iron nutritional status in the United States. In: Filer LJ Jr (ed) Dietary iron: birth to two years. Raven Press, New York, pp 37–61

Chapter 4
Iron Metabolism in Human Body

Abstract Animal and plant sources have haem and non-haem forms of iron. These two forms of iron have variable absorbability in the small intestine. Haem iron is available in animal protein in the forms of haemoglobin and myoglobin. This form of iron is absorbed faster than non-haem iron from vegetarian diet.

Ascorbate, citrate and amino acids act as absorption promoters. Phytic acid, tannins, oxalates, carbonates, phosphates and soil eating retard iron absorption. Duodenal cytochrome b (DCTYB) is a characteristic ferric reductase enzyme of the human body. It is located on the brush border surface of enterocytes. It reduces Fe^{3+} iron into highly soluble Fe^{2+} iron. Ferrous iron is actively transported across the cell membrane of enterocytes. Absorption of haem iron absorption is mediated by haem receptors. These are located on the microvilli of epithelial cells in the small intestine. Iron from the inside of mucosal cells can be exported to plasma by basolateral iron exporter proteins. Ferroportin-1 serves as the chief iron exporter from mucosal cells and hepatocytes. Hepcidin prevents the export of iron in circulation in condition of chronic inflammation like kidney failure and cirrhosis (Ashby et al. 2009). Iron deficiency during 0–3 years of life alters the morphology, physiology and bioenergetics of CNS. Iron is an essential cofactor in the process of myelination. Additional hypothesis related with the effect of iron deficiency on the neurodevelopment is the altered brain bioenergetics. Children who were iron deficient in infancy had comparatively poor physical activity and poor ability to verbalization at the age of 5 years irrespective of the iron supplements which corrected their iron deficiency during infancy.

4.1 Introduction

Iron is the essential trace element of the body. Its daily requirement is less than 100 mg, and it constitutes, even, smaller than 0.01% of total body weight. It is imperatively needed for oxygen transport in circulation. Additionally, iron is elemental for biosynthesis of cytochromes that is critical for redox reactions in mitochondria. It forms the structural part of catalase enzyme that reduces hydrogen peroxide into water and oxygen in body tissues. Iron exists in the body in ferrous (Fe^{2+}) and ferric (Fe^{3+}) states that are interconvertible, hence enabling iron to participate in oxidation-reduction reaction in the body. Although iron is the most abundant

A. Gupta, *Nutritional Anemia in Preschool Children*,
DOI 10.1007/978-981-10-5178-4_4

Table 4.1 Dietary sources of iron

Sources of iron	Iron content
Leafy vegetables	20 mg/100 g
Pulses	10 mg/100 g
Cereals	8 mg/100 g
Nuts, dry roasted, 1 oz (49 nuts)	1 mg
Broccoli, boiled	1 mg
Potato, baked	1 mg
Milk	0.1 mg/100 ml
Liver	5 mg/100 g
Meat	2 mg/100 g
Spleen	6 mg/100 g
Egg, boiled	1 mg
Spinach, boiled, ½ cup	3 mg
Lentils, boiled, ½ cup	3 mg
Jaggery	3 mg/100 g

Sources: IOM/FNB (2001); Ayur times (2014)

element on the earth's crust, it is paradoxical that iron deficiency becomes the chief cause of nutritional anemia, which is the most prevalent form of anemia in children and pregnant women in developing countries, especially in rural population and people residing in slum areas, throughout the world.

4.2 Dietary Sources of Iron

Different dietary sources contain definite amount of iron, as in Table 4.1.

4.3 Nonconventional Sources of Iron

4.3.1 Iron Utensils

I strongly hypothesize that food preparation in iron utensils/pots, irrespective of the size of pot/quantity and quality of food/spices added, definitely fortifies the foods with readily absorbable and bioavailable iron.

To augment my hypothesis, I will cite the works of earlier researchers as follows:

Geerlings et al. (2013a) state that eating food prepared in iron pots increases haemoglobin concentration of anaemic/iron-deficient individuals.

Use of iron pots varies between countries. Further size of cooking pot, quantity of food prepared, duration of food preparation, nature of food prepared, etc. determine bioavailability of iron in the body. Introduction of iron pots or improving their use in communities in developing countries for the preparation of food may be a promising innovative intervention for reducing iron deficiency and iron deficiency anemia. Further research is required to monitor the use and effectiveness of this intervention.

According to Geerlings et al. (2013b), rural Malawian adults, who inhabited high malarial endangered areas, consumed foods cooked in iron utensils for 6 weeks. It was found that these adults showed an uplift of Hb level on resorting to food consumption prepared in iron pots despite having had high malarial transmission.

In rural and remote areas with little medical facility, it becomes difficult to maintain and sustain iron supplementation on a daily basis for long periods to IDA children (Stoltzfus and Dresfus 1998). In view of these difficulties, the use of iron cooking pots could provide an innovative strategy to control iron deficiency anemia, which has fewer logistical difficulties and lower cost than other approaches.

4.4 Iron Absorption

4.4.1 Factors Affecting Iron Absorption

Animal and plant sources have haem and non-haem forms of iron. These two forms of iron have variable absorbability in the small intestine. Haem iron is available in animal protein in the form of haemoglobin and myoglobin. It is absorbed faster than non-haem iron from vegetarian diet.

In duodenum, pH is nearly neutral. Haem iron is not polymerized at alkaline pH and is readily available for absorption. Non-haem iron at alkaline pH has a tendency to become precipitated as ferric hydroxide. Also it chelates with phytates rendering it minimally absorbable (Spiro et al. 1967; Monsen 1988).

Absorption of haem iron is about 20–30% of the total intake, whereas non-haem iron absorption is around 5–10% of the total dietary intake of iron. In Western countries, daily consumption of red meat, a source of haem iron, is predominant in comparison to Indian subcontinent, where the daily diet contains a large amount of cereals, pulses and vegetables. It has been estimated that haem iron constitutes about 75% of the total body iron in a person. Vegetarians are susceptible to iron deficiency due to least absorbability of non-haem iron (Carpenter and Mahoney 1992).

1. Ascorbic acid, citric acid and amino acid

   Absorption of iron occurs in ferrous form. It is a highly soluble form. Ascorbate, citrate and amino acids act as absorption promoters. These exogenous ligands chelate iron and form highly soluble monomeric complexes in the duodenum. Glutathione reduces ferric form into ferrous form, hence favouring its absorption through intestinal mucosa (Conrad and Umbreit 1993; Teucher et al. 2004).

2. Meat factor helps in iron absorption. Presence of a small amount of meat in diet promotes absorption of non-haem iron.

   In a study by Hallberg et al. (1989a), a contrasting feature was observed. Bread meal that contained 88 mg of phytic acid was supplemented with 50 mg of beef by authors (Hallberg et al. 1989b). Iron absorption was not enhanced.

   Later on, in another study by Baech et al. (2003), absorption of non-haem increased by 44% when a dose of 50 g pork was added in a diet rich in phytic acid (220 mg) by authors (Baech et al. 2003).

   It is clearly evident from both studies that an increase in phytic acid in diet in the presence of meat factor resulted into faster non-haem iron absorption. So the effect of meat factor is highly accentuated in the presence of phytic acid. Fish in diet has been shown to uplift non-haem iron absorption (6%–12% rise with 40 g fish) (Aung et al. 1976).

3. The pH of stomach and duodenum favours iron absorption. In stomach, a low pH value promotes disassociation of ferric form of iron that exists in non-haem form bound to plant proteins. Again, gastric pH favours reduction of ferric ion into ferrous ion, hence decreasing polymerization and precipitation potential and increasing solubility of ferric iron (Andrews 2000).

   Further, entry of chyme into the proximal part of the duodenum lowers intraluminal pH, hence enhancing absorption of iron. Iron in monomeric complex form can enter intestinal epithelial cells, but polymeric complexes of iron cannot penetrate the cell membrane of epithelial cells and have a tendency to get precipitated in the duodenum (Jacobs 1967).

   Gastric acid influences absorption of non-haem iron from the food. Achlorhydria severely affects the production of HCl in gastric secretion. Its secretion almost becomes absent. The condition is responsible for malabsorption of non-haem iron, hence causing iron deficiency anemia. Achlorhydria has been noted in children causing IDA. Three cases of achlorhydria causing IDA in children aging 8 months, 13 months and 4 years were reported by Hawksley et al. (1934). Betesh et al. 2015 analysed the effect of gastritis-induced achlorhydria as a cause in iron deficiency anemia. Authors concluded that the absorption of iron from a typical Western diet is reduced in the body of achlorhydric patients. However, iron losses in these patients exceeded the intake of iron; therefore, the process disturbed the iron balance of the body. Nonetheless, a rise in iron absorption by intestinal mucosa in these patients was noted which was an adaptive response of the mucosa in relation of negative iron balance. Overall, it was asserted by authors that gastritis-induced achlorhydria is a definite independent factor of iron deficiency anemia in patients.

4. Iron-deficient individuals have higher ability to absorb iron in comparison to health individuals. This value is 20% in iron deficiency anemia, whereas the value is 10% in healthy individuals.

5. Phytic acid, tannins, oxalates, carbonates, phosphates and soil eating retard iron absorption. Phytic acid and oxalic acid are present in cereals and leaf vegeta-

bles in large quantity. These inhibitors bind with iron and form polymeric complexes which are insoluble in the duodenum. Indian diet contains about 20 mg of iron obtained through cereals and vegetables as main constituents. Nevertheless, a meager quantity of iron (5–10%) is absorbed owing to the presence of large concentration of phytates and oxalates in the diet. Moreover, the presence of bran, cellulose and hemicelluloses in the food interferes with iron absorption. Pectin is a mucopolysaccharide present in the cell wall of plants. It is used as gelling agent in foods. It is supposed to decrease iron absorption. Pectin delays gastric emptying so the food remains in the stomach for longer duration that retards iron absorption since it is absorbed in the duodenum (Hallberg et al. 1987; Di Lorenzo et al. 1988; Sotelo et al. 2010).

Bran, whole-wheat flour, has a high quantity of phytic acid in comparison to white wheat flour. Oats also contain high amount of phytate. To counteract this disadvantage, adequate amount of ascorbic acid should be supplemented in bran and oat flour (Brune 1992).

6. Concomitant use of milk with iron interferes in its absorption. Both elements are necessary for growth and adopt same pathway for their absorption. Calcium in milk or in other dairy products competitively inhibits haem and non-haem absorption. However, calcium cannot be considered as an inhibitor. To overcome inhibition, dose of iron in diet should be increased. So more iron molecules attach to DMT-1 carrier, and its absorption would increase (Hallberg 1991).
7. Tea and coffee also delay iron absorption. These are caffeinated beverages and caffeine decreases iron absorption. Additionally, tea contains tannins, polyphenols and coffee enriched with chlorogenic acid. These organic compounds also limit iron absorption.
8. Heavy metals like lead, cobalt, strontium, manganese and zinc play a role of competitive inhibitor with iron and hinder iron absorption. These metals follow a common absorptive pathway and could have negative interactions affecting absorption of iron. Lead is a highly toxic element and competes with iron for its attachment on the surface of divalent metal transporter (DMT-1). Under normal condition of health and total body iron, iron binding with DMT-1 and, hence, its absorption is faster due to high affinity of iron towards metal transporter. In iron-deficient children, body response increases secretion and release of DMT-1 in the intestine mucosa. So the absorption of lead increases many times leading to lead toxicity (Goyer 1993).

   Lead is taken up by DMT-1 and secondarily blocks iron absorption through competitive inhibition. Lead interferes with another iron-dependent metabolic pathway as haem biosynthesis. Lead has multifaceted impetus on children resulting in iron deficiency anemia and impaired cognitive process.
9. Antacids, possibly including H2 receptor blockers, interfere with the absorption of iron from the gut. Lactoferrin present in human milk is a promoter of iron absorption.
10. Oxidation of iron influences its physical state and chemical property, hence affecting its absorption potential. Absorption of ferrous iron is around more

than twice the absorption of ferric salts. Ferric ions from diet are reduced into ferrous ions in the stomach at low pH. Its solubility is higher than ferric ion. It chelates with ascorbate and citrate that keep ferrous ions in soluble state. Ferric ions have a potential to get precipitated at alkaline pH in the upper part of the duodenum.

4.4.2 Non-haem Iron Absorption

Non-haem iron is abundantly found in green leafy vegetables, cereals, pulses and fruits. In India, jaggery is a good source of iron. In villages, food is prepared in iron utensils that enrich the food with iron. Non-haem or elemental iron is abundant in vegetables, fruits, jaggery, molasses, nuts and grams and is the chief form of dietary iron in developing countries. Haem iron is present in meat, liver, poultry and fish that becomes the main part of daily diet in Western countries (Medline plus 2015).

Although sources of iron are abundant, cheap and economical, however, it is surprising and alarming that iron deficiency anemia is ubiquitous in poor population. Today, its prevalence is obvious in upper socio-economic strata that might arise due to a change in dietary habits. Fast foods and canned foods are rich in calories but deficient in iron contents.

Nutritional anemia in preschool-aged children is associated with poor appetite, diarrhoea, malabsorption syndrome or worm infestation of the alimentary canal. All these factors, collectively, lead to malnutrition that is a potential cause of child morbidity and mortality.

Dietary iron absorption is multifactorial. It depends on the source of iron, presence of other dietary components in the alimentary canal and iron store of the body. Iron absorption occurs through the mucous membrane, in the proximal part of the small intestine, particularly, close to the gastroduodenal junction and duodenum. Its absorption decreases upwards in the small intestine. Acidic chyme favours fast absorption in the gastroduodenal region. Low pH keeps the iron in easily solubilized and least polymerized form which is rapidly taken up by mucosal cells. Rise in pH along the distal part of the small intestine discourages absorption because iron has a tendency to become precipitated at high pH.

Elemental iron exists in plant foods in the form of ferric ion. It remains bound to plant proteins. After dietary intake, iron-protein bonding is hydrolysed by HCl in the stomach. Ferric iron is released in the stomach. Interconvertibility of the oxidative state of iron, in the body, is its unique property. Dietary ingredients like ascorbic acid, citric acid and glutathione oxidize ferric into ferrous iron. It is chelated and solubilized. Ascorbic acid present in the diet, additionally, helps in absorption of elemental iron. Possible pathway is the chelation of iron to form iron-ascorbate complex in the digestive tract. The complex keeps iron in highly solubilized form. It is taken up readily by mucosal cells of the small intestine (Hallberg et al. 1989b).

Duodenal cytochrome b (DCTYB) is a characteristic ferric reductase enzyme of the human body. It is located on the brush border surface of enterocytes. It reduces Fe^{3+} iron into highly soluble, Fe^{2+} iron (Latunde-Dada et al. 2002).

Ferrous iron is actively transported across the cell membrane of enterocytes. Active transportation is facilitated by a transmembrane carrier protein molecule. It is called divalent metal transporter-1 (DMT-1). It is also called as natural resistance-associated macrophage protein 2 (NRAMP-2) and divalent cation transporter-1 (DCT-1). It is encoded in humans by SLC11A2 gene (Vidal et al. 1995; Wu et al. 2015).

It is highly ascribed for its role in transporting ferrous iron across the cell membrane of the intestinal epithelium. DMT-1 expression is induced by the iron store of body. It is a physiological process of iron homeostasis. It carries out proton-coupled transportation of divalent metal ions from the intestinal lumen to the inside of intestinal mucosal cells. DMT-1 is located on the apical surfaces of enterocytes. Additionally, DMT-1 facilitates active absorption of divalent ions like copper, zinc, manganese, cobalt and cadmium (Gunshin et al. 1997; Anderson et al. 2009).

Important endogenous ligand, gastroferrin, helps in iron absorption. It is secreted by oxyntic cells of gastric mucosa. It is a glycoprotein having 15% polypeptide and 85% carbohydrate. It binds with ferrous iron in the lumen of the gut and helps in its absorption (Multani et al. 1970).

Indirect evidence in favour of gastroferrin-induced ferrous absorption was provided by Like et al. (1967). They asserted that secretion of gastroferrin correlated with the iron store of body. In case of iron overload, its synthesis is decreased, whereas its release accelerates during iron deficiency in the body (Like et al. 1967).

Dietary factors like polyphenols, phytates, tannic acids, calcium and soya proteins inhibit the absorption of non-haem iron. Ferrous iron is internalized inside the enterocytes. Within the cytoplasm of enterocytes, Fe^{2+} iron is oxidized into Fe^{3+} form with the help of ceruloplasmin, a major copper-containing protein. It is synthesized in the liver. It contains six copper atoms associated with globulin (Hellman and Gitlin 2002).

It has copper-dependent oxidase activity. It is also called as ferroxidase1. It oxidizes ferrous into ferric iron inside enterocytes. It is synthesized by the liver. It exerts copper-dependent oxidase activity that helps oxidize ferrous iron into ferric form (Song and Dunaief 2013).

Fe^{3+} iron is carried by an intracellular iron carrier (IIC) inside the intestinal mucosal cells. A part of ferric iron is delivered to mitochondria. Another proportion of ferric iron binds to apoferritin, a protein, present inside the cytoplasm of enterocytes. It is synthesized by mucosal cells of the intestine. It stores ferric iron in the form of ferritin within mucosal cells.

Theory of iron absorption by intestinal mucosal cells is called mucosal block theory. It was proposed by Garnick. According to it, iron content of intestinal mucosal cells determines the uptake of iron by intestinal mucosal cells. Whenever ferritin level of mucosal cells is depleted, the rate of iron absorption is increased; otherwise, under iron overload, absorption is blocked. Iron homeostasis is exclusively based on iron absorption by mucosal cells, and it is independent of iron excretion in body.

4.4.3 Haem Iron Absorption

It is an important source of iron in the body. But the mechanism of its uptake is not clearly established. There are two hypotheses regarding haem iron absorption through the intestinal epithelium. Haem iron is taken up by enterocytes as a whole in the form of metalloporphyrin. Current theory proposes that haem iron absorption is mediated by haem receptors. Haem receptors are proteins. These are located on the microvilli of epithelial cells of the intestinal mucosa. Second theory explains the haem iron absorption through the haem transporter (West and Oates 2008). Intestinal haem receptor is located on brush borders of epithelial cells. It has very high affinity for haem, and Grasbeck et al. (1979) demonstrated that receptor was localized on brush borders of the intestine and had high affinity for haem with the use of radioactive haem in porcine small intestine. According to Grasbeck et al. 1979, it could be solubilized with the use of hydrophilic polyethylene oxide (Triton X-100) from the intestine of porcine and human.

Nature of haem receptor is entirely different from albumin, haemopexin, glutathione-S-transferase or aggregated haem as proved by a study by Grasbeck et al. (1979). Further another study by Trenhunen et al. (1980) showed that haem receptor complex had high association as indicated by a dissociation constant of 10^{-6}–10^{-7} mol/L using radioactive ligands (Trenhunen et al. 1980). Haem receptor-mediated uptake by enterocytes in iron deficiency was studied by Roberts et al. (1993) in rat enterocytes. It was found that haem absorption increased with iron deficiency. It was hypothesized that quantity of haem receptor on microvilli of intestinal epithelium and binding capacity of haem determined its absorption. Haem uptake in rat enterocytes increased with iron deficiency.

Above data suggests and augments the endocytosis of haem by the mucosa of the intestine in metalloporphyrin state through the help of haem receptor. Another theory explains haem absorption directly through the cell membrane of intestinal mucosal cells with the help of transporter. Haem carrier protein-1 (HCP-1) has been identified in mammalian intestinal mucosa as haem transporter. Its characterization was done by Laftah et al. (2009) in vivo and in vitro studies in CD-1 mice. It was found that HCP-1 was a proton-coupled folate transporter (PCFT) that had high affinity for folate and limited affinity for haem (Laftah et al. 2009).

HCP-1 was extracted by Shayeghi et al. (2005) from the duodenum by a subtractive hybridization method. Haem carrier transporter is a temperature-sensitive protein responsible for haem transportation across the cell membrane of the intestinal mucosa. Haem transportation is analogous to metal-tetracycline transportation in bacteria. It is located on brush border surface of the duodenum. It is controlled by hypoxia and iron deficiency. Haem is catabolized inside the enterocytes in mammals and humans. This was demonstrated by Brown et al. (1968). Authors administered intragastric (Brown et al. 1968) dose of radiolabelled haemoglobin to dog. Sample of portal blood was collected by Brown et al. (1968) after 3 h. Highest radioactivity was observed in portal blood sample by Brown et al. (1968) that contained non-haem iron. The experiment showed breakdown of haem within enterocytes.

Haem is broken down inside enterocytes with the help of an enzyme. It was first of all demonstrated by Weintraub et al. (1968). A molecular weight of 64,000 Da and kinetic properties of the enzyme pointed to the presence of xanthine oxidase located inside the cytoplasm of enterocytes. It was supposed to degrade haem iron into the mixture of bilirubin isomers and iron (Dawson et al. 1970). Later on, authors (Gray et al. 1972) confirmed through in vivo studies that degradation of haem produced exclusively bilirubin IX-α and not the mixture as supposed by earlier authors in 1970.

Further, research by Raffin et al. (1974) argued in favour of the presence of haem oxygenase enzyme in mucosal cells. The assumption was based on the release of bilirubin IX-α isomer by catabolism of haem through haem oxygenase enzyme. It confirmed the earlier study of Gray et al. (1972) which hypothesized the release of bilirubin IX-α isomer.

Another fact that the activity of haem oxygenase is highest in areas of haem iron absorption in the duodenum was demonstrated by Conrad et al. (1976). Also, iron deficiency in the body correlated with haem iron absorption and haem oxygenase activity, whereas xanthine oxidase activity decreased. Haem oxygenase is a microsomal enzyme contained and synthesized in the smooth endoplasmic reticulum in hepatocytes. It cleavages haem ring at the alpha-methene bridge generating biliverdin IX-α that further undergoes reduction to bilirubin IX-α by biliverdin reductase enzyme. Iron in ferrous form is released along with carbon monoxide (Tenhunen et al. 1969; Kikuchi et al. 2005; Ryter and Choi 2009).

Enzyme haem oxygenase-1 is inducible in the presence of inflammation, nitric oxide, oxidative stress, iron deficiency in the body and heavy metals like Cd, Co, Cr and Cu. Haem iron is catabolized by HO-1 inside the endocytotic vesicle. The entire process needs about 2–3 h. Iron is released inside the vesicle and then transported to the systemic circulation from the enterocyte. Iron export from inside the vesicle within the enterocyte follows the same path as for the non-haem iron after its uptake by the intestinal mucosal cell.

4.5 Iron Transport in Blood Circulation

Iron from the inside of mucosal cells can be exported to the plasma by basolateral iron exporter proteins. Ferroportin-1 is a transmembrane protein encoded by SLC40A1 gene. It serves as the chief iron exporter protein from mucosal cells and hepatocytes. These cells store iron which is released by the lysis of senile RBCs in the body.

Additionally, hephaestin (Vulpe et al. 1999; Ishikawa et al. 1998) is a copper-dependent ceruloplasmin with ferroxidase activity. It facilitates iron export from mucosal cells into the circulation. Possibly, it is also involved in the copper homeostasis in the body (Chen et al. 2006). It is involved in the iron export from enterocytes into blood circulation.

Hepcidin is a 24-amino acid peptide hormone, a key regulator of iron export into plasma. It is secreted by hepatocytes and encoded by HAMP gene (Rossi 2005; Ganz 2003).

Hepcidin is a main iron regulator. Hepcidin binds with ferroportin-1, which is situated on the basolateral surface of enterocytes and cell membranes of macrophages (Jordan et al. 2009).

Hepcidin degrades ferroportin-1, thus antagonizing its action on iron transport and ultimately preventing the egress of iron from hepatocytes and macrophages. Iron is sequestered into macrophages and hepatocytes. Hepcidin also reduces uptake of dietary iron by mucosal cells.

In inflammation and infection, hepcidin level is raised. Therefore, serum iron level falls down, causing anemia. Hepcidin prevents the export of iron in the circulation in condition of chronic inflammation like kidney failure and cirrhosis (Ashby et al. 2009).

Moreover, bioavailability of iron is reduced. Hepcidin synthesis and secretion is regulated by hypoxia, inflammation, infection, iron store of the body and rate of erythropoiesis. Erythroferrone is a peptide hormone, secreted by erythroblasts. It acts on hepatocytes and inhibits the synthesis and secretion of hepcidin; therefore, it increases the absorption and bioavailability of iron for erythropoiesis. Iron in circulation exists in ferric state that binds to apotransferrin and is circulated in the form of transferrin, a chief iron carrier of blood. It helps to deliver iron to cells and tissues; also, it prevents its conversion into ferrous state and averts the synthesis of free radicals that might injure tissues. In blood circulation, a negligible amount of iron remains bound to citrate and albumin.

4.6 Iron Uptake by Cells

Transport of iron from transferrin to the inside of cells of the body is dependent on cytosolic concentration of iron (Harding et al. 1983). It is regulated by transferrin receptor-mediated endocytosis that occurs on the surface of the cell membrane. Transferrin receptor 1 (TFR-1) is highly sensitive to Fe^{3+} bound to transferrin and is ubiquitous in its distribution on the surface of cells, whereas transferrin receptor 2 (TFR-2) exhibits lesser sensitivity to Fe^{3+} and is exclusively located on the surface of hepatocytes. Transferrin, the ligand, attaches to TFR-1 located on the cell membrane and forms a TF-TFR complex (Herbison et al. 2009; Tortorella and Karagiannis 2014). It starts a cellular signal that activates the adapter protein complex, a cytosolic heterotetramer, present inside the cytoplasm of the cell (Nakatsu and Ohno 2003; Tortorella et al. 2014).

It starts sorting of clathrin molecules (Pearse 1976) which are comprised of a network of polypeptide chains. There are three heavy chains and three light chains, wherein each heavy chain ~190 kDa joins to each light chain, ~25 kDa, through its C-terminal (Kirchhausen et al. 2014; Bitsikas et al. 2014).

Clathrin triskelions remain floating in the cytosol. These molecules do not link the cell membrane directly and, however, attach to adapter protein molecules, present on the interior of the cell membrane that, later on, links with another clathrin molecule to form a polyhedral lattice. It stretches and deforms the cell membrane to form vesicles. These are called clathrin-coated vesicles that contain TF-TFR-1 complex and are pinched of and released into the cytoplasm, hence resulting in internalization of TF-TFR1 complex. This process is called as clathrin-coated vesicle endocytosis (McMahon and Boucrot 2011).

ATP-dependent proton pump operates in the cell membrane of these vesicles. It secretes H^+ ions that lower the pH of the interior of vesicles to pH 5.5. Acidification of the clathrin-coated vesicles ensures definite change inside the vesicle. Initially, TF-TFR-1 complex undergoes conformational change and destabilization of the bond between the ligand and receptor. Additionally, acidic medium of vesicles favours the release of Fe^{+3} from transferrin. Later on, vesicle membrane-bound oxidoreductase enzyme reduces the Fe^{+3} iron into Fe^{+2} state, and it crosses the membrane of vesicle and enters the cytosol of the cell. Apotransferrin bound to TFR-1 in vesicles recycles to the surface of plasma membrane and enters the circulation and again transports iron molecule. Within the cytosol of body cells, iron is translocated to different cytoplasmic organelles with the help of a low-molecular-weight carrier that has not been elucidated, till date.

In mitochondria, iron is utilized for biosynthesis of haem. Also, iron is sorted in cells as ferritin.

During iron overload, ferritin is aggregated and engulfed and degraded by lysosomes. Hemosiderin is the by-product of lysosomal digestion and is the agglomerate of denatured protein with iron.

4.7 Regulation of Iron Absorption

Absorption of iron is a thoroughly meticulous and controlled process in humans. It serves to uptake and transport dietary iron but checks iron overload in the body. Iron absorption is regulated by different pathways in the body.

4.7.1 Mucosal Block Mechanism

Iron absorption is regulated at the level of intestinal mucosa. Accordingly, absorption of iron is dependent on the iron store of the body. Depletion of iron store in the body is a stimulus for a rise in absorption of iron through the mucosa of gut. Conversely, adequate iron store discourages iron absorption. This regulation prevents over-accumulation of iron in the body. It is noteworthy that iron is an element with one-way flow in the body (Crosby 1966; Manis and Schachter 1964; Conrad et al. 1964).

4.7.2 *Erythropoietin Stimulus*

Iron deficiency in the body results in diminished Hb saturation with oxygen. The oxygen-carrying capacity of blood is decreased owing to diminished body iron. Kidneys are highly sensitive to hypoxic hypoxia. Juxtaglomerular apparatus releases erythropoietin. It signals to increase iron absorption at mucosa level (Delorme et al. 1992; Dudrick et al. 1985; Eschbach et al. 1987).

4.7.3 *Ferritin–Transferrin Receptor Translational Control*

Iron overload in the body increases ferritin synthesis by mRNA transcription and translation for ferritin. So it stores iron in the body. But transferrin receptor synthesis is aborted through degradation of mRNA for transferrin receptor. It inhibits uptake of iron.

4.7.4 *Hepcidin Level*

DMT-1, a metal ion carrier, and ferroportin help in absorption of iron. Synthesis of these two proteins is regulated by another protein called hepcidin. It is produced in the liver. In high iron load in the body, hepcidin level falls down, in turn decreasing the synthesis of ferroportin and curtailing iron absorption (Ganz and Nemeth 2012; Jordon et al. 2012; Pak et al. 2006).

Conclusively, different proteins regulated the iron absorption through the mucosa.

4.8 Role of Iron in the Growth and Development of Children

Term infants have adequate iron store that may be necessary for the growth and development in the coming 3–4 months of life. However, iron deficiency during infancy would have persistent unfavourable effects on the structure and functioning of the central nervous system. Iron store for adequate brain iron is vital during 0–12 months postnatal life. Extensive research has been undertaken in the past on human and animal models describing the consequences of iron deficiency on the changes in myelination of neurons in white matter, altered biosynthesis of iron-dependent neurohormones in the brain and their impaired functioning and altered functioning of the hippocampus (Todorich et al. 2009).

Oligodendrocytes are iron-rich cells in the central nervous system. These cells are profoundly located in white matter tracts and are focused around cell bodies.

Iron deficiency due to any cause results in iron depletion in oligodendrocytes. Inadequate iron in the body leads to premature death of oligodendrocytes and affects the myelin synthesis (Connor and Menzies 1996). Gratuitous impact of iron deficiency on the growth of CNS resists immediate recovery even after the start of iron supplements and persists for a longer period in the life. Study on rodents (Connor and Menzies 1996) exhibits the long-lasting effects of iron deficiency during gestation and lactation which continued till adulthood period notwithstanding the repletion of iron store.

4.8.1 Hypotheses for Iron Deficiency-Induced Harmful Effects on Neurodevelopment

Infancy, early preschool age and puberty are the periods of rapid growth in the body. The requirement for micronutrients is high during these periods. Endogenous iron store of the body is exhausted, thereby setting a stage for iron deficiency between 6 and 24 months of postnatal life, wherein rapid neurodevelopment requires high brain iron (Rao and Georgieff 2007).

Iron deficiency during 0–3 years of life alters the morphology, physiology and bioenergetics of CNS. Iron deficiency either in intrauterine life and/or early postnatal life, contrarily, influences the neurogenesis and differentiation of brain regions. Studies on rats indicate an alteration in the structure of the hippocampus and striatum in iron deficiency state. Study was done by Rao et al. (2003) on rats to detect a cause for an injury in the hippocampus during perinatal iron deficiency.

A significant change in the concentration of neurochemicals like phosphocreatine, *N*-acetylaspartate, glutamate, γ-aminobutyric acid, aspartate, phosphorylethanolamine and taurine was detected in the hippocampus of iron-deficient rats. These neurohormones might alter resting membrane potential, transmission of nerve impulse and myelination in perinatal iron deficiency (Rao et al. 2003).

Second hypothesis concerning the effect of iron deficiency on neurodevelopment is the functioning of oligodendrocytes in the brain. These are the myelinating cells of CNS. These form myelin sheath over the axons in white matter. Iron is an essential cofactor in the process of myelination. Nonetheless, the precise role of iron in the working of oligodendrocytes is yet to be discovered. Myelin is a lipoprotein. Iron might regulate the transcriptional and/or translational processes (Ortiz et al. 2004). In their study on iron-deficient animal models, hypomyelination was observed characterized by depletion of protein and lipid in myelin composition. Overall, content and compaction of myelin were affected in white matter of the brain.

These changes remain persistent in life and effect adult life. It was observed by Corapci et al. (2006) in their study on Costa Rican children aging 5. Children who were iron deficient in infancy had comparatively poor physical activity and poor ability to verbalization at the age of 5 years irrespective of the iron supplements which corrected their iron deficiency during infancy. Study shows that the effect of iron deficiency in infancy persisted in the form of poor cognition and low level of motor ability during school age.

Still another hypothesis explaining the role of iron deficiency on neurodevelopment is the alteration in the metabolism of monoamines in CNS.

Workers (Beard 2003) examined the level of brain iron, dopamine transporter and dopamine receptors in rats in infancy. Reduction in the level of D2 receptors and brain iron in nucleus accumbens and striatum was observed in iron-deficient rats at the 14th day of postnatal life. Density of dopamine transporter also reduced to half in the nucleus accumbens of iron-deficient rats. Dopamine and norepinephrine are mainly concerned with the psychomotor activity, learning, memory and sleep cycles. Dopaminergic neurons in the nigrostriatal tract and mesolimbic pathways as well as the noradrenergic neurons in the midbrain regulate most of the psychomotor activities.

Alteration in biosynthesis, secretion and/or catabolism of these neurohormones is responsible for impaired cognitive, motor and sociobehavioural pattern in children. However, evidences for alterations in dopamine or norepinephrine level in humans are limited.

Researchers (Voorhess et al. 1975) analysed urine samples of 11 iron-deficient children and detected high levels of norepinephrine, epinephrine and dopamine which normalized after intramuscular iron therapy. Activity of monoamine oxidase is dependent on brain iron level. Inadequate brain iron level inactivates MOA which is responsible for degradation of monoamines in the body. It might be the cause of high secretion of catecholamines in the urine.

Additional hypothesis related with the effect of iron deficiency on the neurodevelopment is the altered brain bioenergetics. The brain is a highly sensitive organ to partial pressure of oxygen. It is solely dependent on available glucose in circulation for biosynthesis of ATP through glucose oxidation. De Deungria et al. (2000) studied the activity of enzyme in the brain responsible for bioenergetics through utilization of glucose. Cytochrome c oxidase is a haem-conjugated enzyme. It is the terminal enzyme in the electron transport chain in mitochondria responsible for ATP synthesis by oxidative phosphorylation. Activity of cytochrome c oxidase is mainly dependent on iron content of circulation; hence, in iron-deficient state, brain iron distribution is restricted, which further diminishes activity of cytochrome c oxidase in brain areas. It was observed that its activity decreased in subareas of the hippocampus, dentate gyrus, piriform cortex, medial dorsal thalamic nucleus and cingulate cortex. These areas of the brain are mainly linked with higher cognitive functions and memory. Therefore, disrupted energy balance in highly sensitive areas of the brain in infancy results in impaired cognitive development.

References

Anderson GJ, Frazer DM, McLaren GD (2009) Iron absorption and metabolism. Curr Opin Gastroenterol 25:129–135

Andrews NC (2000) Intestinal iron absorption: current concepts circa. Dig Liver Dis 32(1):56–61

Ashby DR, Gale DP, Busbridge M, Murphy KG, Duncan ND, Cairns TD, Taube DH, Bloom SR, Tam FW, Chapman RS, Maxwell PH, Choi P (2009) Plasma hepcidin levels are elevated but responsive to erythropoietin therapy in renal diseases. Kidney Int 75(9):976–981

Aung TB, Thein T, Thane T (1976) Iron absorption from Southeast Asian rice-based meals. Am J Clin Nutr 29:219–225

Ayur times+ (2014) Jaggery nutritional value, nutritional facts and analysis. Available at: https://www.ayurtimes.com/jaggery-nutritional-value-nutrition-facts-analysis/

Baech SB, Hansen M, Bukhave K, Jensen M, Sorensen SS, Kristensen L, Purslow PP, Skibsted LH, Sandstrom B (2003) Nonheme-iron absorption from a phytate-rich meal is increased by the addition of small amounts of pork meat. Am J Clin Nutr 77:173–179

Beard J (2003) Iron deficiency alters brain development and functioning. J Nutr 133:1468S–1472S

Betesh AL, Santa Ana CA, Cole JA, Fordtran JS (2015) Achlorhydria a cause of iron deficiency anemia? Am J Clin Nutr 102(1):9–19

Bitsikas V, Correa IR Jr, Nichols BJ (2014) Clathrin-independent pathways do not contribute significantly to endocytic flux. eLife 3:e03970

Brown EB, Hwang YF, Nicol S, Ternberg J (1968) Absorption of radiation-labeled hemoglobin by dogs. J Lab Clin Med 72:58–64

Brune M (1992) Iron absorption from bread in Humans: inhibiting effects of cereal fiber, phytate and inositol phosphates with different numbers of phosphate groups. J Nutr 122:442–449

Carpenter CE, Mahoney AW (1992) Contributions of heme and nonheme iron to human nutrition. Crit Rev Food Sci Nutr 31:333–367

Chen H, Huang G, Su T, Gao H, Attieh ZK, McKie AT, Anderson GJ, Vulpe CD (2006) Decreased hephaestin activity in the intestine of copper-deficiency mice causes systemic iron deficiency. J Nutr 136(5):1236–1241

Connor JR, Menzies SL (1996) Relationship of iron to oligodendrocytes and myelination. Gila 17(2):83–93

Conrad ME, Umbreit JN (1993) A concise review: iron absorption – the mucin-mobilferrin-integrin pathway. A competitive pathway for metal absorption. Am J Hematol 42:67

Conrad ME, Weintraub LR, Crosby WH (1964) The role of intestine in iron kinetics. J Clin Invest 43:963–974

Conrad ME, Weintraub LR, Sears DA, Crosby WH (1976) Absorption of hemoglobin iron. Am J Phys 211:1123–1130

Corapci F, Radan AE, Lozoff B (2006) Iron deficiency in infancy and mother child Interaction at 5 years. J Dev Behav Pediatr 27(5):371–378

Crosby WH (1966) Mucosal block. An evaluation of concepts relating to control of iron absorption. Semin Hematol 3(4):299–313

Dawson RB, Rafal S, Weintraub LR (1970) Absorption of hemoglobin iron: the role of xanthine oxidase in the intestinal heme-splitting reaction. Blood 35:94–103

de Deungria M, Rao R, Wobken JD, Luciana M, Nelson CA, Georgieff MK (2000) Perinatal iron deficiency decreases cytochrome c oxidase (CytOx) activity in selected regions of neonatal rat brain. Pediatr Res 48:169–176

Delorme E, Lorenzini T, Giffin J, Martin F, Jacobsen F, Boone T, Elliott S (1992) Role of glycosylation on the secretion and biological activity of erythropoietin. Biochemistry 31:9871–9876

Di Lorenzo C, Williams CM, Hajnal F, Valenzuela JE (1988) Pectin delays gastric emptying and increases satiety in obese subjects. Gastroenterology 95:1211–1215

Dudrick SJ, O'Donnell JJ, Raleigh DP, Matheny RG, Unkel SP (1985) Rapid restoration of red blood cell mass in severely anemic surgical patients who refuse transfusion. Arch Surg 120:721–727

Eschbach JW, Egrie JC, Dwoning MR, Browne JK, Adamson JW (1987) Correction of the anemia of end-stage renal disease with recombinant human erythropoietin. N Engl J Med 316:73–78
Ganz T (2003) Hepcidin, a key regulator of iron metabolism and mediator of anemia of inflammation. Blood 102(3):783–788
Ganz T, Nemeth E (2012) Hepcidin and iron homeostasis. Biochim Biophys Acta 1823(9):1434–1443
Geerlings PD, Brabin BJ, Omar AA (2013a) Food prepared in iron cooking pots as an intervention for reducing iron deficiency anemia in developing countries: a systematic review. J Hum Nutr Diet 16(4):275–281
Geerlings PP, Brabi B, Mkumbwa A, Broadhead R, Cuevas LE (2013b) The effect on haemoglobin of the use of iron cooking pots in rural Malawian households in an area with high malaria prevalence: a randomized trial. Tropical Med Int Health 8(4):310–315
Goyer RA (1993) Lead toxicity: current concerns. Environ Health Perspect 100:177–187
Grasbeck R, Kouvonen I, Lindberg M, Tenhunen R (1979) An intestinal receptor for heme. Scand J Haematol 23(1):5–9
Gray CH, Nicholson DC, Tipton G (1972) Degradation of haem compounds to bile pigments. Nat New Biol 239:5–8
Gunshin H, Mackenzie B, Berger UV, Gunshin Y, Romero MF, Boron WF, Nussberger S, Gollan JL, Hediger MA (1997) Cloning and characterization of a mammalian proton-coupled metal-ion transporter. Nature 388(6641):482–488
Hallberg L (1991) Calcium: effect of different amounts on nonheme-and heme-iron absorption in Humans. Am J Clin Nutr 53:112–119
Hallberg I, Brune M, Rossander L (1989a) The role of vitamin C in iron absorption. Int J Vitam Nutr Res 30:103–108
Hallberg L, Brune M, Rossander L (1989b) Iron absorption in man: ascorbic acid and dose-dependent inhibition by phytate. Am J Clin Nutr 49:140–144
Hallberg L, Rossander L, Skanberg AB (1987) Phytate and inhibitory effect of bran on iron absorption in man. Am J Clin Nutr 45:988–996
Harding C, Heuser J, Stahl P (1983) Receptor-mediated endocytosis of transferrin and recycling of the transferrin receptor in rat reticulocytes. J Cell Biol 97(2):329–339
Hawksley JC, Lightwood R, Bailey UM (1934) Iron-deficiency anemia in children: its association with gastro-intestinal disease, achlorhydria and haemorrhage. Available at: http://adc.bmj.com/content/9/54/359.full.pdf
Hawksley JC, Lightwood R, Bailey UM (Retrieved July, 2017) Iron deficiency anemia in children: its association with gastrointestinal disease, achlorhydria and haemorrhage. group.bmj.com. Available at:http://adc.bmj.com
Hellman NE, Gitlin JD (2002) Ceruloplasmin metabolism and function. Annu Rev Nutr 22:439–458
Herbison CE, Thorstensen K, Chua AC, Graham RM, Leedman P, Olynyk JK, Trinder D (2009) The role of transferrin receptor 1 and 2 in transferrin-bound iron uptake in human hepatoma cells. Am J Physiol Cell 297(6):C1567–C1575
Institute of Medicine/Food and Nutrition Board (2001) Dietary reference intakes for vitamin A, vitamin K, arsenic, boron, chromium, copper, iodine, manganese, molybdenum, nickel, silicon, vanadium and zinc: a report of the panel on micronutrients. National Academy Press, Washington, DC
Ishikawa K, Nagase T, Suyama M, Miyajima N, Tanaka A, Kotani H, Nomura N, Ohara O (1998) Prediction of the coding sequences of unidentified human genes. X. The complete sequences of 100 new cDNA clones from brain which can code for large proteins in vitro. DNA Res 5(3):169–176
Jacobs A (1967) Effect of gastric juice and pH on inorganic iron in solution. Nature 216:707–708
Jordan JB, Poppe L, Haniu M, Arvedson T, Syed R, Li V, Kohno H, Kim H, Schnier PD, Harvey TS, Miranda LP, Cheetham J, Sasu BJ (2009) Hepcidin revisited, disulphide connectivity dynamics and structure. J Biol Chem 284(36):24155–24167
Kikuchi G, Yoshida T, Noguchi M (2005) Heme oxygenase and heme degradation. Biochem Biophys Res Commun 338(1):558–567

Kirchhausen T, Owen D, Harrison SC (2014) Molecular structure, function, and dynamics of clathrin-mediated membrane traffic. Cold Spring Harb Perspect Biol 6(5):a016725–a016725

Laftah AH, Latunde-Dada GO, Fakih S, Hider RC, Simpson RJ, Mckie AT (2009) Haem and folate transport by proton-coupled folate transporter/haem carrier protein 1 (SLC46A1). Br J Nutr 101(8):1150–1156

Latunde-Dada GO, Van der Westhuizen J, Vulpe CD, Anderson GJ, Simpson RJ, Mckie AT (2002) Molecular and functional roles of duodenal cytochrome B (Dcytb) in iron metabolism. Blood Cell Mol Dis 29(3):356–360

Like CG, Davis PS, Deller DJ (1967) Change in gastric iron-binding protein (gastroferrin) during iron-deficiency anemia. Lancet 1(7496):926–927

Manis J, Schachter D (1964) Active transport of iron by intestine: mucosal iron pool. Am J Physiol 207:893–900

McMahon HT, Boucrot E (2011) Molecular mechanism and physiological functions of clathrin-mediated endocytosis. Nat Rev Mol Cell Biol 12:517–533

Medline Plus (2015) Iron in diet. Available at: https://www.nlm.nih.gov/medlineplus/ency/article/002422.htm

Monsen ER (1988) Iron nutrition and absorption: dietary factors which impact iron bioavailability. J Am Diet Assoc 88(7):786–790

Multani JS, Cepurneek CP, Davis PS, Saltman P (1970) Biochemical characterization of gastroferrin. Biochemistry 9(20):3970–3976

Nakatsu F, Ohno H (2003) Adaptor protein complexes as the key regulators of protein sorting in the post-Golgi network. Cell Struct Funct 28(5):419–429

Ortiz E, Pasquini JM, Thompson K, Felt B, Butkus G, Beard J, Connor JR (2004) Effect of manipulation of iron storage, transport, or availability on myelin composition and brain iron content in three different animal models. J Neurosci Res 77:681–689

Pak M, Lopez MA, Gabayan V, Ganz T, Rivera S (2006) Suppression of hepcidin during anemia requires erythropoietic activity. Blood 108:3730–3735

Pearse BM (1976) Clathrin: a unique protein associated with intracellular transfer of membrane by coated vesicles. Proc Natl Acad Sci U S A 73(4):1255–1259

Raffin SB, Woo CH, Roost KT, Price DC, Schmid R (1974) Intestinal absorption of hemoglobin iron-heme cleavage by mucosal heme oxygenase. J Clin Invest 54:1344–1352

Rao R, Georgieff MK (2007) Iron in fetal and neonatal nutrition. Semin Fetal Neonatal Med 12:54–63

Rao R, Tkac I, Townsend EL, Gruetter R, Georgieff MK (2003) Perinatal iron deficiency alters the neurochemical profile of the developing rat hippocampus. J Nutr 133:3215–3221

Roberts SK, Henderson RW, Young GP (1993) Modulation of uptake of heme by rat small intestinal mucosa in iron deficiency. Am J Phys 265(4 Pt 1):G712–G718

Rossi E (2005) Hepcidin – the iron regulatory hormone. Clin Biochem Rev 26(3):47–49

Ryter SW, Choi AMK (2009) Heme oxygenase-1/carbon monooxide from metabolism to molecular therapy. AM J Respir Cell Mol Biol 41:251–260

Shayeghi M, Latunde-Dada GO, Oakhill JS, Laftah AH, Takeuchi K, Halliday N, Khan Y, Warley A, McCann FE, Hider RC, Frazer DM, Anderson GJ, Vulpe CD, Simpson RJ, McKie AT (2005) Identification of an intestinal heme transporter. Cell 122(5):789–801

Song D, Dunaief JL (2013) Retinal iron homeostasis in health and disease. Front Aging Neurosci 5:24

Sotelo A, Gonzalez-Osnaya L, Sanchez-Chinchillas A, Trejo A (2010) Role of oxate, phytate, tannins and cooking on iron bioavailability from foods commonly consumed in Mexico. Int J Food Sci Nutr 61(1):29–39

Spiro GT, Page L, Saltman P (1967) The hydrolytic polymerisation of ferric citrate. I. The chemistry of the polymer. J Am Chem Soc 89:5555–5559

Stoltzfus RJ, Dresfus ML (1998) Guidelines for the use of iron supplements to prevent and treat Iron deficiency anemia. International Life Sciences Institute, ILSI Press, Washington, DC

Tenhunen R, Grasbeck R, Kouvonen I, Lundberg M (1980) An intestinal receptor for heme: its partial characterization. Int J BioChemiPhysics 12:713–716

Tenhunen R, Marver HS, Schmid R (1969) Microsomal heme oxygenase. Characterization of the enzyme. J Biol Chem 244:6388–6394

Teucher B, Olivarres, Cori H (2004) Enhancers of iron absorption: ascorbic acid and other organic acids. Int J Vitam Nutr Res 74(6):403–419

Todorich B, Pasquini JM, Garcia CI, Connor JR (2009) Oligodendrocytes and myelination: the role of iron. Gila 57(5):467–478

Tortorella S, Karagiannis TC (2014) Transferrin-receptor mediated endocytosis: a useful target for cancer therapy. J Membr Biol 247:291–307

Vidal S, Belouchi AM, Cellier M, Beatty B, Gros P (1995) Cloning and characterization of a second human NRAMP gene on chromosome 12q13. Mamm Genome 6(4):224–230

Voorhess ML, Stuart MJ, Stockman JA, Oski FA (1975) Iron deficiency anemia and increased urinary norepinephrine excretion. J Pediatr 86:542–547

Vulpe CD, Kuo YM, Murphy TL, Cowley L, Askwith C, Libina N, Gitschier J, Anderson GJ (1999) Hephaestin, a ceruloplasmin homologue implicated in intestinal iron transport, is defective in the sla mouse. Nat Genet 21:195–199

Weintraub LR, Weinstein MB, Huser HJ, Rafal S (1968) Absorption of hemoglobin iron: the role of a heme-splitting substance in the intestinal mucosa. J Clin Invest 47:531–539

West AR, Oates PS (2008) Mechanisms of heme iron absorption: current questions and controversies. World J Gastroenterol 14(26):4101–4110

Wu W, Song Y, He C, Liu C, Wu R, Fang L, Cong Y, Miao Y, Liu Z (2015) Divalent metal-ion transporter 1 is decreased in intestinal epithelial cells and contributes to the anemia in inflammatory bowel disease. Sci Rep 5:16344

Chapter 5
Aetiology of Iron Deficiency in Children

Abstract Iron deficiency is the commonest nutritional deficiency disorder prevalent worldwide. It is the main cause of anemia during pregnancy and lactation among women and infancy and preschool age in children. In the developing countries, the prevalence is usually highest in infants and preschool-aged children, whereas in developed countries, iron deficiency and IDA are observed among women owing to higher physiological need of elemental iron during pregnancy, lactation and menstrual cycles. Chief aetiology of iron deficiency and IDA in poor population in villages, remote settings and slum dwellers is the low dietary iron intake.

Loss of blood from gastrointestinal mucosa is ascribed to the presence of anal fissure. It is a common cause of anemia in infants. *Helicobacter pylori* is responsible for primary gastritis in children and leads to nutritional anemia. Hookworm infection is the globally prevalent predisposing factor of iron deficiency and IDA among preschool children. Chronic haemorrhage is another sequel of helminthic infestation in children. Hookworms result in chronic intestinal loss of blood and anemia. Severe malarial anemia is the malarial infliction in childhood characterized by an intense reduction in HB concentrations <5 g/dl.

Environmental enteropathy is an inflammatory condition of the intestinal mucosa prevalent in population residing in crowded places and among those individuals who adopt poor hygiene and improper sanitation practices. Environmental enteropathy is associated with increased epithelial permeability, poor absorption and defunct localized immunity of GIT. This condition decreases the absorption of vitamin B_{12}, folate and iron from the dietary food by the intestinal mucosa. These micronutrients are necessary for haemoglobin synthesis and erythropoiesis in bone marrow. Malabsorption syndrome could be a reason for iron deficiency with or without anemia. Coeliac disease is an autoimmune disorder of the mucosa of the small intestine. Its prevalence is observed in individuals of all ages from infants to old age persons. Children suffers from diarrhoea, anorexia, abdominal discomfort and weight loss. Geophagy is the habit of eating inedible substances like chalk, soil, clay and pencil. These are passed into the stool unchanged. These unnatural substances do not provide any energy to the body. Clay particles may have the tendency to withhold dietary iron in complexes. So it is unavailable for absorption in intestinal lumen. Ingested soil particles may damage intestinal mucosa resulting in impaired absorptive capacity.

A. Gupta, *Nutritional Anemia in Preschool Children*,
DOI 10.1007/978-981-10-5178-4_5

5.1 Introduction

Highly prevalent dietary deficiency disorder is the iron deficiency and IDA which are frequently observed among women, infants and preschool-aged children. Iron deficiency is pandemic in its distribution and is the exclusive dietary inadequacy found among women in developed nations as well as among infants, preschool-aged children and women in developing nations. Onset of puberty among girls is a predisposition towards higher need of elemental iron. Further, consistent growth during infancy and among under-5 aged children imposes additional iron demand which is to be fulfilled through diet rich in iron. Failure to enrich body iron store by dietary iron is one of the causes of iron deficiency and IDA in preschool children (De Macyer and Adiels-Tegman 1985).

Additionally, iron deficiency anemia in underprivileged population coexists with nutritional disorders like protein/energy malnutrition, vitamin A deficiency, folate deficiency and infection (Florentino and Guirriec 1984; Fleming 1982; Masawe 1981; Lechner and Raja 1984).

In tropical regions of the world, parasitic infections and haemoglobinopathy are frequently prevalent. In the developed countries, iron deficiency is the main nutritional health problem. Environmental factors, food availability and medical facility are the intervening factors that decide the prevailing disorder in any country of the world (UNICEF 1995; Nazir et al. 2011; Ubino and Ijaz 2010; CDC 2011a).

Adequate dietary constituents, macronutrients and micronutrients maintain nutritional status of the body. Iron deficiency affects infants and preschool-aged children severely. Reduction of daily supply of dietary iron, exaggerated iron loss from body in the form of diarrhoea and occult blood and increased physiological demand for iron are responsible for iron deficiency among infants and preschool-aged children. Moreover, iron deficiency aggravates in case of a preterm baby (Olivares et al. 1999).

Milk is a poor source of iron (Cashman 2006). A diet purely based on milk is not advisable to supply iron to infants and preschool children. Infants who are provided cow milk early in life are more vulnerable to iron deficiency and IDA. Iron in cow milk is poorly absorbed (Leung and Sauve 2003; Fomon 2001). Formula-fed infants, who are not enriched with iron supplements, have comparatively higher probability of IDA. Further, iron deficiency among infants is amplified in case infants have premature birth and have low weight. These physiological conditions are associated with low iron store of body (Rao and Georgieff 2010).

It has been observed by workers that majority of infants who are born prematurely have comparatively higher tendency for iron deficiency and IDA (Halliday et al. 1984; Lundstorm et al. 1977; Borigato and Martinez 1998; Dauncey et al. 1978). Full-term infants have adequate body iron store, and deprivation of supplementary foods to full-term term infants results in iron deficiency between 6 and 12 months. Contrarily, natural body iron store in preterm infants is deficient at the time of birth, and iron deficiency in preterm infants appears earlier (Iwai et al. 1986; Franz et al. 2000).

After age of 5, children's diet is diversified. It lessens the burden of iron deficiency. School-aged children have comparatively lesser demand for iron owing to decreased body growth (Mayo Clinic 2016).

Decreased dietary iron supply, increased iron loss from the body and decreased iron bioavailability in the body compromise the iron pool of body and bioavailability of iron to tissues. It results in latent iron deficiency (LID). It is also called as iron-deficient erythropoiesis (Yehuda and Mostofsky 2010). This is regarded as a condition in which iron deficiency persists in the body without the presence of anemia. Latent iron deficiency has a diagnostic value. It is an indication that the person with LID will develop anemia in the near future when the iron deficiency is not corrected by appropriate iron supplements. Latent iron deficiency manifests in three stages. In stage 1, there is a reduction in the iron level in the bone marrow. Haemoglobin level and iron in serum are normal.

This condition results in compensatory increase in the iron absorption and serum transferrin level. Iron-binding capacity rises and serum ferritin concentration drops down. In stage 2, serum iron and transferrin saturation are decreased. Consequently, erythropoiesis is decreased. In this stage, serum iron level sinks to below 50 μg/dL, and transferrin saturation is below 16%. In stage 3, haemoglobin level is decreased, which manifests as anemia. In peripheral blood smear, erythrocytes appear normal in size and shape. Further, stage 4 is characterized by microcytic and hypochromic erythrocytes (Lichtin 2016). In the last stage, body tissues are negatively affected as a sequel of iron deficiency. Therefore, stage 1 and 2 are characterized by reduction of iron store of the body.

Concentration of serum transferrin receptor (TfR) increases, and it is followed by an increase in free erythrocyte protoporphyrin (FEP) coupled with a decline in transferrin saturation (Sat). It has been assessed by Olivares et al. (1999) that all these factors are indicative of early fall in haemoglobin concentration. If iron deficiency continues, it perpetuates into iron deficiency anemia.

Amount of iron absorption from dietary sources is variable and dependent on multiple factors. Quantity of iron in the foods is the prime factor. Iron-rich foods provide, comparatively, higher amount of iron to the body unlike foods deficient of iron. Another factor is the form of iron, whether ferrous or ferric, present in the dietary foods. The form of iron determines its solubility and rate of absorption in the gut. The last factor is the concomitant presence of iron absorption inhibitors and/or enhancers in the foods (Charlton and Bothwell 1983; Olivares et al. 1993).

Forms of iron as haem iron and non-haem iron have different rates of absorption from the mucosa of the small intestine. Population who rely predominantly on the non-vegetarian diet has lesser chances of developing iron deficiency and IDA (Olivares et al. 1999). Term infants are protected from iron deficiency by their endowment at birth and iron present in breast milk which is provided to infants in the first half of infancy (Arvas et al. 2000; Raj et al. 2008; WHO 2001; ACP 1997).

Absorption of iron from milk is rapid and underlying mechanism is uncertain (Bhatia 2007; Kleinman 2004). Awareness among population in developed nations has led to the widespread use of iron-enriched formula-fed and a continuous use of iron-rich supplementary foods during weaning periods. These dietary practices have reduced the incidence of iron deficiency among infants (Olivares et al. 1999; Yeung 2016; Rees et al. 1985; Rees et al. 1985).

According to workers (Olivares et al. 1999), it is customary to nourish infants and children with cereal- and legume-based supplementary foods in developing nations. These dietary foods contain high quantity of inhibitors like phytic acid which hinder the absorption of iron. According to workers (Olivares et al. 1999; Badamosi et al. 1995), it was found that children in school age are provided with staple foods like wheat, potato, corn and/or rice. It is the practice throughout the world. As a consequence, children in school-age group suffer from iron deficiency anemia.

It has been observed by Suharno et al. (1993) that vitamin A deficiency and nutritional anemia, generally, occur together. Workers have documented a strong correlation between retinol plasma concentration and iron deficiency anemia in preschool-aged children (Suharno et al. 1993). It is customary to feed infants in their first half of infancy on cow milk in some portions of world. This practice, if continued for a prolonged period, may result in damage of mucosa of the small intestine of infants. It is the chief cause of occult bleeding (Ziegler et al. 1990; Walter et al. 1990, Jiang et al. 2000; Axelsson II et al. 1999; Jiang et al. 2000). Allergy to cow milk is a common cause of damage to the gut mucosa due to IgE-mediated immunity against cow milk protein (Vandenplans et al. 2007; Host 2002, Hill et al. 1986a; Harvey et al. 2014).

Loss of blood from GIT is mainly caused by anal fissure in infants. It is responsible for iron deficiency anemia. It may result in blood-stained stool and/or blood spots in the undergarments of infants. It is the manifestation of apparent neonatal milk protein allergy. Ductus arteriosus in intrauterine life links pulmonary artery to descending aorta. It is an important passage for the entry of deoxygenated blood into the placenta via the descending aorta and arteries in the umbilical cord. Its persistence in neonates is considered as an abnormality. Indomethacin, a nonsteroidal anti-inflammatory drug, is used for patent ductus arteriosus in postnatal life. The drug helps in the closure of patent ductus arteriosus. It acts by blocking the cyclooxygenase enzyme and inhibits prostaglandin synthesis. However, drug in neonates might predispose to GIT bleeding. Indomethacin is responsible for vasoconstriction and platelet dysfunction (Mehta et al. 2003).

Helicobacter pylori is responsible for primary gastritis in children. It is the frequent cause of gastritis (Hill et al. 1986b; Rowland et al. 2004; Wolfrom 2016).

In tropics, parasitic infestation is another common cause for blood loss from the gut which may predispose to iron deficiency anemia. *Necator americanus* and *Ancylostoma duodenale* are common endoparasites responsible for mucosa damage, blood loss, and iron deficiency anemia. It has been assessed by workers (Baker 1981; Stolzfus et al. 1997a; Roche and Layrisse 1966; CDC 2014; NHS 2015) that amount of loss of blood from the gut is linked directly to the damage and number of parasites in the gut lumen.

Schistosomiasis is another common cause of haematuria. It is a predisposing factor for IDA (Prual et al. 1992; CDC 2014; NHS 2015). Acute or chronic inflammatory disorders might be implicated in the pathogenesis of nutritional anemia in preschool children (Yip and Dallman 1988; Inoue et al. 2016; Agarwal and Prchal 2009).

Reduced haemoglobin is attributed to interference in iron release from the reticuloendothelial system and a reduction in intestinal absorption of iron. Iron inadequacy in the body results in reduced iron availability for erythropoiesis, inhibition of erythropoiesis, inappropriate erythropoietin production and reduction of erythrocyte survival. Immunoactivation releases cytokines that are mainly responsible for the changes on iron metabolism, inhibition of erythropoiesis and lower erythropoietin production observed in inflammation or infection (Means and Krantz 1992). Acute infections are very frequent in childhood, especially in subjects of low socioeconomic strata of developing countries. Even mild infections that do not warrant medical consultation induce a significant decrease in Hb, Fe and TIBC (Olivares et al. 1989).

Multiple laboratory tests for detection of iron status are necessary to reach a proper diagnosis. Widespread infections in endemic areas might interfere with the laboratory tests employed for screening of anemia. In case of rampant infections, laboratory tests, according to workers (Ferguson et al. 1992; Olivares et al. 1995; Cook et al. 1993; Joosten et al. 2002; Punnonen et al. 1997), may not determine the accurate prevalence of nutritional anemia in the population in inflicted with infection. In such conditions, TfR assay can be beneficially used to assay the iron status of population. According to Olevares et al. (1999) and Jansonn et al. (1986), either acute or chronic infection does not influence the TfR estimation.

5.2 Chronic Helminthic Infection

Iron deficiency anemia (IDA) is affecting the lives of more than one billion people in the world. It is more prevalent in population afflicted with poverty, illiteracy and inaccessible medical facility. This scenario is widespread in rural areas and slum areas, especially in developing countries (Gupta 2015a, b).

Helminthic infestation is another consequence of illiteracy and ignorance towards hygiene and sanitation. Children in young age and preschool-age group are susceptible to parasitic infestation owing to poor personal hygiene practices. Children become weak, emaciated and anaemic (Odebunmi et al. 2007; Cappello 2004; Sorensen et al. 2011). Soil-transmitted helminthic infections account for high disease burden in tropical countries and constitute around 40% of the major diseases affecting children in tropical nations apart from *Plasmodium*-induced infection (Cappello 2004).

According to Odebunmi et al. (2007), susceptibility of preschool children to parasitic infections is due to immature immunity in comparison to grown-up individuals and poor hygiene and sanitary which favour the growth of endoparasites inside body of hosts. Iron deficiency anemia frequently manifests in round worm and tapeworm infections. Anemia is a secondary manifestation due to associated symptoms developed in worm-infested children. Children develop anorexia, diarrhoea and frequent episodes of upper respiratory infections.

Chronic haemorrhage is another sequel of helminth infestation in children. Hookworms result in chronic intestinal loss of blood and anemia. Authors (Hotez and Molyneux 2008) reported a blood loss of about 1/5th of a millilitre (Hotez and Molyneux 2008) per day due to hookworm infection in children.

Hookworm infection is acquired through the use of soil-contaminated edibles, utensils and/or following unhygienic practices. Children as well adults could get infected when they walk barefoot in soil contaminated with larvae. So it is a soil-transmitted helminthiasis. It is a disease thriving in tropical and subtropical regions of the world, because of the humid and hot climate in these countries that protect and nourish the parasites. Ancylostomiasis is the predominant cause of iron deficiency anemia in preschool children (CDC 2011b; Haburchak et al. 2016; de Siva et al. 2003; Hotez et al. 2004).

Hookworms are smaller than round worms. They inflict more injury to intestinal walls than causing physical obstruction. Hookworm infection causes chronic haemorrhagic anemia and loss of body iron and protein owing to persistent injury in the intestinal wall. Blood loss is not noticeable in stools. Hookworm infection affects preschool children and pregnant women, globally (Brooker et al. 1999).

It is the main cause of child morbidity resulting in impaired intellectual, cognitive and growth potential (Chan 1997; Bleakley 2003). In women, it could lead to premature birth of babies. These features are common in population in developing countries. However, in developed countries, according to Haburchak et al. (2014), hookworm infection is hardly ever fatal, but it definitely causes iron deficiency anemia in preschool children and pregnant ladies. Helminthic infection in children under age of 5 years affects the academic performance of children and has a negative consequence on the economic health of children in the future (Miguel and Kremer 2003). Worm infections are the predisposing factors for secondary infections like AIDS, malaria and/or acute upper respiratory infection (Finchman et al. 2003; Le Hesran et al. 2004; Horton 2003).

Iron deficiency anemia caused by hookworm infection affects nearly one billion people in tropical and subtropical countries (Harhay et al. 2010; Milman 2011; WHO 2005). According to Roche and layrisse (1966), hookworms injure wall of the intestine. They feed on the blood from capillaries. Adult hookworm infestation of intestinal mucosa is responsible for nearly 0.2 ml blood loss per day. According to authors (Foy and Kondi 1960), blood losses on account of hookworm infection depend on the worm load in the intestine. Authors reported a blood loss of 180 ml per day for a worm load of 834 *Ancylostoma duodenale* in the body. Authors posited that such a heavy blood loss stimulates bone marrow activity and in turn leads to enhanced iron absorption from diet by the intestinal mucosa. However, this compensatory mechanism of the body could not replenish depleted iron store of the body and, according to authors, might progress into iron deficiency anemia in children. According to Tasker (1961), blood loss varies from 2 to 90 ml and relies on worm load.

Ancylostoma duodenale harbours proximal part of small intestine, whereas ascaris is an enodoparasite of the entire small intestine small, and trichuris is a parasite in the large intestine. Infection occurs by ingestion of developed eggs of ancylostoma and trichuris (Crompton 2001). Chronic ancylostomiasis has severe repercussion on physical and mental health of preschool children. According to Yu and Shen (1990), retarded growth, anemia and malnutrition are common in children. Worm infection has been noted in neonates. Transmission occurs through suckling of breast milk. It resulted in intense haemorrhage from the intestine of neonates which could be life threatening.

Warmth and moisture hasten the development of larva in soil (Brooker et al. 2006; Brooker and Michael 2000). Hookworms secrete anticoagulant that facilitates bleeding from the gut mucosa. The feeding mechanism of hookworms is analogous to other blood-sucking parasites on human body like leeches and ticks. The authors (Cappello et al. 1993) identified anticoagulant activity in the secretory products of hookworms.

Exact chemical nature of these anticoagulants had not been ascertained. Moreover, these parasites have an effective mechanism that interferes with the host clotting system. Cappello et al. (1993) reported an anticoagulant in the secretion of adult *Ancylostoma caninum* that is capable of inhibiting human clotting factor Xa. High prevalence of infection by *Trichuris* and *Ascaris* has detinfections have beeimental consequence on the health of children.

The authors (Le et al. 2007) depicted in their study, high prevalence of Trichuris infection among school children in Vietnam. These children had almost twice the probability of anemia as per authors. But anemia was not as a consequence of iron deficiency and concluded by the authors that the association requires an intensive study to explore the correlation between *Trichuris* infection and anemia.

Another study was performed by Kaminsky et al. (2015). The authors observed more than 50% of the trichuris infection in children in age group between 2 and 12 years. Further, the authors insisted that children belonged to poor families. According to Kaminsky et al. (2015), all children in above age range were detected to have chronic infection and suffered episodes of dysentery frequently. Children had Hb concentration below 11 g/100 ml and were severely stunted (Kaminsky et al. 2015).

Parasitic infection and its sequela in the form of anemia, malnutrition and stunting are a common scenario and public health problem in the children especially in age group in between 2 and 5 years in developing nations. A study by Tsuyuoka et al. (1999) was conducted on school children in Northeast Brazil. The authors observed around 42% intestinal parasitic prevalence in children in age below 8 years. Infection by *Ascaris* predominated among intestinal parasites. Poor environmental factors like unacceptable sanitation and poor personal hygiene were associated with parasitic infection in children. The authors commented that parasitic infection was not followed by anemia; however, the nutritional status of children was severely compromised with stunting as the main negative health indicator.

In a study by Osazuwa et al. (2011) in rural areas of Edo state, Nigeria, it was found that ascariasis was the highly prevalent helminthic infection in children.

These children were malnourished. As per authors (Osazuwa et al. 2011), anemia was present in 38.6% of the children, and anemia in worm-infected children was significantly related to ascaris and hookworm infection. This descriptive study clearly delineates the interrelationship among helminthic infection, anemia and malnutrition. It is always an interplay among factors, wherein a cause resulted in an effect, which in turn precipitates the cause. Helminthic infection weakens the children as manifested in terms of stunting, underweight, wasting and anemia; hence, compromised immune response of children hastens exposure and expression of infection in the body.

Malaria and helminthic infestation are the main public health problems affecting preschool and school-aged children in sub-Saharan Africa (Adegnika and Kremsner 2012). Population residing in tropical and subtropical regions of the world is living in hot and humid conditions. In these areas, intestinal parasitic infections are common in children. *Plasmodium falciparum* infection causing malaria is another health hazard responsible for high child morbidity. Coexistence of helminth infection and malaria in children is responsible for high burden of diseases (Snow et al. 2005). *Plasmodium* parasite and soil-transmitted helminth together are the major public health problem among children (Brooker et al. 2012; Pullan et al. 2011; Mazigo et al. 2010).

Malarial fever in sub-Saharan Africa is associated with high mortality in children. Helminth infection is not as fatal as malaria but has high morbidity in children (Hotez and Kamath 2009). Nematodes, cestodes and trematodes contribute to helminth infection in children. However, distribution of helminth infection is largely determined by immunity of host and environmental factors like sanitation, hygiene, drinking water and hand wash practice (Brooker and Clements 2009).

Poverty, lack of medical facility, defaecation in the open, poor hygiene, contaminated drinking water and malnutrition individually and collectively affect the prevalence of helminth infection in children (Booth 2006). *Plasmodium falciparum* infection and helminth infection have common predisposing factors (Adedoja et al. 2015; Brooker and Clements 2009; Booth 2006; Ouf et al. 2012).

Co-endemic existence of *Plasmodium* parasitic infection and soil-transmitted helminthic infection in children, especially in developing countries, is inundated with similar environmental and geographical conditions and is associated with anemia, diarrhoea and malnutrition (Naing et al. 2013; Snow et al. 2005). Further, these infections have similar distribution pattern in tropical and subtropical regions of the world. Widespread consequence of this coinfection is the iron deficiency and iron deficiency anemia in preschool and school-aged children (Naing et al. 2013).

Iron deficiency anemia as a coexistence of malarial and of helminth infections in children is, additionally, worsened when dietary intake is inadequate and poor in quality with low iron and folic acid as in staple foods like rice, maize and cassava.

Worms feed on the digested food present in the intestinal lumen of children, like tapeworm and also injure and feed on the capillary blood, as hookworm. Weight of children is reduced. Absorption of minerals is affected. Metabolism is perturbed. Child growth and development are severely obstructed (Papier et al. 2014).

Hookworm infection is responsible for persistent blood loss from the body of children and critically ruins the physical, mental and cognitive health of preschool children (Miller 1966). According to Lwambo et al. (1992), a number of worms are linked to haemoglobin concentration in patients. Large number of endoparasites in lumen of the small intestine can deviate the normal absorptive process and, hence, are responsible for lowering of Hb conc. below 11 g/dl.

However, the authors suggest that the generalization is not absolutely true. The type of species of worm is also accountable for blood loss from injured intestinal mucosa. *Ancylostoma duodenale* is more injurious than *N. americanus*, and hence, the blood loss is comparatively more in case of *A. duodenale* (Hotez et al. 2004; Albonica et al. 1998; Miller 1966).

Soil-transmitted helminthic infection is responsible for high disease burden in children and pregnant women. Hookworm infection is a potential threat to the health of children and women in pregnancy. There are enough evidences provided by workers to support the hypothesis. According to the work by Stolzfus et al. (1997b), Navitsky et al. (1998) and Christian et al. (2004) in Africa and Asia, it was proved that large proportion of moderate to severe anemia was prevalent in pregnant women and was caused by hookworm infestation. Intervention studies by Brooker et al. (1999) posited that administration of antihelminthic drugs during pregnancy has profound positive effect on health of mothers and children.

The attempt resulted in increase of Hb level of pregnant women and birth weight of infants. According to Bethony et al. (2006), hookworm infection is responsible for severe anemia in preschool children. The anemia resembles to that caused by iron deficiency. Other clinical manifestations of hookworm infection are similar to that of iron deficiency anemia in children.

Despite health hazards of hookworm infection in children, a little is known about the mechanism which hookworms undertake to inhabit in intestine and modify the local and systemic immunity of the host. In the body of the host, helminth exposure-related immunity does not develop. Hookworm infection should be crucially reckoned for the health of preschool children. As stated by Loukas et al. (2005), vaccination of children in preschool and school age against the hookworms infection is the novel approach to end the deadly menace of this soil-transmitted endoparasite. Further, they stated that gene sequencing has been done for more than 10,000 hookworms which has provided cDNA sequences. It would be possible to develop an efficient and cost-effective vaccine against hookworm infections in the near future (Diemart et al. 2008; Loukas et al. 2006).

Soil-transmitted helminthic infestation is fatal in terms of malnutrition, iron deficiency anemia, poor cognitive development, and compromised immunity. Infection is rarely life threatening. Moreover, burden of disease hinges on the chronic blood loss from the intestinal wall leading to iron deficiency anemia. Heavy hookworm population in the alimentary canal of a child depletes the iron store of the body and according to Hotez et al. (2005) is called as hookworm disease.

Further, Stoltzfus et al. (2004a) conducted a placebo-controlled, randomized and double-blind trial using 10 mg of iron per day and (Stoltzfus et al. 2004b) antihelminthic drug, mebendazole in a dose of 500 mg (Stoltzfus et al. 2004a) once in

3 months time period over 459 Zanzibari children in age group of 6–71 months. It was concluded (Stoltzfus et al. 2004b) that iron alone could not improve anaemic condition of children. However, after administration of antihelminthic drug, overall improvement in wasting and anemia was observed. According to Stoltzfus et al. (2004a), helminthic infection progressed into malnutrition and anemia and affected the protein metabolism in the body of children.

Ancylostoma duodenale, adult worm, secretes biologically active peptides that enable worm to feed on blood. One such anticoagulant is the serine protease inhibitor. It inhibits the clotting factor Xa. Adult hookworms secrete pharmacologically active peptides that facilitate blood feeding. According to Harrison et al. (2002), most potent are novel serine protease inhibitors that anticoagulate blood by inhibition of factor Xa.

Another anticoagulant (Lee and Vlasuk 2003) secreted by hookworm is nematode anticoagulant protein c2 (NAPc2). This factor inhibits factor VIIa/tissue factor. It is a powerful factor. It has high affinity to activated factor X. *Ancylostoma* secretes a factor, according to Del Valle et al. (2003), that is responsible for inhibition of platelet aggregation. This platelet inhibitor attaches to surface receptors, glycoprotein Iib/IIIa and GPIa/Iia.

Hookworm disease manifests as iron deficiency anemia and protein energy malnutrition owing to heavy hookworm load in the body. Iron deficiency anemia associated with hookworm disease is microcytic and hypochromic (Beaver et al. 1984).

Cutting organs of hookworm help it invade deeper into the mucosa and submucosa of the intestine. Adult worm penetrates its anterior end of the body deeper into the tissues of the mucosa and sucks blood from capillaries. Anterior glands of parasites secrete protease and anticoagulants which the parasites release to digest the host body wall. Endoparasites are endowed with teeth in *A. duodenale* and cutting plates in case of *N. americanus* parasite. These armed structures of parasite enable it to anchor itself to host body wall (Loukas and Prociv 2001; Cappello et al. 1995; Atoda et al. 1991). It damages capillaries. Anticoagulants secreted by hookworm maintain oozing of blood at the point of worm attachment on the wall. Liberated blood enters the alimentary canal of hookworm. Red blood cells are broken down inside its alimentary canal. Released Hb is digested with the help of metallohaemoglobinases. A few RBC remain intact and pass out of the gut of parasite. The attachment of hookworms' cutting organs to the intestinal mucosa and submucosa and the subsequent rupture of intestinal capillaries and arterioles cause blood loss. The secretion of factor Xa and VIIa/TF inhibitors and antiplatelet agents by the parasite helps to maintain continuous oozing of blood at the hookworm attachment site and the free flow of blood through the parasite alimentary canal (Williamson et al. 2003; Ledizet et al. 2005; Loukas and Prociv 2001; Cappello et al. 1995; Atoda et al. 1991).

Blood loss in a *Ancylostoma duodenale* infection is the chief and long-term morbidity in the body of preschool children. Chronic and persistence oozing of minute droplets of blood for prolonged periods might terminate into iron deficiency anemia (Albonico et al. 1998).

Chronic blood loss causes haemorrhagic anemia, and it resembles to that of pernicious anemia Suarez (1931). Hookworm population (Stoltzfus et al. 1997a) of 40–160 worms in the gut is responsible for iron deficiency anemia. It further depends on the iron store of the host body (Stoltzfus et al. 1997b).

Trichuris trichiura infection is heavy and alarming in rural population in developing world affecting about 1049 million population (Ramdath et al. 1995). High burden of infection by *Trichuris* results into stunting, anemia and dysentery in children. High-intensity infection causes lesions in colon, blood loss and malnutrition. These manifestations are the consequence of chronic infestation of worm in the body (Stephensom et al. 2000).

Ascaris lumbricoides- and *T. trichiura*-induced infections are highly prevalent in children in age group between 10 and 15 years. Intensity of infection declines with advancing age. According to Galvani (2005), acquired immunity towards same antigen from same species in adolescence and adulthood is the main factor that limits the worm infection.

Soil-transmitted helminths like ascaris, trichuris, and hookworm are the frequent worms that either at least one or three together harbour the small intestine of poor children, especially in developing countries. These endoparasites are the source of potential injury to the small intestine and are responsible for impaired physical, mental, intellectual and cognitive development (Bethony et al. 2002). These endoparasites have the ability to regulate the host immune response. Helminths survive in lymphatics, GIT or blood stream. These parasites have large body and complex life cycle in the body of host. Parasites pass through different developmental stages. Endoparasites can prevent host immune attack. They create a niche for their feeding, reproduction and survival (Maizels et al. 2014).

Clear pathophysiology regarding the initiation of immune response in soil-transmitted helminth infection is lacking (Taylor et al. 2012). Parasite and host interact which determines the survival of parasites in the host body for a long period. Parasites modulate host immune response by suppression, whereas host immunity that develops over sustained exposure to parasitic antigens limits the spread of parasite infection in the host body (McSorley and Maizels 2012). Soil-transmitted helminth infection is involved in multiple of physiological processes like eosinophil-induced lysis of parasitic larva, synthesis of immunoglobulin E, degranulation of mast cells and activation of mucous-secreting cells in GIT (Bradley and Jackson 2004). Despite the T helper −2 cell response, worms thrive in the body of host for long time. Possibly, a balance is obtained between the parasitism and immune response of the host.

This induces the activation of T helper-2 cells (Yazdanbakhsh et al. 2001), which in turn results into production of cytokines and activation of mast cells, eosinophils, macrophages (Ouyang et al. 2011; Gause et al. 2013). Prominently, IL-10 and transforming growth factor-beta are associated with diminished responsiveness of the host body to helminthic infection. It is responsible for sustainability of helminths in the host body. Exact mechanism is not understood (Maizels et al. 2014). Other theory holds that secretions produced by worms enable their survival in the host body. Natural killer cell binding protein is secreted by adult *N. americanus*. This secretion binds particularly with NK cells of host. This complex induces the synthesis of interferon-gamma. This macromolecule limits the activation of T helper-2 cells and maintains the chronicity of worms (Hsieh et al. 2004).

Ascaris secretes pepsin inhibitory substance through its body wall. This secretion protects the worms against the proteolytic enzymes in the gut of the host. It is believed to secrete phosphorylcholine molecule attached with glycoprotein or glycolipids (Lochnit et al. 2000). This complex impair the cell signalling pathway in the host body (Deehan et al. 2002). *Trichuris trichiura* releases a protein which forms pores in the phospholipid layer of the host. Ion-conducting pores in lipid bilayers (Drake et al. 1994), thus enabling parasite to invade GIT of host.

5.3 *Helicobacter pylori* Infection

Helicobacter pylori is the gram-negative, spiral, microaerophilic bacilli which were discovered by Warren and Marshall (1983) as the pathogen for gastritis and peptic ulcer.

It is the pathogen in the gut of human being, but it has been isolated from the stomach only portraying the fact that it is highly adherent to the mucosa of the stomach with a mechanism which has not been explained properly. Great majority of world population is infected with *H. pylori* (Yamaoka 2008).

Its prevalence is higher in developing nations in comparison to developed nations owing to deteriorated hygiene and sanitation and poor dietary intake. Generally, it infects the stomach in early childhood. Its infection occurs by faeco–oral route (Brown 2000; Sabbi et al. 2005; Razak and Ragab 2014). Human being is the reservoir of *H. pylori*. So poor hygiene, improper human excreta disposal, use of contaminated drinking water and food preparation and serving without hand wash are the main reasons for its transmission through faeco–oral route (Brown 2000). Large population of preschool children in slums and villages become the high-risk population for its transmission (Parente et al. 2006; Bhuiyan et al. 2009; McColl et al. 1998; Sarkar et al. 1997).

Oral to oral route is another means of its spread in human beings. This mode of infection becomes the sole ground for its spread in children in institutions and crowded places. Presence of *H. pylori* DNA in the vomitus and saliva conforms this assumption.

So it can be said with certainty that almost all factors of poverty and malnutrition in children are responsible for the acquisition and spread of *H. pylori* infection. Knowledge about the epidemiology of *H. pylori* would help to understand its mode of transmission. Also study delineates population who has higher susceptibility towards its infection and consequences.

H. pylori remains attached to the gastric mucosa. It has high urease activity and hence produces ammonia in the gastric lumen. It neutralizes the gastric acidity. *H. pylori* decreases gastric secretion and results in a state of hypochlorhydria. It has been proposed to be a cause of iron deficiency anemia in preschool children (Windle et al.2007; Seo et al. 2002; Kostaki et al. 2003; Marignani et al. 1997; Yip et al. 1997; Gold et al. 2001; Bahu Mda et al. 2003; El-omar et al. 1997).

Hypochlorhydria is the cause of accentuated enteric infection. It is attributed to decreased antimicrobial activity in the stomach lumen. Higher chances of diarrhoea, nutritional anemia and malnutrition are the commonest manifestations in children (Dale et al. 1998; Passaro et al. 2001).

Helicobacter pylori could be responsible for iron deficiency anemia in children, despite the absence of peptic ulcer. Researchers studied the aetiology of refractory iron deficiency anemia in children who had concomitant infection by *H. pylori*. It was observed by authors that Hb concentration normalized once the infection was treated successfully with anti-*Helicobacter pylori* drugs. So it can be concluded that *H. pylori*-induced gastritis was the chief cause for iron deficiency anemia (Barabino et al. 1999; Annibale et al. 1999; Choe et al. 1999; Ashorn et al. 2001; Capurso et al. 1999; Akcam et al. 2007; Barabino et al. 1999; Annibale et al. 1999; Choe et al. 1999; Ashorn et al. 2001).

Iron oxidation and absorption are determined by gastric acidity. Infection by *Helicobacter pylori* raises the gastric pH. It might impair the absorption of dietary iron in the gut. It was proved by Annibale et al. (1999) and Capurso et al. (2001) that gastric acidity was decreased to pH 7 in patients who suffered from *H. pylori* gastritis and associated iron deficiency anemia. Hypochlorhydria-induced reduction in the absorption of iron in gut is responsible for IDA in *Helicobacter pylori* infection.

Cobalamin is attached to protein in food. The pH of the gastric lumen decides the proteolysis and release of cobalamin in the stomach lumen. According to Stopeck (2000) and Kaptan et al. (2000), infection caused by *H. pylori* disturbs the normal pH of stomach in causing hypochlorhydria. It might be a reason in the failure of dietary cobalamin absorption and its deficiency in the body and, ultimately, appearance of clinical signs of pernicious anemia in children.

Conversely, authors (Cohen et al. 2000) depicted a slight opposite observation on the *H. pylori* infection and iron absorption. The authors said that hypochlorhydria is not always responsible for failure of cobalamin absorption. Malabsorption can be managed by appropriate antibiotic therapy.

Probably, prolonged use of antacids, proton pump inhibitors and/or H2 receptors antagonist in patients of *H. pyloric* infection can induce hypochlorhydria and may be the cause of dietary cobalamin malabsorption (Howden 2000). Infection from *H. pylori* is responsible for malnutrition, anemia and impaired development in children (Franceschi et al. 2014). Its infection creates a state of hypochlorhydria in gastric lumen. Physiological functions of gastric secretion are affected. It facilitates the entry of other pathogens in gut.

It culminates into a vicious cycle of impaired iron absorption, diarrhoea and halting of growth and development and leading to stunted growth. According to authors (Franceschi et al. 2014), early acquisition of *H. pylori* infection in children might be responsible for growth retardation.

Compromised environmental proponents, further, accentuate the negative synergistic effect of coexistence of diarrhoea and micronutrient deficiency in children. So the sanitation, hygiene and water quality in developing countries determine the prevalence of *H. pylori* infection in children and associated consequences on their health.

Irrespective of the economic and industrialization status of countries, about one half of the total population of world is inflicted with *H. pylori* infection. Its onset of infection is quite early in life around the first 5 years of life which is a significant threat to the health of children (Suerbaum and Michetti 2002). However, according to Megraud et al. (1989), *H pylori* infection is acquired, in developing countries by the age up to 10 years across most of the world population.

Again, regarding prevalence rate, according to Rothenbacher et al. (1998), it varies between 1.2 and 12.2% in developed countries (Shaman et al. 2009). Authors (Rothenbacher et al. 1998) conducted a cross-sectional study on preschool children(5–8 year) in Southern Germany to find out manifestation and prevalence of *H. pylori* infection.

The authors (Rothenbacher et al. 1998) observed a prevalence of 13.2% of *H. pylori* infection in children. In developing countries, the prevalence, comparatively, is much higher. Prevalence of seropositive cases of *H. pylori* infection as observed by authors (Sullvian et al. 1988) in infants was 15% in age under 20 months. Further, authors (Sullvian et al. 1988) found an increase in prevalence to 46% in age group between 40 and 60 months. They compared the seropositive cases of *H. pylori* infection of infants with age-matched healthy controls. Around 50% (Sullvian et al. 1990) of the seropositive cases suffered from persistent diarrhoea and malnutrition.

The authors (Parsonnet et al. 1992) studied a cohort of epidemiologist to find out the associated symptoms and prevalence of *H. pylori* infection. They (Parsonnet et al. 1992) tested the serum samples of the participants for presence of antibodies to *H. pylori*. For the sake of comparison, authors used the stored serum samples (Parsonnet et al. 1992) of participants and on the other side, a fresh serum samples were tested for comparison. The authors observed the decline of infection between a time interval of 1969–1988 in participants (Parsonnet et al. 1992). *H. pylori* infection, generally, persists for decades having had a low rate of complete eradication.

Some cases of protein losing enteropathy have been reported by researchers in children. The disorder is manifested by a loss of serum protein from the alimentary canal owing to a defect in the tight junctions of epithelial cells of the small intestine. Frequent causes of PLE is the association of inflammatory bowel disease, environmental enteropathy and of endoparasites in gut and rarely infection by *H. pylori* (Seidman 1995; Gleason 1993; Landzberg and Pochapin 2001).

Studies by (Urganci et al. 2002; Cohen et al. 1991; Hill et al. 1987) observed a relation between *H. pylori* infection in children and PLE. Probably, the manifestation is a result of intense inflammation in the mucosa of stomach which disrupts tight junctions and increases the permeability of gastric epithelium. *H. pylori* infection is responsible for the inflammation of the gastric mucosa in children. Colonization of *H. pylori* in gastric mucosa is followed by infiltration of inflammatory cells in the mucosa of the stomach in preschool children (Uc and Chong 2002).

Helicobacter pylori infection is responsible for gastric ulcer and duodenal ulcer among preschool children (Huang et al. 1999).

5.4 Severe Malarial Anemia in Children

5.4.1 Overview

Severe malarial anemia is the malarial infliction in childhood characterized by (WHO 2004) an intense reduction in HB concentrations <5 g/dl associated with *P. falciparum*-induced parasitaemia, according to WHO (2004), (>10,000 parasites/μl) with or without cerebral malaria in children.

Severe malarial anemia is the major cause of child mortality and morbidity in child population under 5 years of age residing in endemic areas (Scott et al. 2014). Worldwide mortality of around 20% in under 5 aged children is caused by severe malarial anemia (WHO 2013; UN MDG 2012; Kassabaum et al. 2014; Stolzfus et al. 2004b).

Plasmodium is the causative agent of malaria in human population (Collins and Jeffery 2007; Madabushi et al. 2005). It is a unicellular sporozoan and is a cellular parasite. *Plasmodium* needs two hosts, namely, female anopheles mosquitoes and man to complete its life cycle (CDC 2016). Sporozoites are spore-like infective organisms, a stage in life cycle of plasmodium, remain lodged inside the salivary glands of female mosquitoes and are injected into blood circulation by mosquitos during their attempt to suck blood (Prudencio et al. 2006).

Female mosquitos, though harbour sporozoites, are not affected by *Plasmodium*. Sporozoites migrate to the liver, invade hepatocytes and commence a new stage of parasitic life inside the liver (NIH 2014). Haploid forms, called merozoites, are produced inside liver cells. These are haploid forms of plasmodium which move into the blood stream and invade RBCs (Karapelou and Despommier 1987).

Merozoites grow and proliferate inside erythrocytes and are called erythrocytic stage of plasmodium, wherein infected erythrocytes rupture and release toxin, haemozoin, which is phagocytosed by macrophages and monocytes and induces signs and symptoms of malaria (Casals-Pascual et al. 2006; Skorokhod et al. 2010; Barrera et al. 2011).

A few merozoites from schizont are transformed into gametocytes inside the human body, which are responsible for transmission of parasitic forms into the body of female *Anopheles* mosquitos, while other merozoites continue with asexual phase of *Plasmodium* (Bruce et al. 1990; Smith et al. 2000; Silvestrini et al. 2000).

Female *Anopheles* mosquitoes are the vector of *Plasmodium* parasites (CDC 2015) and are wholly responsible for transmission of malarial endoparasites and spread of malaria. When a female mosquito bites an infected person, it ingests a large number of gametocytes. Within its midgut, gametocytes are released and undergo maturation into male and female gametes ready for sexual phase of its life cycle (Bousema et al. 2014; Collins and Jeffery 2007; Wang et al.). Gametes fuse and form diploid zygote and get transformed into an ookinete. It is highly motile and active form and penetrates inside the walls of the stomach of mosquito and further gets converted into encysted form, called as oocyst, and it grows and proliferates within a female mosquito to form thousands of diploid sporozoites. The outer wall of oocyst ruptures and releases a stream of sporozoites. Mature diploid sporozoites enter haemolymph and, ultimately, invade and lodge the salivary glands. Organisms are ready for their transmission into the blood stream of a healthy person (Vanderberg and Rhodin 1967; Sinden 2002; Wang et al. 2005; Bousema et al. 2014; Collins and Jeffery 2007).

According to WHO (2008) report, malaria is widespread in Asian and African countries. As per (WHO 2008), around 247 million population of the world is infected with malaria round the year. African children (WHO 2008) under the age of 5 years are highly vulnerable to malarial infection and is accompanied by high mortality among one million population of children.

Generally, four species of *Plasmodium*, namely, *P. falciparum*, *P. ovale*, *P. malariae* and *P. vivax* are the causative sporozoans for malarial infection in humans. According to Sabbatani et al. (2010), a new species, *P. knowlesi*, has been labelled to cause malaria in humans in remote cases.

According to a report by WHO and UNICEF (2008), *P. falciparum* is the most virulent species of *Plasmodium* and is the cause of fatal malaria (White 2008; Snow et al. 2005). It has the highest prevalence in African regions. Sub-Saharan Africa faces a very high mortality and morbidity as a result of *P. falciparum*-induced malarial fever and is strongly associated with severe anemia (Genton et al. 2008; Eibach et al.2012; Ketema and Bacha 2013).

The other three species of *Plasmodium*, namely, *P. vivax*, *P. ovale* and *P. malariae*, are responsible for less severe form of malaria whose prevalence is around 10% worldwide. Severe malarial anemia (SMA) in children has been defined by World Health Organization (WHO 2000). Accordingly, SMA is characterized by a sharp fall in Hb concentration below 5.0 g/dL, according to WHO (2000), in a child in the presence of any density of parasitaemia.

Severe malarial anemia is a serious and potential threat to the life of children under 5 and is highly accountable for high mortality and morbidity in children. SMA is a key public health issue, especially in developing countries. According to English et al. (2004), SMA is accountable for around 3–46% of mortality in children admitted to hospitals.

SMA is contributing to high disease burden in children under age of 5 years in African countries. Brabin et al. (2001a) adopted a cross-sectional study approach to find out the relation between anemia and child mortality. Research depicted that SMA mortality is much higher in comparison to iron deficiency anemia. Most of the data points were based on hospitalized patients and were those who had anemia due to severe malarial infection. Intensive efforts are needed to ameliorate the prevalence of SMA and iron deficiency anemia in children in malarious regions of the world.

The study by Reyburn et al. (2005) was conducted over hospital-based patients located in areas with varying rate of malarial transmission in population in north-eastern Tanzania. Reyburn et al. (2005) observed an inverse relation between transmission intensity of malaria and age of children. In high transmission areas, median age of patients was 1 year (Reyburn et al. 2005), whereas age of patients increased by a decrease in transmission severity of malaria.

5.4.2 Aetiopathogenesis of Severe Malarial Anemia in Children

Enormous factors are accountable for the pathogenesis of anemia in children in malaria endemic regions. Due to a number of biochemical and metabolic changes that the parasites induce in the body, it is difficult to isolate a single factor for severe malarial anemia in children. Transmission intensity of malaria and genetic traits and dysregulation of innate immune response influence other causes of malaria-induced anemia like suppression of erythropoiesis, destruction of uninfected erythrocytes, bone marrow suppression and synthesis of inflammatory mediators.

Multiple factors contribute to aetiopathogenesis of SMA. Erythrocytes which are invaded by parasites undergo sequestration in the body. Innate immune system reacts to the pathogens and their toxins, and hence, a number of pharmacologically active substances are synthesized in circulation, which definitely influence erythropoiesis in the bone marrow. Further, the host becomes frail, partly, as a consequence of decreased intake of food, which is again negatively correlated with the erythropoiesis. Nevertheless, factors act in totality to produce clinical and biochemical manifestation of SMA in children.

A reduction in Hb concentration is the key clinical sign which intensifies the mortality rate of children under 5 years of age, especially in malaria endemic regions of the world, where the transmission intensity of malaria is very high. According to Menendez et al. (1997), *P. falciparum*-induced malaria and iron deficiency anemia are chiefly responsible for rising mortality and morbidity among preschool children, especially in malaria endemic regions. Data is available to depict the pathogenesis of severe malarial anemia. However, scarce knowledge is available to show the effect of chemoprophylaxis on occurrence of new cases of malaria and severe malarial anemia in children in endemic regions. A placebo-controlled clinical trial after randomization (Menendez et al. 1997) was conducted to find out efficacy of antimalarial drugs for the prevention of malaria in infants and role of iron supplements on malaria susceptibility and control of anemia in infants. It was observed by Menendez et al. (1997) that oral administration of iron supplements reduced the frequency of anemia in infants; however, it had no impact on the incidence of malaria in infants. Menendez et al. (1997) further observed that antimalarial regimen reduced the incidence of malarial episodes and anemia in infants.

Another placebo-controlled clinical trial was conducted by Schellenberg et al. (2001). Both the groups of researchers clearly posit that malaria control programs can effectively reduce the occurrence of malarial episodes and associated anemia in children in malaria endemic regions. Hence, it can be concluded that malaria caused by *Plasmodium* is the root cause of severe anemia in children.

Apart from other causes for disturbed erythropoiesis, impairment of innate immunity is the most important factor underlying defective erythroid response in children suffering from SMA. Additionally, malarial parasites release endotoxin, called haemozoin (Hz). The pigment is engulfed by macrophages and neutrophils and is the key substance for initiation of synthesis of vast variety of chemical compounds which induce drastic changes in the body including impaired innate immune response.

Plasmodium falciparum-derived haemozoin (PfHz) is central to suppression of erythropoiesis in the bone marrow by means of its potential to persuade imbalance between pro-inflammatory and anti-inflammatory compounds in the body.

P. falciparum proliferates very rapidly within body tissues. Erythrocytes infected with parasites produce a protein called as *Plasmodium falciparum* erythrocyte membrane protein 1 (PfEMP1), as described by Brown and Brown (1965), first of all.

Infected RBCs and PfEMP-1 are the potential risk factor for the host. The parasite protein is exported to the outer surface of infected erythrocytes. PfEMP-1 has the ability to bind to (Senczuk et al. 2001) host body tissues. It can remarkably bind to intercellular adhesion molecule-1 (ICAM-1) (Carlson et al. 1988) and chondroitin sulphate A (NLM 2016) present on the vascular endothelium of host.

ICAM-1 is a glycoprotein which is encoded by ICAm 1 gene in humans. It is also called as CD54 (Carlson et al. 1988; Katz et al. 1985).

ICAM-1 protein is a member of a diverse group of proteins called as immunoglobulin superfamily, holding a pivotal role in cell recognition and cell–cell adhesion. ICAM-1 is located on the endothelium and immune system cells. It is a transmembrane protein possessing extracellular as well as cytoplasmic domains. Amino terminal is located on the cell surface, whereas carboxy terminal lies inside the cytoplasm (Bella et al. 1998).

ICAM-1 protein is present in low concentration under normal condition on the cell membranes of endothelial cells and lymphocytes and macrophages. Pro-inflammatory substances enhance its synthesis, thereby raises its concentration under inflammation and infection. Probably, interleukin-1(IL-1) (Pelletier and Girard 2007) and tumour necrosis factor-alpha (TNF) (Xu and Zhong 1997) induce expression of ICAM-1. Recently, its central role in the pathogenesis of malaria has been fully explained.

PfEMP-1 has high affinity for ICAM-1 protein located on the surface of endothelial cells. So it is responsible for cytoadhesion. Infected erythrocytes, expressing PfEMP-1, bind with endothelial cells and, consequently, undergo sequestration in the high endothelial venules. Brain microvasculature also captures the infected RBCs, which is the cause of cerebral malaria. *P. falciparum*'s ability to proliferate tremendously and infect fresh RBCs is the basis for high parasitaemia, impairment of immune system of host and frequent spells of disease relapse, which altogether holds for the chronicity of malarial infection (Hbiid 2005).

According to Urquhart (1994), severity and fatality of malarial infection by *P. falciparum* were, previously, thought to be associated with the virulence of parasite. The author insisted that exaggerated innate immune response of the host is equally accountable for the pathogenesis of severe malarial anemia, and it is uncertain whether the antigenic load given by the parasites determines the intensity of immune response of host. The author said that it is clear that the parasitic toxin, mainly, haemozoin, induces the synthesis of cytokines in the body which have dual effect, partly, protective against the parasites that limit the spread of infection and, partly, deleterious on the body of host owing to its impact on the innate response of host.

Overall, it is well established that pro-inflammatory and inflammatory substances produced in response to parasitic toxins modulate the innate immune response of host. According to Weidanz (1982), In the early phase of infection, inflammatory substances activate the innate immunity of the host. However, in the later stage of infection, immune response is suppressed and/or has a markedly reduced response intensity. According to Schwarzer et al. (1998) and Zuckerman (1977), the prominent feature is probably a decline in the number of T cells in blood circulation. Nevertheless, Ho et al. (1986) observed that T helper and T suppressor cell population decreased in blood, and their number was restored after the complete removal of parasites from circulation. The authors said that the feature is an adaptive mechanism of parasite by which it persists in blood circulation by suppressing the immunity of host. *P. falciparum* releases toxic metabolite, mainly haemozoin and others such as glycosylphosphatidylinositol into the circulation.

In a study by Rodriguez-Morales et al. (2006) over aetiology of anemia and thrombocytopenia due to *Plasmodium vivax* infection among preschool children, they observed that patients developed anemia and thrombocytopenia along with fever and shivering and the patients needed hospitalization (Zuckerman 1977; WHO 2000, 2004, 2008; CDC 2016; Sabbatani et al. 2010; WHO and UNICEF 2008; English et al. 2014; Barbin et al. 2001a, b; Reyburn et al. 2005).

Multicentre studies conducted by Taylor et al. (2012) indicated that severe malarial anemia involves around 7.5–34% (Taylor et al. 2012) of the African children. It is associated with case fatality rate of 8.4% (Taylor et al. 2012) in the affected children population. Haemozoin pigment is the key factor for the stimulation of innate immune response. *Plasmodium* feeds upon the haemoglobin of host and derives nutrients for its survival. During the process, parasite degrades the host Hb into haem (Pandey and Tekwani 1996; Menendez et al. 1997; Schellenberg et al. 2001; Senczuk et al. 2001; Hviid 2005; Rodriguez-Morales et al. 2006; Shio et al. 2010; Romani et al. 1997; Kachroo 2004; Fitch et al. 1983).

According to Francis et al. (1997), host Hb is digested inside the digestive vacuole of parasite.

According to Slater et al. (1991), as more of the Hb is catabolized, concentration of haem accumulates within the parasite. It is toxic to the parasite. Nonetheless, parasite can neither excrete haem nor can it convert into free iron and pyrrole rings. In due course of time, free haem builds up inside parasite body and is converted into an insoluble, brown-coloured crystalline compound. Authors studied the chemical structure of haemozoin and posited that it is made up of polymer of haems molecules.

According to Rosenthal and Meshnick (1996), *P. falciparum* digests around 80% of the host haemoglobin during its asexual phase inside the erythrocytes. According to Fitch et al. (1983), its proteolysis results into formation of monomeric alpha-haematin. Alpha haematin is a pro-oxidant and can generate reactive oxygen species inside erythrocytes. As per authors, haematin can disrupt the cell membrane of erythrocytes of host.

Under condition of stress and oxidative damage, organisms resort to the process of biocrystallization, wherein biomacromolecules inside their body are converted into insoluble crystals, a unique adaptive response to sequester the toxic compounds (Kachroo 2004).

Plasmodium converts alpha-haematin into beta-haematin insoluble crystals, a distinctive process of confining a undesirable substance in the body (Hempelmann 2007).

According to Pagola et al. (2000), clear structure of haemazoin produced by Plasmodium is still uncertain. It is a assumed that haemozoin structure resembles to that of a synthetic compound beta-haematin. In *Plasmodium*, according to Sullivan (2002), haemazoin crystals accumulate in food vacuoles. Each one is made up of around 80,000 haem moieties.

Infected RBCs are captured by monocytes, polymorphonuclear cells (PMN), macrophages and dendritic cells. Infected erythrocytes burst inside these immune cells and release haemozoin along with other contents of *Plasmodium*. According to Shio et al. (2010), monocytes and macrophages are the most prominent immune cells that can internalize the haemozoin. Further, Shio et al. (2010) described that around 30% of the total cell volume of monocytes may be occupied by haemozoin, and the malarial pigment may remain there for long periods in time. Haemazoin is produced during malarial parasite infection by any one of the four species. However, amount of haemazoin is variable depending on the type of *Plasmodium* species. According to Oliver et al. (2014), concentration of haemazoin in the host body during cerebral malaria caused by *P. falciparum* is much higher in comparison to amount of haemazoin that was produced by other species of parasite, namely, *P. malaria*, *P. vivax* and *P. ovale*.

Haemozoin concentration within monocytes and neutrophils can be a valuable tool to assess the prognosis of malarial infection. According to Nguyen et al. (1995), haemazoin concentration that is released into the blood circulation is a marker of the parasitic burden in *P. falciparum* malaria. The authors observed that malarial fatality in patients was associated with high count of peripheral neutrophils containing haemazoin. They observed remarkably low neutrophil count in those patients who were infected but survived the fatal malarial attack.

Furthermore, in another study conducted by Amodu et al. (1998), where it was tried to observe whether percentage of leucocytes containing haemazoin was related to the severity of malaria in children in age group between 6 months and 14 years (Amodu et al. 1998). Severity of malaria was assessed by Amodu et al. (1998)

through categorization into four groups, namely, no malaria, asymptomatic malaria, mild malaria and cerebral malaria. Amodu et al. (1998) detected the highest percentage of neutrophils having malarial pigment in peripheral blood in children who were suffering from cerebral malaria. However, haemazoin containing neutrophil count was least in children (Amodu et al. 1998) who suffered from 'no malaria'. Therefore, peripheral neutrophil count harbouring haemazoin may be a valuable tool in assessing the severity of malaria in comparison to peripheral parasitic count (Amodu et al. 1998).

Haemazoin can induce innate immune response in children through its affinity to toll-like receptors. These are protein molecules which are located on cell membranes of immune cells like macrophages and dendritic and natural killer cells. These receptors play a central role in the recognition of microbe-generated antigenic substances in the body and initiation of innate immune response. Toll-like receptors, additionally, activate T cells and help in the development of adaptive immunity in host. Toll-like receptors have been observed in the lymphocytes which can induce the synthesis of cytokines (Chiron et al. 2008; Medzhitov et al. 1997).

Haemozoin binds with the Toll-like receptors located on macrophages and dendritic cells, and the complex, thereafter, transduces the signal for activation of adapter protein. Within cytosol, other proteins in sequence are activated by phosphorylation process. Sequestered parasites inside the infected erythrocytes are phagocytosed.

Irrefutable evidence is provided by the work of Coban et al. (2005). Coban et al. (2005) demonstrated in their work that haemazoin from *P. falciparum* infection can bind with toll-like receptor 9 and can initiate the production of cytokines (Coban et al. 2005). They also worked on the capability of synthetic haemazoin, in vivo, to induce cytokine production through its binding with toll-like receptor 9. Their work depicts that haemazoin can stimulate innate immune response through interaction with TLRs.

5.4.2.1 Role of Pro-inflammatory Biomacromolecules

In an intial phase of parasitic infection, pro-inflammatory biomolecules like tumour necrosis factor-alpha, interleukin-12 and interferon-γ are produced. Each molecule has its unique role in the suppression of parasitic proliferation and to limit the spread of infection in the host. Innate immune cells participate in response to these pro-inflammatory substances. Innate response is short lived and is devoid of generation of memory for the future attack of same microbes (Romani et al. 1997; Fearson and Locksley 1996; Clark 1978).

Interleukin-12 is a heterodimer, composed of two dissimilar monomeric units which are covalently linked. It is synthesized by the dendritic cells, neutrophils and macrophages after the invasion of a virulent microbe in the host body (Trinchieri 1998). One monomeric unit of IL-12 is a light chain of 35 kDa, labelled as p35, whereas other monomeric unit is a heavy chain of 40 kDa, designated as p40. These chains are linked by disulphide linkage (Podlaski et al. 1992). IL-12 is necessary for the development of T helper 1 cells. According to Locksley (1993), IL-12 plays a significant function in induction of interferon-gamma. This process is initiated by the presence of intracellular parasites and is carried out without the involvement of T lymphocytes.

According to Takenaka et al. (1997), synthesis of IL-12 by macrophages and dendritic cells is initiated by two different pathways. Pathogenic microbes and/or their toxins induce the synthesis of IL-12 (Takenaka et al. 1997), and this pathophysiology is independent of the participation of T cells. Other pathway involves the activation of T cells for IL-12 production in the body (Takenaka et al. 1997).

IL-12 has innumerable functions in the body. It is linked with the differentiation of immature T lymphocytes into T helper 1 cells (Hsiech et al. 1993).

Hence, it can be regarded as T lymphocytes growth factor. IL-12 is additionally involved in the synthesis of TNF-alpha and interferon-gamma through the stimulation of T cells and natural killer cells. IL-12 has remarkable immunomodulatory property. It has pro-inflammatory potential. IL-12 induces the production of cytokines by T cells and natural killer cells. These cells actively proliferate and exert cytotoxic activity in the presence of IL-12 (Gately et al. 1994).

A study performed by Malaguarnera et al. (2002) clearly depicted a role of IL-12 in the stimulation of T helper 1 cells and regulation of innate immune response to the intracellular parasites as *Plasmodium falciparum*. Malaguarnera et al. (2002) posited that this cytokine also exerts its protective and disease-limiting effect on the parasite in blood stage. Malaguarnera et al. (2002) observed a rise in IL-12 level in children who suffered from mild *P. falciparum* infection, whereas its level was depressed in severe form of malarial infection.

Pathogenic microbes, endotoxins and intracellular parasites like *Plasmodium* are the potent antigenic characters which can induce the synthesis of IL-12 or natural killer cell stimulatory factor (D'Andrea et al. 1992) through monocytes. According to Chan et al. (1991) and Chan et al. (1992), IL-12 stimulates T lymphocytes and NK cells to synthesize lymphokines for essential cytotoxic effect against pathogens. IL-12 brings about differentiation of T helper 1 cells and synthesis of interferon-gamma in the body.

Thus, IL-12 acts as a pro-inflammatory cytokine in response to infection. Furthermore, IL-12 behaves as a pro-inflammatory mediator. It induces the production of other immunomodulators like tumour necrosis factor-alpha and granulocyte-macrophage colony-stimulating factor in the host body. Granulocyte-macrophage colony-stimulating factor (Francisco-Cruz et al. 2014) enhances the production of IL-12. Hours after the exposure to pathogenic organisms, NK cells produce interferon-gamma. It potentiates the activity of AP cells and helps in the lysis of pathogens.

TNF-α is another competent immunoregulator during the parasitic infection which stimulates innate immunity. Studies by authors have hypothesized the involvement of TNF-alpha in the pathogenesis of malarial infection. It was the study of Grau et al. (1989) where authors observed the increased levels of tumour necrosis factor-alpha in children who suffered from *P. falciparum* malaria. The authors confirmed its raised levels in the severe form of malaria and cerebral malria in children.

In the study by Kwiatkowski et al. (1990), plasma concentration of TNF-alpha was observed in Gambian children with uncomplicated malaria and compared by Kwiatkowski et al. (1990) with children who suffered and hospitalized for cerebral malaria. Kwiatkowski et al. (1990) spotted two times higher levels of TNF-alpha in children who survived after an attack of cerebral malaria. TNF-alpha levels were observed to be ten times higher (Kwiatkowski et al. 1990) in children who died in hospital.

TNF-alpha is partly responsible for symptoms of malarial fever. Kwiatkowski et al. (1989) studied TNF-alpha concentrations in malaria children and healthy adults. They induced the monocytes with malarial endotoxins in vitro and observed its level comparable to that secreted in malarial children. Kwiatkowski et al. (1990) concluded that TNF-alpha is released in concomitance with rupture of schizonts, and the phenomenon is responsible for malarial-related symptoms.

TNF-alpha and IL-6 were assayed and found to be higher in concentration in patients of malaria before the start of treatment. Kern et al. (1989) conjectured that a rise in cytokine concentration is a predictor of severity of malaria. It was hypothesized whether an elevated level of TNF could result in symptoms severity and organ damage in malarial infection. Study was conducted by Clark et al. (1989). The authors administered, parenterally, recombinant human TNF/cachectin into the body of healthy and mildly *Plasmodium vinckei*-infected mice. Kern et al. (1989) scrutinized neutrophilic ingress into pulmonary vessels and liver damage in mildly infected mice. It was concluded by the authors that a high level of TNF is associated with parasitaemia and organ injury in the infected individuals.

TNF-α induces the production of macrophage migration inhibitory factor (Calandra and Bucala 1995). According to Nishihira (2000), macrophage migration inhibitory factor (Calandra and Roger 2003) is a pleiotropic protein which has inflammatory and immunostimulatory response. According to the author, MIF is a lymphokine participating in delayed hypersensitivity reaction and impart phagocytic, tumouricidal function to macrophages.

MIF counteracts the anti-inflammatory action of glucocorticosteroids in the body.

TNF-alpha can stimulate inducible nitric oxide synthase enzyme and can induce synthesis of nitric oxide in the host body. It was observed by Rockett et al. (1991) in vitro that oxidation products of nitric oxide can kill *Plasmodium* parasites. So NO in the body can minimize parasitaemia. TNF-alpha has a protective role in malarial infection through pro-inflammatory action. However, it can intensify inflammation by the stimulation of cyclooxygenase (COX-2) and subsequent induction of pharmacologically active metabolites like prostaglandins in the host (Perkins and Kniss 1997).

Prostaglandins along with TNF-alpha can lead to expression of malaria-associated clinical manifestations like fever, shivering, headache, nausea, vomiting, anorexia, diarrhoea, myalgias and thrombocytopenia (Schwartz et al. 1989).

IFN-γ and TNF-alpha are synthesized by natural killer in the initial phase of malarial infection by *P. falciparum* (Hensmann and Kwiatkowski 2001). The authors showed that infected erythrocytes produced these cytokines within 18 h of infection with *P. falciparum*.

Artavanis-Tsakonas and Riley (2002) concluded that natural killer cells, probably, induce synthesis of IFN-gamma in the early stage of infection. This cytokine is immensely involved in the induction of protective molecules against malarial parasites.

Malarial parasitic-infected individuals exhibit heterogeneity in the IFNg response in the body of host. D' Ombrain et al. (2008) conducted a cohort study covering 206 school-aged children in New Guinea country in Papua, for 6 months. They observed a high-level IFNg response among children against *P. falciparum* malarial infection and was concluded that early production of IFNg is associated with protection against parasites.

According to Kremsner et al. (1995), tumour necrosis factor-alpha and interferon-gamma are important pro-inflammatory cytokines. These are produced in the body of children during intial stage in response to *Plasmodium* infection. These cytokines activate the monocytes and macrophages against malarial parasites and phagocytose the pathogens and help in limiting the proliferation of parasites.

Further, Lyke et al. 2004 and Perkin et al. (2000) observed the cytotoxic role of these pro-inflammatory cytokines in the body during advancing stage of malarial infection. These cytokines exhibit dual characteristic in different stages of infection, but their precise function in the body is still under investigation.

Malaria-induced immune mechanism is still not fully recognized. Cytokines like TNF-α, IFN, and IL-12 have a proven protective role against the malarial parasites. But cytokines as IL-4, IL-5, IL-10 and TGF-β can aggravate severity of infection.

Interleukin-10 is an effective anti-inflammatory cytokine. It is produced in the later stage of malarial infection. According to Ho et al. (1998) and Lyka et al. (2004), anti-inflammatory substance, effectively, regulates the synthesis of pro-inflammatory cytokines in response to malarial parasites. IL-10 is associated with innate immune response of the body in *P. falciparum* infection. It prevents the hyperactivity of pro-inflammatory cytokines and, hence, is protective in nature against the pro-inflammatory cytokines. As per a study by Ho et al. (1998), a high concentration of TNF-alpha and a low concentration of IL-10 in *Plasmodium* infection is indicative of advancing infection.

According to Nathan and Xie (1994), nitric oxide synthase catalyzes the conversion of L-arginine. Hence, nitric oxide is produced which in turn modulates the innate immune response and is importantly involved in the pathogenesis of severe malarial anemia. Later on, Perkins et al. (1999) postulated that inducible nitric oxide synthase (iNOS) is responsible for the synthesis of nitic oxide during *Plasmodium* infection. Polymorphonuclear cells, monocytes and macrophages harbour iNOS. This form of enzyme is induced by cytokines in the body, as per Perkins et al. (1999). Further, research by Gyan et al. (2002) emphasized on the role of nitric oxide in the body during malarial parasite infection. Tumour necrosis factor-alpha,

IFN- γ and IL-12 prominently participate in the activation of iNOS, which enhances the production of nitric oxide. This molecule has malarial parasitic activity and, hence, prevents the spread of parasite in circulation. Therefore, NO has a protective role in the malarial pathogenicity. As per authors (Gyan et al. 2002), NO has a dual function. It is implicated in the pathogenesis of cerebral malaria. Chronic production of NO results into suppression of bone marrow activity. It has deleterious effect on the erythropoiesis, and sustained release of NO by iNOS activity in the splenic macrophages endorses phagocytosis of RBCs. Overall, a chronicity of malarial infection and NO synthesis enhances the severity of anemia in children. Anti-inflammatory cytokines like IL-10 help to suppress the expression of iNOS and minimize the release of NO and offer a host cytoprotective role.

Falade et al. (2007), Reddy et al. (1996) and Romand et al. (1994) reported cases of congenital malaria as a result of transmission of *P. falciparum* through the placenta. However, malaria is seen in all age groups.

Although dietary deficiencies are widespread in malaria endemic regions, the influence of reduced folate and iron levels is not thought to be a major contributor to dyserythropoiesis seen during SMA (Abdalla 1988; Wichramsinghe and Abdalla 2000).

During human malarial infection, sequestration of infected and uninfected erythrocytes is a prominent feature of *P. falciparum* infection. Looareesuwan et al. (1987) worked on the animal model to assess the splenic filtration of RBCs. Looareesuwan et al. (1987) utilized heated 51Cr-labelled autologous erythrocytes and observed (Looareesuwan et al. 1987) the clearance in acute *P. falciparum* malarial patients and controls. It was observed by the authors that in patients who had splenomegaly, clearance of labelled RBCS was very fast in comparison to patients who did not have spleen enlargement (Looareesuwan et al. 1987).

They concluded that the hyperactivity of the spleen in malarial infection is directed against infected RBCs and whether the uninfected erythrocytes are also removed from the circulation is not still clear. Further, research is needed to elicit the exact role of the spleen in malarial infection.

Further study by Price et al. (2001) and Looareesuwan et al. (1987) characterized some risk factors which predispose to *P. falciparum* infection. Probability of enlarged spleen and age below 5 years are the important determining factors for infection. Children under 5 age have higher propensity towards anemia in malarial infection in comparison to adult patients.

Jaleman et al. (1999) postulated that destruction of uninfected erythrocytes occurs by phagocytosis, and dyserythropoiesis does not have a significant role in the cause of anemia in children suffering from *P. falciparum* infection.

Lamikaran et al. (2007) concluded by their work on human and mice that anemia in case of *P. falciparum* infection is caused by multiple factors that have a synergistic role. As per authors, infected RBCs undergo sequestration in the spleen, and uninfected erythrocytes are cleared from circulation as a consequence of hypersplenism. The authors observed that the body of the host is unable to produce erythroid tissues and this factor is equally important in inducing anemia in *Plasmodium* infection.

Malarial infection induces proliferation of macrophages in the body and may contribute to hyperactivity of macrophages and increased removal of uninfected cells. Brown et al. (1990) assessed the macrophage activation in malaria-infected individuals. They measured urinary neopterin and serum interferon-gamma levels and were found to be elevated in most of patients. Neopterin is synthesized by monocytes/macrophages after their stimulation by interferon-gamma. According to authors (Murr et al. 2002; Cesur 2005; Wirleitner et al. 2005; Fuchs et al. 1992), assessment of neopterin concentrations in serum, cerebrospinal fluid and/or urine is an important parameter to determine activation of t helper cell-1 in host. Its level is increased in viral infection like HIV, intracellular parasites like *P. falciparum* and in autoimmune disorder. As per authors, neopterin is indicative of the synthesis of reactive oxygen species and hence is a marker of the level of cellular immunity.

The study of Mohan et al. (1995) and Jenkins et al. (2006) revealed a relation between monocytic-induced damage of RBCs in malaria. As per authors, monocytes generate reactive oxygen species and, as a consequence, result in lipid peroxidation of membranes of erythrocytes and, therefore, exhibit a clear role in erythrocytic lysis in malarial infection.

In the study by Ladhani et al. (2002), high degree of leucocytosis was observed which was associated with high rate of mortality, according to authors. *Plasmodium falciparum* infection is detrimental to erythrocytes which are destroyed due to vivid causes. But malarial parasites also affect the lymphocytes, which was studied by Abdalla (1988), and WBCs undergo profound changes after malarial infection in children living in high transmission areas. According to Berkley et al. (1999), high WBC count might be attributed to superimposed bacterial infection. Another researcher (Modiano et al. 2001) did not assign any specific cause for leucocytosis in malarial infection.

Exaggerated oxidation of the membrane of uninfected erythrocytes has been observed by authors (Dondorp et al. 2003; Omodeo-Sale et al. 2005) in the *Plasmodium falciparum* infection in children. Synthesis of pro-inflammatory cytokines and the injury caused by inflammatory mediators are responsible for oxidation of RBCs membranes. Also the metabolites produced by the parasites might be the factor for the increased clearance of uninfected erythrocytes.

Increased oxidation of membranes of uninfected erythrocytes has been reported among preschool-aged children who were infected with *Plasmodium falciparum* malaria. Acute inflammatory cytokines and/or toxic metabolic products of parasites impair the erythrocyte deformability.

Parasite products that may be a part of immunoglobulin-antigen complexes are deposited on the membranes of uninfected RBCs. The major parasitic metabolite is the *P. falciparum* ring surface protein 2 (RSP-2). This protein expresses shortly after merozoite invade RBC. Hence, protein mediates adhesion of iRBCs to endothelial cells of hosts, which are captured by immune cells (Pouvelle et al. 2000; Douki et al. 2003).

The authors (Douki et al. 2003; Lekana Douki et al. 2002) identified that RSP-2 is the key component of RBCs sequestration involving both infected erythrocytes and non-infected erythrocytes. RSP-2 is deposited onto the membranes of uninfected and infected RBCs. Hence, coated RBCs undergo process of opsonization and are phagocytosed by macrophages. Ring surface protein-2 is responsible for cytoadhesion of infected erythrocytes to endothelial cells and early ring stage-infected erythrocytes (rIEs) to host endothelial cells and epithelium of placental villi (Scherf et al. 2001; Costa et al. 2006; Lekana et al. 2002).

Abdalla (1990) examined bone marrow of Gambian children who suffered from *P. falciparum* infection and furnished a detailed account of haematological changes in malarial anemia in children. The author asserted that hyperplastic erythroid tissues and dyserythropoiesis were of common occurrence in histological findings in malarial patients. These patients exhibit severe anemia and low parasitic count, as per author. Other prominent features described by the author (Abdalla 1990; Wickramasinghe and Abdalla 2000) were lymphocytosis in bone marrow and increased count of monocytes, neutrophils and lymphocytes in blood circulation.

According to the author, severe malarial anemia is characterized by either normochromic, normocytic or hypochromic, microcytic appearance of erythrocytes. Low serum ferritin, low iron-binding capacity and high concentration of inflammatory proteins are the prominent features of severe malarial anemia (Abdalla 1990). These features are characteristics of anemia of chronic inflammation. Cytokines like interleukin-1 and interleukin-6 have been detected in patients of SMA (Issifou et al. 2003; McDevitt et al. 2004). Cell lines in bone marrow are suppressed. Erythropoiesis is inhibited, and young erythrocytes are haemolyzed by cells of reticuloendothelial cells (Edington and Gilles 1976).

Persistent hypoferrinaemia in malarial anemia is attributed to temporary intracellular sequestration of iron, decreased level of transferrin and chelation of iron by iron binding proteins other than transferrin in the presence of inflammation(Ghosh 2007). Overall, bioavailability of iron to host tissues and parasite is reduced. Bone marrow functioning is affected (Oppenheimer 1989). Hypersplenism is noticed in severe malarial anemia (Looareesuwan et al. 1987).

Non-infected erythrocytes are sequestered by the macrophages and the reticuloendothelial system (Ghosh 2007). Splenomegaly syndrome in repeated or chronic *P. falciparum* infection could lead to reduced RBC count and anemia.

Dyserythropoiesis is a feature of malarial anemia in children. It is characterized by the synthesis of morphologically and/or functionally defunct erythrocytes in bone marrow. These RBCs have fragmented nucleus and vacuoles in cytoplasm and may show the presence of intercytoplasmic bridge. Erythropoiesis is disturbed in *Plasmodium* infection. According to Abdalla et al. (1980), children with malarial anemia exhibited erythroid hyperplasia and dyserythropoiesis.

A study was done by Abdalla et al. (1980) in three groups of Gambian children who suffered from *Plasmodium falciparum* infection. According to authors (Abdalla et al. 1980), children with acute infection developed anemia after treatment. In the second group, children suffered from chronic *Plasmodium* infection. These children had anemia during the duration of infection. Dyserythropoiesis was observed by

Abdalla et al. (1980) in the bone marrow aspiration in these children. The third group of children showed evidence of intermediate haematological changes, as described by Abdalla et al. (1980). This variation in haematological change in children with different duration of malarial infection is possibly attributed to body response to different stages of malaria (Abdalla et al. 1980).

A fall in Hb and subsequent reduction in oxygen tension should stimulate elevated levels of erythropoietin (Epo) in patients with SMA. The clinical evidence for appropriately raised levels of Epo in malaria is somewhat contradictory. Studies were conducted on adult population inhabitant of Thailand and Sudan, and it was suggested by Burgmann et al. (1996) and el Hassan et al. (1997) that Epo concentration was elevated in plasma of infected persons. But its level is not responsible for the manifestation of anemia.

However, several studies of malaria in African children suffering from malarial anemia have shown appropriately raised erythropoietin (Epo) concentrations (Kurtzhals et al. 1997; Verhoef et al. 2002). In fact, the Epo levels in malarial anemia are more than threefold higher when compared with anaemic children without malaria (Casals-Pascual et al. 2006). It is possible that ineffective or inadequate Epo synthesis does contribute to malarial anemia in some settings, possibly related to age, ethnic origin or presentation of the patient.

Contrary to elevated Epo level, bone marrow is suppressed leading to reduced haemoglobin concentration. According to Sandau et al. (2001), high concentration of tumour necrosis factor-gamma in malaria is responsible for hypoxia. TNF induces the activation of hypoxia-inducing factor. This transcription factor has serious impact on pathogenesis of malaria and suppression of bone marrow. Research indicates that abnormal activity of HIF is the cause of distorted internal environment of the cells and helps in the promulgation of inflammation. Consequently, an equilibrium is disturbed in cytokines, growth factor and reactive oxygen species (Smith et al. 2008). Hypoxia and ischaemia lead to raised Epo level. In malaria, proliferation of erythroid progenitors as burst forming colonies and colony forming units-erythroblast is inhibited due to the decreased level of interleukin-12 and interleukin-10 in *Plasmodium*-induced malaria in children. These cytokines are the leading stimulator of haemopoietic precursor cells. These cytokines are responsible for differentiation and proliferation of these precursors. Inadequacy of IL-12 and IL-10 and bone marrow suppression are together the cause of severe malarial anemia in children (Angela O'Donnell et al. 2007).

Nitric oxide synthesis is increased in malaria (Clark et al. 1991). This potent vasodilator inhibits the Na+/K+ ATPase enzyme located in cell membrane of erythrocytes. NO also causes lipid peroxidation and distorts the structure of cell membrane. Coenzymes like NAD are depleted in the RBC which leads to inhibition of glycolysis and deficiency of ATP (Clark and Cowden 2003). In response to these metabolic changes, erythrocytes are sequestered by the spleen. According to Xie and Wolin (1996), nitric oxide injures the mitochondria of erythroid precursors and suppresses the bone marrow.

Haemozoin contains iron in ferrous state. This free iron catalyzes the formation of superoxide free radicals. Consequently, phospholipids in RBC membrane produce 15-hydroxyeicosatetraenoic acid (15-HETE) (Takahashi et al. 1985) and 4-hydroxy nonenal (4HNE) (Awasthi et al. 2004). These two are the lethal lipid peroxidation metabolites and cause cell function impairment. These molecules enhance the release of lysozymes from monocytes. Mitogen-activated protein kinase (Pearson et al. 2001) and nuclear factor(NF)-κB are involved in the release of lysozyme (Pollimeni et al. 2013). As a result, erythrocytes become rigid and lose their flexibility and are removed by cells of RE system (Arese and Schwarzwer 1997).

Anyona et al. (2006) have demonstrated the presence of spherocytes in blood circulation in malaria. Another group of authors (Gamain et al. 2001; Oquendo et al. 1989) suggested that fatty acid translocase which is also called as CD-36 is an important cell surface protein located on the endothelial cells which has a role in the sequestration of infected erythrocytes. This protein has high affinity to malarial parasite-infected RBCs. These stick to endothelial cells and are removed from circulation. This condition might lead to acute anemia in malaria.

Pathophysiology of severe malarial anemia is regulated by an imbalance between pro- inflammatory and anti-inflammatory molecules. Overall intense decline in Hb concentration in *Plasmodium*-induced malaria in children, especially in malaria endemic regions of the world, is partly attributed to the phagocytosis of infected and uninfected erythrocytes by macrophages and monocytes and partly due to the suppression of bone marrow leading to defunct erythropoiesis and consequently leads to anemia in children.

5.4.3 Clinical Manifestations

Severe malarial anemia is characterized by nausea, vomiting, loss of appetite, fever, jaundice and difficulty in breathing. Severe anemia is followed by a haematocrit value below 15% and Hb concentration less than 5 g/dL. Tachycardia, hepatomegaly and splenomegaly are the perceptible signs. Anemia in malaria is marked by normochromic and normocytic appearance of RBCs. However, marked deficiency of folate is responsible for macrocytic anemia in malaria in children. Differential count of monocytes is raised along with granules in the cytoplasm of monocytes (Edington and Gilles 1976). Bone marrow undergoes erythroblastic hyperplasia (Wiwanitkit 2008, Karunaweera et al. 1998).

5.5 Environmental Enteropathy and Anemia in Children

Environmental enteropathy is an inflammatory condition of intestinal mucosa prevalent in population residing in crowded places and among those individuals who adopt poor hygiene and improper sanitation practices (Fagundes-Neto et al. 1994; Korpe and Petri 2012).

These prevailing conditions are common in developing countries. Poor environmental conditions like contaminated drinking water, deprived hygiene and sanitation expose the individuals to faecal pathogens. Endotoxins induce inflammatory changes in the intestinal mucosa; its chronicity results in architectural and hence functional abnormality of mucosa of the intestine. Environmental enteropathy is associated with increased epithelial permeability, poor absorption and defunct localized immunity of GIT (Ramakrishna et al. 2006).

Persistent exposure to pathogens through contaminated food and drinking water as a result of non-adoption of personal hygiene practices perpetuates to consistent gastroenteritis. Decreased length of villi, hyperplasia of crypts of Lieberkuhn, lymphocytic infiltration of lamina propria and presence of intra-epithelial lymphocytes are the common features in the structure of intestinal mucosa. Primarily, the absorptive capability of mucosa is impaired (Humphrey 2009; Prendergast and Kelley 2012).

According to a study by Martin et al. (1994), infants inhabiting in slum areas in Brazil were found to have abnormal length of mucosal villi. These infants were normal at birth, and inflammatory condition of intestinal mucosa developed within 12 weeks postnatal life. These infants had impaired D-xylose absorption pattern when compared with infants belonging to middle-class families who were served with clean drinking water.

Pathogens invade GIT through the use of contaminated water, food and/or preparation, and serving of food without hand washing, invariably, induces immunity-mediated inflammation in the intestinal mucosa. It might be an adaptation of the gut mucosa to survive in hostile environmental condition (Fagundes-Neto et al. 1984).

Environmental enteropathy is also called as tropical sprue. It deteriorates the absorptive capacity of mucosa. This condition decreases the absorption of vitamin B_{12}, folate and iron from the dietary food by the intestinal mucosa. These micronutrients are necessary for haemoglobin synthesis and erythropoiesis in the bone marrow (Humphrey 2009).

Anemia in preschool children who are living in slum areas and in those regions of world where pitiable condition of WASH prevails is multifactorial. Inhabitants suffer from nutritional anemia due to either inadequate and poor quality of dietary intake or impaired absorption from inflamed intestinal mucosa or the coexistence of these conditions acting in negative synergy (Rosenberg and Bowman 1982).

Defaecation in open field is a malpractice, largely, adopted by population in developing countries. According to an estimate by UNICEF and WHO (2014), open defaecation is practiced by around 1 billion (UNICEF) population of the world and is common in South Asia. Further analysis shows that majority, (90%) of all the total population which defecates in open, resides in rural areas. Multilevel approach by government agencies, NGOs and spread of education among masses have resulted in apparent decline in open defaecation in almost all parts of the world. Globally, it decreased from 24% prevalence in 1995 rate to 15% prevalence rate in 2011.

Open defaecation is one of the determinants of environmental condition in which a child is reared and brought up. It is related to the health of the child (Haghighi et al. 1997; Lin et al. 2013).

The greater the number of families who defecate in the open, the larger would be the prevalence of malabsorption, faltered growth, anemia and malnutrition in children, especially in preschool-age group. Open defaecation and no hand wash practiced by the family members during either food preparation, food serving or handling a child transmit harmful entero-parasites via faecal-oral route. The infection initiates a number of after-effects in the body. Malabsorption of micronutrients, diarrhoea, induction of cell-mediated immunity, protein energy malnutrition and iron deficiency anemia coupled with anemia of chronic infection are the noteworthy manifestations of the invasion of faecal pathogens in GIT. Nutritional anemia is the cause of poor absorption of iron, vitamin B_{12} and/or folate (Cairncross 2003).

It was hypothesized by Walker (2003) that open defaecation renders children susceptible to bacterial infection and invasion of worms (Ziegelbauer et al. 2012).

It leads to decreased absorption of micronutrients vital for the synthesis of haemoglobin in the bone marrow. Nutritional anemia is commonly considered to be the effect of inadequate and poor quality diet. However, environmental factors like hand wash, potable drinking water, hygiene and sanitation have empirically, theoretically and clinically same implication on nutritional anemia as those of dietary factors. WHO and UNICEF have incorporated poor hygiene and sanitation as the contributing factors to anemia in preschool and school children.

5.6 Malabsorption Syndrome

Malabsorption syndrome comprises various clinical manifestations arising due to inability of the small intestine to absorb diffusible form of nutrients from the lumen of the small intestine. Dietary food substances are digested in the gut (Jensen 2007; Bernstein et al. 2003). These are transformed into easily diffusible form by the action of hydrolytic enzymes. Any abnormality in the structure or functioning of mucosa of the small intestine manifest into malabsorption syndrome (Bernstein et al. 2003). Common are diarrhoea, flatulence, anorexia and emaciation over a period of time. Prolongation of malabsorption syndrome results into deficiency of iron store of the body, protein deficiency and depletion of necessary micronutrients in the body (Sleisenger 1969).

Overall, children suffer from nutritional anemia, varying degrees of wasting and stunting and compromised immunity of the body. Any cause of malabsorption would invariably lead to the following outcome:

Deviation in Body Built Due to sustained loss of nutrients from the alimentary canal, there is defunct absorption. The body of children suffers from inadequate calorie intake. It results into wasting, and chronicity of malabsorption would cause stunting and underweight in children. Adult undergo weight loss, fatigue and loss of productivity (Pietzak and Thomas 2003).

Chronic diarrhoea. It is the passage of three or more loose stools per day for more than 2–4 weeks and is called chronic or persistent diarrhoea (CDC 2014). Steatorrhoea is another clinical manifestation of the inflammatory condition of the mucosal cells of the small intestine. Fats are digested and absorbed. Excess fats are excreted in stools which are clay coloured and oily in appearance and have foul smell (Cheifetz et al. 2011a, b).

Other deficiency associated with malabsorption could be iron deficiency with or without anemia, vitamin B_{12} or folate deficiency anemia and protein energy malnutrition. Preschool children in villages and slums can be easily diagnosed for malabsorption symptoms in the absence of any physicians by trained health workers. Common symptoms are malodorous stools, chronic diarrhoea, weight loss, stunted growth and definitely pallor (Ravikumara et al. 2006). Moreover, in advanced stage of malabsorption, children develop oedema over the feet and face and distention of the abdomen (pot belly) and have a tendency for bruising. These children should be referred to physicians where he will determine the cause of malnutrition and anemia. A detailed history of children would reveal the cause of malabsorption syndrome.

A syndrome was observed in children and adults about 35 years back. It was characterized by chronic diarrhoea, loss of appetite, increased susceptibility to acute infection and failure to growth and development in children. According to Savilahti (2000), the syndrome was the consequence of consumption of cow milk by infants. It caused the damage of the mucosa of the jejunum. Manifestations was decrease in absorption of D-xylose. Infants suffered from steatorrhoea, iron deficiency anemia and hypoproteinaemia. Infants developed cell-mediated immunity to cow milk. Intraepithelial count of lymphocytes was increased. It resulted in inflammation of the surface epithelium and lamina propria of the jejunum. Length of villi decreased and microvilli became shorter.

It was found that IgA and IgM level in serum was increased, but IgE level was normal. Eosinophil count was increased in infants who were exposed to cow milk. All these immunological changes in the body of infants indicated the T cell-mediated immunity to the cow milk protein in the small intestine, as was described by Savilahti (2000). It is suggestive of food-induced enteropathy in the infants. The child's body would be deprived of nutrients and essential minerals due to defective absorption process. It can materialize into harmful effects on the body and mind. Coeliac disease is an autoimmune disorder of the mucosa of the small intestine (Sheldon 1959).

Coeliac disease is a hypersensitivity to gluten protein present in wheat and proteins in barley and rye. The disease is seen in genetically predisposed individuals (Green et al. 2001). Its prevalence is found in individuals of all ages from infants to old age persons. These individuals have genetic predisposition to develop coeliac disease. Disease affects children early in age after the feeding diet containing gluten. Children suffers from diarrhoea, anorexia, abdominal discomfort and weight loss. These manifestations arise due to inability of intestinal mucosa to absorb nutrients from the diet (van der Windt et al. 2010; Fasano 2005).

Nutritional anemia is also seen in coeliac disease in children (Farrell and Kelley 2002; Green and Celler 2007; NIH 2004). A 33-amino acid residue of gliadin has been identified which escaped digestion by protease enzymes secreted in the gastric, pancreatic and intestinal juices in the body. This large peptide entered circulation and is capable of inducing innate immunogenic response in the body (Shan et al. 2002).

Host immunity is directed against gluten. It is made up of gliadins and glutenins. These are plant storage proteins in wheat, rye, barley and oat. Gluten has high content of proline and glutamine amino acids. Tissue transglutaminase is an enzyme that has calcium-dependent catalytic activity in the body. It cross-links proteins in the body between an ε-amino group of a lysine residue and a γ-carboxamide group of glutamine residue. It results in high resistance of protein to proteolysis. tTG is mainly implicated in the pathogenesis of coeliac disease. This enzyme modifies gluten peptides. It causes deamidation of glutamine present in ingested gliadin. It produces glutamic acid. Deamidation confers higher antigenicity to glutamine because it has higher affinity for DQ2 and DQ8 molecules. It is presented by antigen-presenting cell (APC) to T lymphocytes and initiates cell-mediated immunity in intestinal mucosa. Pro-inflammatory cytokines and interferon (IFN)-γ are produced in the mucosa and results into structural changes in the intestinal mucosa (Molberg et al. 2001).

Another modification of gluten protein by tissue transglutaminase enzyme is through the process of transamidation. Glutamine residue of gluten protein is cross-linked to lysine residue of tTG enzyme. This reaction causes a permanent covalent linkage between gluten and tTG enzyme (Fleckenstein et al. 2002). Finally a new complex is formed that is more immunogenic in comparison to gluten alone. The new antigenic epitope can trigger the formation of antibodies against tTG (Koning et al.2005).

According to Thomas et al. (2006), gliadin induces antigenic response in the body through innate mechanism. Peptides of gliadin, 33-mer and p31-43, have the potential to induce synthesis of zonulin (Fasano 2011), a glycoprotein responsible for altering the permeability of cell membrane in GIT. Overall, increase in permeability of intestinal epithelium and release of cytokines bring about structural and hence functional abnormality in the mucosa of the small intestine and increase in intestinal permeability and macrophage pro-inflammatory gene expression and cytokine secretion. Gliadin-induced zonulin release, increased intestinal permeability and cytokine production were dependent on myeloid differentiation factor 88 (MyD88), a key adapter molecule in the TLR/IL-1R signalling pathways but were neither TLR2- nor TLR4-dependent.

According to Maiuri et al. (2003), activated innate immunity generates interleukin-15 through the intestinal epithelial cells. Apart from innate immunity, there is a role of adaptive response of the intestinal mucosa in susceptible individuals to coeliac disease.

Interleukin-15 increases the permeability of intestinal epithelium. Toxic peptides like 33-mer can transgress into the epithelium and reach the lamina propria, wherein dendritic cells present these peptides to T lymphocytes. IL-15 can also induce apoptosis of intestinal epithelial cells which further aggravate the permeability of mucosa (Rossi and Young 2005).

Conclusively, it is well established that gliadin, a component of wheat, can trigger activation of innate immunity along with the activation of adaptive immune response in GIT in genetically determined individuals and children. Anemia is a frequent symptom in any disorder affecting digestive system. Defective absorption of micronutrients from the dietary food by the intestinal mucosa is the root cause of nutritional anemia in digestive disorders.

5.7 Habit of Geophagy in Children and Anemia

5.7.1 Overview of Geophagy

Geophagy is the persistent and compulsive behaviour involving consumption of nonedible and harmful substances (APA 1994). It involves the habit of eating substances like chalk, soil, clay and pencil. These are passed into the stool. These non-nutrients do not provide energy to the body (Lopez, et al. 2004). Precise aetiology of geophagy is uncertain. Iron deficiency anemia predisposes children to the habit of geophagy. Probably, it is an attempt of the body to recover micronutrients like iron, calcium, zinc and copper from nonedible food substances (Lanzkowsky 1959).

Children below age of 5 years and pregnant women have been observed to practice soil eating (Njiru et al. 2011). Moreover, children in higher age group who suffered mental retardation have been observed (Herguner et al. 2008) to consume soil. In particular regions of the world, geophagy is endorsed by customs and beliefs (Vermeer and Frate 1979). It is associated with worm infestation in the gut (Lumsden et al. 1979). It is considered as compulsive behaviour (Herguner et al. 2008) in children. However, the researchers have varied opinions regarding the aetiology and ill effects of pica.

Geophagy is construed to be a consequence of psychiatric disorder in children (Herguner et al. 2008). Geophagy is the premeditated ingestion of clay, soil, ice and other non-nutritive substances. Since time immemorial, it has been practiced by humans in all continents of the Earth by people of different culture, race and ethnicity (Anell and Lagercrantz 1958).

Maravilla and Berk (1978), radiologists, observed that eating a small amount of clay or soil is possibly harmless. They insisted that this habit has serious consequences on health. Consumption of soil which is contaminated by toxic substances like lead could be harmful to the individual. They posited that physical obstruction of airway is another concern regarding soil eating.

Soil is contaminated with metals like lead, zinc, copper, arsenic and microbes and worms. Geophagy has high repercussions on the health of intestinal mucosa; thereby, its absorptive capability is impaired in children (Rawlins et al. 1991).

Pica is associated with anemia and followed during pregnancy (Adam et al. 2005). In an explanation by Aufreiter et al. (1997), geophagy is assumed to be the consequence of either increased iron demand in the body or a deficiency of iron in the body; thereafter, an individual satisfies one's craving for iron through perverted dietary habit of eating earth.

Pica, a millennium old tradition and practice of ingesting nonedible substances, ubiquitous around the Earth with age and gender predilection, hitherto, stands unresolved.

The authors (Hagopian et al. 2011) observed the practice of pica between the second and third pica years of life. Its continuation beyond age of 3 year is considered a psychosomatic disorder in children. Tendency to pica decreases with advancing age of children. Mentally deranged and intellectually compromised adults have been observed with the practice of soil eating.

Clay particles may have the tendency to withhold dietary iron in complexes. So it is unavailable for absorption in intestinal lumen. Ingested soil particles may damage intestinal mucosa resulting in impaired absorptive capacity (Young 2010). Another empirical evidence was given by Sayers et al. (1974a). Soil particles could bind with mucin of the gut lumen and may form a physical barrier that might impede with the dietary iron absorption through intestinal enterocytes. So pica can cause iron deficiency in the body, which, in turn, replicates with supplementary ingestion of clay.

Youdim (1977) posited brain enzymes in the hypothalamus for regulation of hunger and might be responsible for perverted appetite for non-food substances that is linked to micronutrient deficiency. The authors (von Bonsdorff 1977) asserted that the hypothalamus regulates appetite. It is possible under normal serum iron level. Hypoferraemia restricts normal functioning of the hypothalamus; hence, the outcome is geophagy.

Helminthic infestation after soil eating is a common aftermath in most of the children with habit of geophagy. Mucosa of the small intestine is inflamed and may undergo ulceration due to repeated friction with the ingested soil and food particles. Predominantly, *Trichuris trichiura*, *Ascaris lumbricoides*, hookworm and *Strongyloides stercoralis* are the intestinal worms along with prominent protozoans as *Giardia lamblia* and *Entamoeba histolytica*.

Soil eating induces a sense of satiety due to filling of the stomach by non-nutritive substance, wherein the individual is deprived of necessary calories and minerals. Geophagy is implicated in child malnutrition. Geophagy is correlated with iron deficiency anemia among children under age of 5 years. Health is determined by multiple factors. According to FAO (2005), nutritional status of a child is the physiological level of nourishment as established by adequate level of macro- and micronutrients in the body. Further, it can be assessed by child's anthropometric parameters, dietary factors, environmental factors, clinical signs and symptoms and social history.

Cause and effect relationship between habit of geophagy and micronutrient deficiency in the body is controversial. It is difficult to ascertain whether a child develops habit of soil eating as a consequence of iron deficiency or the former is responsible for iron depletion in the body (Sayers et al. 1974b).

A little systematic research is available in literature that depicts the association between geophagy and serum micronutrients level of iron, zinc, copper and calcium in the body. Further it is not universally acceptable that the habit of pica is due to deficiency of micronutrient deficiency. Singhi et al. (2003) conducted an observational case–control study comprised of 31 children with a habit of geophagy and 60 children as control. Both the group of participants were age and sex matched. Mean serum iron level was lower by 20% in geophagy children in comparison with controls.

In another study by Acharya (2007), they describe low serum iron level in children with geophagy. The study included 50 children with pica in age range between 1.5 and 6 years and another 50 children with age and sex matched who were normal. There was a significant decrease in plasma level of iron of pica children in comparison to normal children. Lanzkosky (1959) was the first who proposed that the iron deficiency results into the habit of geophagy.

Iron deficiency in the body may debauch dietary behaviour of an individual. Soil eating, although, non-nutritive, can satisfy the perverted appetite. According to Coltman (1969), pagophagia is the habit of eating ice for a period more than 2 months. It was monitored that pagophagia was given up after the iron deficiency was fulfilled. Further it was observed by that correction of pagophagia and iron deficit coincided with the haemoglobin synthesis and mucosa cell repair in the body.

The work of Coltman (1969) substantiates the cause and effect relationship between iron deficiency and geophagy in children. Ingested clay and soil particles meddle with the iron absorption through the intestinal mucosa. This effect is attributed to cation exchange property (Coltman 1969) of clay. It is the number of cations that can be exchanged in a 100 g sample of dry soil. It depends on the type of clay, particle size, surface area, environment of exchange reaction and many more factors. Cation exchange occurs between cations of clay and iron ion. It causes formation of insoluble complexes in the gut lumen rendering dietary iron unavailable for absorption (Minnich et al. 1968).

Poverty is the most notable, irreversible, limiting factor in the acquisition of good quality and adequate food to nourish and nurture and satiate children. Hunger is a drive to perverted appetite. Children get on with the consumption of non-food substances (Laufer 1930).

Children who eat soil, clay or other nonedible substances are exposed to the risk of worm infestation in the gut lumen. Ingestion of clay contaminated by eggs of *Ascaris* and *Trichuris* is responsible for ill effects of worm infection and anemia in children (Antelman et al. 2000).

In a study by Robischon (1971), iron deficiency anemia was prevalent in children who were indulged in the habit of clay eating.

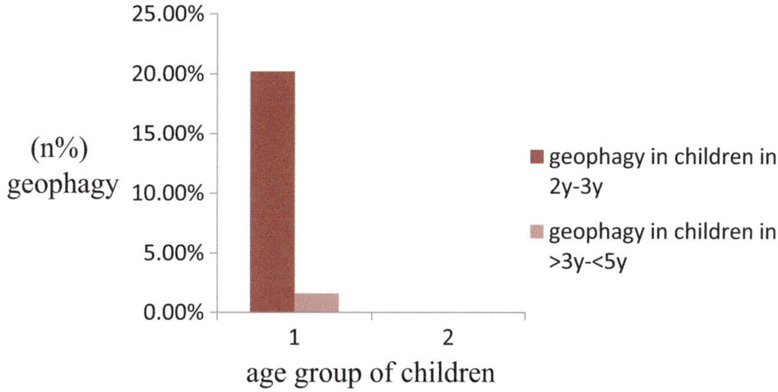

Graph 5.1 Showing prevalence of geophagy among children

5.7.2 Cross-sectional Research Study (1)

Cross-sectional research study was conducted by Gupta (2015a), wherein the study was focused to find out the effect of soil eating on nutritional level of children below 5 years of age.

The author adopted a cohort, descriptive and cross-sectional research plan. Through random and two-stage sampling procedure, the author selected (n = 440) children which was in age group >2 years to <5 years. Sample of participants was drawn from schools, anganwadi (childcare centres) and slums.

Geophagy is the behaviour related to unrelenting eating of soil by children in a specific age group. According to Gupta (2015a), the habit of geophagy is age dependent. It deviates the absorption and assimilation of digested food in the body owing to an abnormality in the structure and function of mucosa of the small intestine.

In the study by Gupta (2015a), it was observed that age of children was significantly ($p < 0.0001$) associated with the habit of geophagy. As in Graph 5.1, children (n = 7) out of total (N = 271) children in age group of (>3 year to <5 year) exhibited the habit of geophagy with a marginal prevalence of 1.6%. Contrary to it, around children (n = 34) out of total (N = 169) children in younger age group (2–3 year) were observed with the habit of geophagy having had a significantly higher prevalence (20.2%). In the present research work, Gupta (2015a) observed overall prevalence (9.4%) of the habit of geophagy among children between 2-year and below 5-year age group in the city of Fazilka in Punjab in India. Increase in age of children resulted in decrease in the habit of geophagy. Further, odds ratio (OR = 9.5) between above age groups of children showed that younger children had much higher probability of contracting the abnormal habit of soil eating.

Previous work by Hagopian et al. (2011) on the age of children and geophagy was reviewed by the author which in turn supplemented the data observed by Gupta (2015a). Both studies revealed a negative linear relationship between geophagy and age of children. It is considered physiological in age group between the 18th month

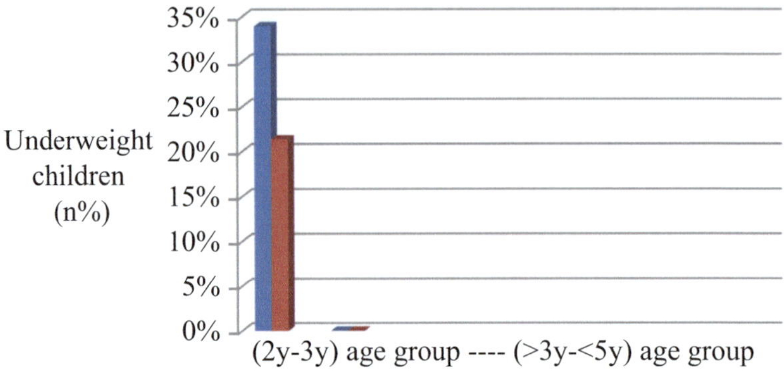

Graph 5.2 Showing prevalence of underweight among children

and 2nd year of life; thereafter, habit of eating non-nutritive substance is viewed as abnormal. Further, data by Hagopian et al. (2011) yielded a prevalence of 25–33% soil eating among children in younger age group, whereas in a study by Gupta (2015a), its prevalence was 20.2% in 2–3-year age group. Although prevalence of geophagy differs, however, both studies very well authenticate the prevalence of geophagy in children in early childhood period ranging between the first year and third year of life. Variation in prevalence might be due to difference in customs, culture, housing condition, socio-economic status and number of participants selected in studies.

Researchers are ambiguous about the aetiology of geophagy. Moreover, ill effects of geophagy on health of children in tender age group are tremendous. In a study by Gupta (2015a), highly significant ($p < 0.0001$) correlation was observed between children who were underweight and simultaneously and had the habit of soil eating as shown in Graph 5.2. An odds ratio of (OR = 15.7), in a study by Gupta (2015a), was observed between geophagy and underweight and indicated very high probability of undernutrition (underweight) among children who develops soil eating habit. Further, odds ratio was statistically proved by significant chi square value of (68.7).

Previous study by Diouf et al. (2000) on geophagy, intestinal parasites and malnutrition supplemented the work of Gupta (2015a), and both studies collectively established the impeding effect of soil eating over dietary status of preschool children. Geophagy habit derails the availability of nutrition to the body. Children become susceptible to nutritional anemia, predominantly iron deficiency anemia. Further research on these issues involved in child health will be discussed in the forthcoming chapters of the book.

Statistically significant ($p < 0.0001$) association has been observed among age, underweight and habit of soil eating in children below five in a study by Gupta (2015a).. Significantly high prevalence of underweight and geophagy has been estimated among children in the second to third year of life in contrast to children in greater than the third to below the fifth year of life. Early childhood period is marked by rapid growth and development which necessitates high demand for energy and micronutrients. Improper and inadequate weaning, poverty, prolonged exclusive breast-feeding, lack of supplementary semi-solid diet, disease and/or soil eating together are responsible for young age-related undernutritional status. Findings in a study of Karoui and Karoui (1993) also supplemented above correlation. Children who are provided scarce food or food with poor calorific value resort to soil eating to achieve satiety. According to Prasad et al. (1983), components of soil damage the intestinal mucosa. Toxic compounds of soil like lead induces immunity-initiated inflammation, called environmental enteropathy. Consequently, absorptive function of the mucosa is decreased. Overall, a child in the period of growth and development bears underweight due to poor assimilation of nutrients. Another study by Geissler et al. (1998) insisted that soil eating is the chief cause of helminthic infection by hookworms, ascaris and whip worms which in turn slacken nutritional status of children by inducing diarrhoea, anemia and malnutrition.

Soil eating habit, even though an adaptive response to micronutrient deficiency, hunger and food scarcity, significantly spoils health of children, especially in early childhood period. Parents, guardians and child caregivers should be educated about unhealthy habit and motivated towards seeking medical help to intercept the habit of geophagy and to prevent from ill health effects.

5.7.3 *Cross-sectional Research Study (2)*

Pica signifies craving for non-calorific materials like soil, clay, hair, wood, chalk and ice. There is ambiguity regarding definition of pica among researchers. Researchers and academicians have a fragmented views about the aetiology of pica and types of pica together its treatment and/or interceptive modalities. Geophagy is a subtype of pica which is characterized by Cynthia (2014) as ingestion of soil, clay and/or mud. In human beings, this behaviour is most noticeable during early childhood and pregnancy. Harbouring geophagy by children older than 3 years is indicative of developmental disability in children. It is firmly established that pica poses great health concerns in children. The habit of geophagy is linked with malnutrition, hygiene practice, poverty and anemia. It is a public health hazard. According to Halsted (1968), geophagy is largely observable phenomena in low socio-economic population; however, pica is a worldwide eating disorder. The author asserted that pica is in practice since 10 B.C. and was practiced for medicinal use.

Chatterjee and Gettman (1972) studied a population of 136 children and asserted that ingestion of lead as pica was prevalent aggressively among black children, which they claimed to be a permissive socialization of oral behaviour.

The study of Hook (1978), on the basis of personal interviews with women after delivery, concluded that pica is a craving of pregnancy. Further Bothwell (1972) insisted on the nutritional deficiency theory of geophagy. Iron deficiency in children is responsible for geophagy. The author poised that most likely inadequate diet or loss of iron, independently or together, results into iron deficiency and ultimately a craving for soil eating develops.

Bothwell insisted that anemia is more prevalent in children who eat clay or starch.

Vyas and Chandra (1984) described that geophagy is either the cause or result of iron deficiency in the body. Nevertheless, initiation of iron supplements has positive effects and often helps in the management of geophagy.

Lackey (1982), a dietician, pointed out different hypotheses for the origin of pica. She presented a psychological origin, cultural cause, physiological cause, nutritional necessity and microbiological origin of pica.

Soil eating is widespread in rural areas in India, some regions of United states and in African tribal population. According to Abraham (2003), geophagy is associated with the customs and traditions, prevalent in their families.

It is believed in some tribal belts that soil eating might sustain fertility where it is ingested to treat nausea and vomiting, while others believe that clay eating gratifies the calcium demands in pregnancy.

A few children, generally, eat soil to overcome mineral deficiency. Soil particles interefere in absorption of iron in the small intestine. Hence, children suffer from anemia and malnutrition.

Gupta (2015b) hypothesized that geophagy habit could lead to detrimental effects on health of children and executed a descriptive and cross-sectional research study that was comprised of total (n = 440) children in age group between 2 and 5 years as in Graph 5.3. Sample population was selected by two-stage random selection method from the city of Fazilka in India.

The study revealed overall prevalence of around 9.4% geophagy among children in the city of Fazilka in India. This abnormal eating habit is due to different causes. Knowledge regarding its effects on the body are inconsistent. Generally, nutritional deficiency of minerals like iron, zinc and/or calcium inclines to the habit of geophagy. An individual takes to soil eating, probably to accomplish deficiency.

In an earlier study by Gupta and Gupta (2005), prevalence of pica in children was 32%. Still another study by Ismail and Hill (1996) on children in three districts of Kazakhstan, they reported a prevalence 23.8% of geophagy in children.

Another previous study by George et al. (2015) described 18% prevalence of geophagy in children under 5 in Bangladesh. Reasonable explanation for variation in prevalence of geophagy in under 5 children could be sample size, literacy rate, sampling site and socio-economic status of parents and geographical differences. Above hypothesis is augmented by Njiru et al. (2011). They stated that prevalence of geophagy differs between and within countries and is in range between 10 and 75%.

Another reason for flexible prevalence of geophagy was given by Young (2010). The author stated that underestimation of geophagy in children might be the result of hiding information due to humiliation, improper understanding and interpretation of the questionnaire by respondents.

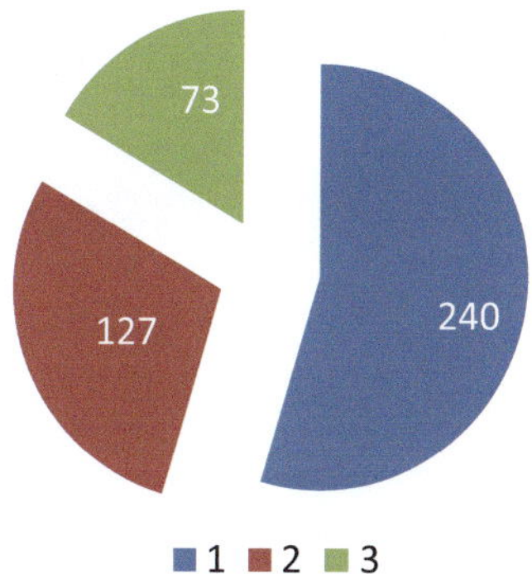

Graph 5.3 Showing distribution of population sample

Despite wide difference in prevalence of geophagy between present study and earlier study, a literature available in NLM, USA, accorded the prevalence 10–32% of geophagy in children between the age of 1 and 6 years.

In another study by Vermer and Frate (1979), pica observed in 57% of the women and 16% of the children, but it was not found among adult males or adolescents.

Conclusively, geophagy is dependent on multiple factors like socio-economic status of families, type of houses, literacy rate, weaning habits and medical facilities.

However, geophagy is injurious to the health of children. Gupta (2015b) in the present study observed that (n = 115) children suffered from pallor of nail beds, whereas another (n = 325) children were normal, as shown in Graph 5.4. Moreover, the author (Gupta 2015b) observed odds ratio (OR = 82.8) between group of children with pallor and geophagy habit and other group of children with pallor and without geophagy habit. The author posited that children with the habit of soil eating have around 82 times higher odds of developing pallor (anemia) in comparison to those who were not addicted to the habit of geophagy. Correlation was found statistically significant ($p < 0.0001$) with ($\chi^2 = 111.46$).

Earlier study by Lanzkowsky (1959) involving 12 children from the department of child health, Red Cross war memorial children's hospital, Cape Town, South Africa, were found to have deficiency of iron. Correlation between soil eating and anemia was again supported by the previous work by Moore and Sears (1994).

Miao et al. (2015) asserted 2.4 times higher probability of pica with soil eating in participants. In present study, the author assessed anemia among children by observation of nail beds in investigation. The author assumed the method to be simple, economical and rapid for anemia screening in preschool children, especially in rural population, slum areas and remote settings where laboratory facility is negligible.

It was observed by Sinha et al. (2008) that pallor has a sensitivity 73% and specificity 98% in detection of anemia as compared with WHO colour scale for Hb estimation.

Sahli's haemoglobinometer is a simple and cost-effective method of determination of anemia. However, its use comprises aspiration of capillary blood, and

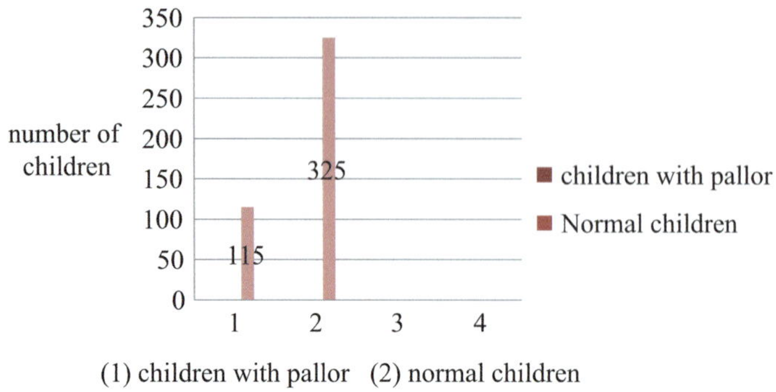

Graph 5.4 Showing prevalence of pallor among children

children under 5 are highly apprehensive about the needle prick, so its utility is overmarked by the non-invasive observational method involving the nail beds, palm, conjunctiva and mucosa of the mouth (Sheth et al. 1997).

Generally, physicians clinically observe anemia by the first sign of pallor which is further verified by lab estimation. The clinical assessment method of anemia is more appropriate in surveys and remote settings or doctors; moreover, the author has his opinion that invasive method for estimation of anemia is not commendable to follow which in turn depends on time, severity and cost factors.

Plausible explanation is that children eating soil are negatively affected in terms of absorption and dietary intake apart from other ill effects of soil eating on their body. Soil particles contain toxic elements which damage the epithelium of the small intestine. So absorption of nutrients is impaired. Children, as a consequence, suffer from energy and micronutrient deficiency (Prasad et al. 1983). Infants and toddlers have a natural tendency to concept formation by placing objects in their mouth. It is a physiological process. But a few children develop habit of eating nonedible substances during the period (Advani et al. 2014).

Finding in present research is authenticated by Roselle (1970). The author claimed that additional cause of geophagy may be iron, zinc or calcium deficiency in children. Nevertheless, geophagy is observed invariably in children having iron deficiency. Probably, clay and/or starch particles chelate iron and render iron unavailable for absorption in the small intestine. But exact interrelation between iron deficiency and geophagy is uncertain.

Say et al. (1969) reported severe hepatomegaly, bleeding tendency and delayed wound healing in children along border of Iran and Turkey. Children were observed to have iron deficiency and chronic soil eating tendency. Possibly, iron deficiency associated with zinc deficiency as per authors which resulted in secondary manifestations in the body.

5.7.3.1 Conclusion

Good health is an outcome of balanced diet and good hygiene habits. Geophagy is ubiquitous in a poverty-dependent environment. It divests health of child, especially during formative period of child age. Educative, interceptive, corrective and remedial measures should be adopted to defeat it to save our future of mankind.

5.8 HIV-Induced Anemia

Anemia is a frequent haematological complication among children suffering from HIV infection (Moore 1999: Moore 2000). Anemia negatively affects the prognosis of HIV disease (Moyle 2002). It mars the clinical outcome of disease (Belperio and Rhew 2004). Studies by Kuller et al. (2008), Ferruci et al. (2005) and Rawat et al. (2013)) strongly favour a relation between pathogenesis of anemia in patients suffering from HIV. The authors asserted that a number of pro-inflammatory cytokines are synthesized in the body of HIV patients which predispose to the development of anemia (Kreuzer and Rockstroh 1997; Scadden 1997: Coyle 1997).

Anemia is decided as the predictor of HIV infection among adults and preschool children. A decline in the concentration of Hb is accompanied by the probability of progression HIV infection (Volberding et al. 2004). Anemia worsens the mortality of HIV-infected children (Volberding 2000, 2002; Mocroft et al. 1999a: Clark et al. 2002: Sullivan et al. 1988: Osterbauer et al. 2012; Malaba et al. 2007).

Anemia has been observed in patients who are undergoing combination antiretroviral therapy (cART) for HIV and also in patients who have yet to start the treatment (Redig and Berliner 2013). The hypothesis suggests that the virus itself may be responsible for the inhibition of erythropoiesis resulting in anemia.

According to Blanc et al. (1968), for the diagnosis of anemia in HIV patients, a sharp cut-off has been suggested, and the value is Hb <10 g/100 ml. This cut-off has been applied for the study in Asia and Africa in HIV patients, as has been asserted by Blanc et al. (1968).

Anemia in HIV is caused by drug toxicity. In a study performed by workers (Moore et al. 1992) on patients undergoing cART and patients who not following active treatment, it was observed that cART helped in recovering of anemia in HIV patients. The work is supported by another study (Anude et al. 2013), where it was proved that anemia was associated with those HIV patients who failed to comply with drug regimen and could not achieve viral suppression. So it was assumed that anemia in these patients is an indicator of treatment failure.

Other studies (Moore et al. 1992; Russell et al. 2010; Mocroft et al. 1999b; Sullivan et al. 1990) support that anemia in HIV is a marker of the prognosis of disease. It refers to the progression of morbidity and mortality associated with HIV. These studies posited that virus has a dominant role in the cause of anemia.

Anemia in HIV is multifactorial in origin. Each factor has its role in individuality and totality. Secondary infections in HIV patients can also cause anemia. HIV patients living in endemic regions where malaria and helminthic infections are common have higher susceptibility of developing signs of anemia (Redig and Berliner 2013). It is due to compromised immunity of these patients. Moreover, deficiency of micronutrients like iron, folate and cobalamin in HIV patients could be the additional cause of anemia. Infection of HIV definitely leads to abnormal immune response. This condition exacerbates the pathogenesis of anemia. Molecular similarity has been advocated between erythropoietin (EPO) (Elliott and Sinclair 2012) of host and HIV-1 p17 protein (De Francesco et al. 2002). This mimicry could be the cause of synthesis of autoantibodies against endogenous EPO in HIV+ patients. It might cause dysregulation of erythropoiesis at the level of the bone marrow (Tsiakalos et al. 2011: Kordossis et al. 1997: Jubault et al. 2000: Zandman-Goddard and Shoenfeld 2002: Kreuzer et al. 1997).

According to Nyesigire et al. (2012), anemia in infected children is a major health problem especially in African regions. The study was conducted among HIV-infected children in age between 3 months and 18 years. Nyesigire et al. (2012) observed around 57.6% prevalence of anemia in children with about 62% of the children had mild anemia, whereas approximately 5% of children suffered with severe anemia. They concluded that anemia was associated with HIV infection among children in Uganda clinic.

A descriptive study was conducted by Enawgaw et al. (2015) among HIV-infected children in age between 6 months and 14 years in the University of Gondar Hospital. They observed that around 1/5th of the total participants suffered from mild anemia and (Enawgaw et al. 2015) only 2% of the children had severe anemia, as per WHO cut-off for Hb.

In developing countries, children are inflicted with nutritional anemia. Moreover, infection with HIV, additionally, deteriorates the iron store in the body, and the prevalence of iron deficiency anemia, according to Semba (2003), Brabin et al. (2001a, b) and Dhillon (2000), is exacerbated.

Sipsas et al. (1999) reported the presence of autoantibodies in plasma against the endogenous erythropoietin. Therefore, the authors (Tsiakalos et al. 2010) posited the role of autoimmunity in the pathogenesis of anemia in HIV children.

Autoimmunity has been explained by Stratton et al. (2009) and Benchley et al. (2006). Impaired control of T and or B lymphocytes might be implicated in the progression of autoimmunity against endogenous erythropoietin in HIV.

Molecular similarity in sequence and/or structure between the host and pathogen could initiate an autoimmune reaction (Godschalk et al. 2004). Viral peptides can also initiate autoimmune response.

A study by Calenda and Chermann (1992) reported the effect of human immunodeficiency virus on the erythroid cells. In vitro study demonstrated a direct inhibitory effect on the erythropoiesis by affecting the differentiation at the level of BFU-E and CFU-E. In another study by Calis et al. (2010), HIV-infected children suffered with anemia, and the number of CD34+ cells and erythroid progenitor cells in HIV-positive children with anemia was reduced in comparison to HIV− children with anemia.

The authors have different views regarding the susceptibility of CD34+ cells towards AIDS virus. A study by Josefsson et al. (2012), Shen et al. (1999) and

Weihold et al. (1998) reported that these cells are non-responsive to AIDS virus. Contrarily, a study by Mcnamara and Collins (2011) and McNamara et al. (2013) have detected AIDS virus in the CD34+ haematopoietic progenitor cells, whereas the viral load in blood circulation was untraceable. Anemia in HIV sufferers may have varied aetiology, like deficiency of micronutrients, initiation of cART and AIDS virus; however, any cause of anemia in HIV is essentially accompanied by normochromic and normocytic erythrocytes, normal iron stores, low reticulocyte count and dysregulated erythropoiesis (Redig and Berliner 2013). These characteristics are also observed in anemia of inflammation (Vanasse and Berliner 2010). Therefore, it can be asserted that pro-inflammatory cytokines, additionally, have profound effect on the pathogenesis of anemia in HIV.

Administration of zidovudine in HIV-infected patients is commonly accompanied by anemia. Richman et al. (1987) conducted placebo-controlled, double-blind clinical trial to find the effect of oral administration of azidothymidine in HIV patients. They (Richman et al. 1987) observed adverse effects of the drug in patients. Nausea, headache, muscle ache and insomnia were frequent symptoms. Bone marrow suppression was posited by the authors (Richman et al. 1987) in patients. Further, they (Richman et al. 1987) noticed decline in Hb to 7.5 g/100 ml, characteristic of anemia in patients.

Studies by Volberding et al. (1990) and Collier et al. (1990) posited that the severity of anemia in HIV patients receiving oral zidovudine is proportional to the dose of zidovudine and severity of HIV disease. Oral zidovudine has suppressive effect on the bone marrow and is predisposed to anemia (Fischl et al. 1990; Walker et al. 1988; Gill et al. 1987; Dainark et al. 1988; Falguera et al. 1995).

Recent trend in the use of combination ART against HIV infection is associated with fewer adverse drug reactions in comparison to zidovudine alone. A study was conducted by Eron et al. (1995) involving the use of lamivudine and zidovudine in comparison to use of individual drugs for treatment of HIV infection. The authors confirmed that combination ART have high efficacy in combating HIV infection and lowering the plasma HIV-1 RNA levels and raising the CD4+ cells. The cART has least possible adverse effects than the use of either drug.

Immunocompromised patients are prone to infection by *Mycobacterium avium* complex. AIDS patients in the advanced stage of disease are easy target for this type of complicated infection.

These patients have high grade involving the bone marrow, hence suppressing erythropoiesis. A study conducted by Jacobson et al. (1990) at San Francisco General Hospital posited the direct correlation between MAC-induced bacteraemia and high probability for the need of blood transfusion. According to authors (Mata-Marin et al. 2010), profound cause of anemia in HIV infection is the chronic blood loss in these patients resulting in haemorrhagic anemia.

Erythropoiesis in HIV disease is decreased leading to anemia. It might be attributed to bone marrow infection (Hambleton 1996), use of drugs for marrow suppression (Levine 1999), AIDS virus(Coyle 1997) and decreased synthesis of erythropoietin (Spivak et al. 1989).

Exaggerated haemolysis in HIV infection is another major cause of anemia. It is the consequence of increased sequestration of erythrocytes in the spleen, synthesis of autoantibodies against erythrocytes, thrombocytopenia or deficiency of glucose-6-phosphate dehydrogenase (McGinniss et al. 1986; Rules et al. 1991; Blanche et al. 1994; Sasadeusz et al. 1990; Rarick et al. 1992).

Still, another important cause for anemia in AIDS patients is deficiency of folic acid, vitamin B_{12} and iron. Malabsorption as a consequence of secondary infection of GIT in AIDS disease is the frequent cause of cobalamin and folic acid deficiency resulting in nutritional anemia (Harriman et al. 1989). A study by Griffin (1990) confirmed a reduced localized immunity in the gut mucosa due to HIV infection. It becomes a favourable ground for the expression of opportunistic enteric organism. However, a clear understanding of the compromised intestinal immunity in HIV patients is lacking. HIV-associated nutritional anemia and macronutrients deficiency lead to malnutrition in children (IPISG 1993; Schwenk et al. 1993; Salomon et al. 2002; Kotler 1995; Ehrenpreis et al. 1994). According to Griffin (1990), autonomic neuropathy of jejunum could cause diarrhoea in AIDS patients. Malabsorption of fat and sugars might result in emaciation. The authors (Lambi et al. 1996; Asmuth et al. 1994; Kotler and Orenstein 1998; Mor et al. 2009) observed the presence of microsporidia in HIV-infected patients who had chronic diarrhoea.

5.9 Iron Deficiency Anemia in Coeliac Disease

Coeliac disease is an autoimmune response to the mucosa of the small intestine, also known as gluten-induced enteropathy. Mucosal villi undergo atrophy with hyperplasia of crypts (Catassi and Fasano 2008).

Nutritional anemia is a common symptom associated with coeliac disease. Deficiency of vitamin B_{12} and folate are the frequent consequence of coeliac disease; however, iron deficiency has also been reported with coeliac disease. It has been observed by Jones et al. (2006) that 46% of the patients suffering from subclinical coeliac disease were associated with iron deficiency anemia. It implies that the absorptive capability of intestinal mucosa is diminished in those patients. Coeliac disease had not expressed clinically. But T cell-mediated immunity against the mucosa has caused degenerative changes in its structure and function. Iron deficiency anemia was the first discernible symptom of the inflammatory changes taking place within the intestinal mucosa.

Intact intestinal mucosa is a prerequisite for absorption of iron from diet. It occurs in proximal part of the intestinal mucosa. However, inflammatory changes derange normal architecture of the mucosa. Atrophy of mucosal villi is the prominent feature. Hence absorption of iron and other micronutrients is impaired in coeliac disease. Iron deficiency anemia is the easily and essentially noticeable symptom in these patients (Halfdanarson et al. 2007). Therefore, iron deficiency among coeliac disease patients arises from its impaired absorption through the mucosa of the small intestine as a consequence of atrophy of the villi. Its manifestation is iron deficiency anemia.

Additional cause of iron deficiency anemia in coeliac disease (CD) can be the loss of blood in faeces. It is not visible, but blood in faeces can be tested by faecal occult blood test. This assumption is proved by the work of Fine (1996), wherein occult blood test was conducted in 34 patients who suffered from coeliac disease. CD patients were grouped as partial atrophy of the villi and complete atrophy of the villi. Occult blood test was positive in 25% of the patients with partial villi atrophy, whereas 54% of the patients with complete villous atrophy had symptom of occult blood loss. Persistent and prolonged blood loss from the damaged intestinal mucosa invariably results into iron loss from the body. Occult blood loss in coeliac disease is negatively supplemented with defective iron absorption. These factors, collectively, contribute to iron deficiency anemia in CD patients.

Still another cause of anemia in coeliac disease is the autoimmune-mediated chronic inflammation in the mucosa of the intestine. It is called as anemia of inflammation. Coeliac disease is an autoimmune-mediated disorder of the intestinal mucosa affecting genetically predisposed individuals with equal age predilection.

Anemia of inflammation also known as anemia of chronic disease is the consequence of chronic infection, inflammation and/or malignancy in the body and is characterized by low serum iron despite the presence of normal iron body store. Haemoglobin level in anemia of inflammation is usually 8 g/100 ml. It is associated with normal shape and normal Hb content of erythrocytes (normocytic normochromic). Iron deficiency anemia is associated with hypochromic microcytic RBC. Clinically, it is difficult to delineate both the types of anemia, and they exist together in any chronic inflammation (Cartwright and Lee 1971).

Nutritional anemia, predominantly, iron deficiency anemia, is commonly observed in coeliac disease as an outcome of impaired absorption of micronutrients by intestinal mucosa; nevertheless, scarce research has been done to find out the presence of anemia of inflammation/anemia of chronic disease in patients suffering from coeliac disease.

It is hypothesized that anemia of chronic disease should be prevalent in coeliac disease with or without the presence of iron deficiency anemia. The assumption is based on the fact that coeliac disease is characterized by the T cell-mediated intense inflammation of the intestinal mucosa in response to the antigenicity of gluten in diet.

Bergamaschi et al. (2008) conducted a study over 152 patients affected with coeliac disease. it was detected that (Bergamaschi et al. 2008) 42% of the patients with coeliac disease were suffering from anemia. Further analysis (Bergamaschi et al. 2008) yielded that 81% of the patients out of 42% of anaemic patients had iron deficiency anemia. Another 16% (Bergamaschi et al. 2008) of the patients suffered from low serum erythropoietin level and raised serum interferon-gamma level. This observation could be indicative of anemia of chronic inflammation in coeliac disease patients. Patients of both the types of anemia were given gluten-free diet(Bergamaschi et al. 2008). Anaemic condition improved, and response to gluten-free diet was equally effective in iron deficiency anemia and anemia of chronic disease.

Inflammatory changes of the intestinal mucosa regressed following gluten-free diet. Hence, absorption of micronutrients enhanced, and pro-inflammatory cytokines level in the blood circulation was normalized.

5.9.1 IL-6-Hepcidin-Hypoferraemia

Hypoferraemia is the main characteristic (Elizabeta et al. 2004) sequel of anemia of chronic inflammation. Iron remains locked in the storage sites. There are three prominent iron storage tissues in the body. The first is the intestinal mucosal cells. The second is the hepatocytes, and the third is the macrophages which contain iron that was sequestered and recycled from the senile erythrocytes. Hence, the main cause of low serum iron in anemia of chronic inflammation is the impaired iron export from storage site to circulation, thereby restricting the availability of iron to bone marrow for erythropoiesis. Hepcidin is the principal regulator of iron in the body. It is a peptide hormone secreted by the liver. During inflammation, the liver increases its secretion. In coeliac disease, interferon-γ and interleukin-6 level is raised. These cytokines induce the hypersecretion of hepcidin. It interferes with iron transport.

5.9.2 Hepcidin-Induced Ferroportin Degradation

Hepcidin binds to ferroportin, a peptide responsible for iron export from within the cells to the outside into circulation. Ferroportin undergoes conformation changes in response to binding with hepcidin. Ferroportin is degraded. Export of iron is prevented and in turn results into low serum iron level in anemia of chronic inflammation (Nemeth et al. 2004).

5.9.3 Erythrophagocytosis

Additional assertion in anaemic of infection is the premature phagocytosis of erythrocytes. It was studied by Freireich et al. (1957) with the help of radioactive iron in patients with juvenile rheumatoid arthritis. It was found that RBC survival was higher in healthy patients in comparison to RBC in anemia of inflammation.

Possibly, chronic inflammation provokes the synthesis and release of interleukin-6, interleukin-1 and interferon-γ. These peptides stimulate macrophages. So macrophage-mediated phagocytosis of erythrocytes is the common feature in anemia of inflammation. Low RBC count due to excessive destruction and premature lysis is responsible for anemia in infection and inflammation as is seen in coeliac disease.

5.9.3.1 Suppression of Erythropoiesis by Inflammation

Furthermore, erythropoiesis in inflammation and infection is inhibited causing anemia. Pro-inflammatory cytokines like TNFα, IL-1 and IFN-γ, principally, suppress the erythropoiesis in bone marrow. These cytokines suppress the activity of erythroid progenitors. Libergts et al. (2011) developed a mouse model to assess effect of IFN-γ on erythropoiesis. It was observed that hyper-synthesis of IFN-γ in mice resulted in development of anemia. It was due to decrease in lifespan of RBC owing to macrophage-mediated phagocytosis in the spleen. Also, suppression of erythroblasts and committed stem cells (BFU-e) was a prime feature in mouse model.

Nutritional anemia, predominantly, the iron-deficient anemia, is a common feature in patients with coeliac disease. Low serum iron is a manifestation of iron deficiency anemia associated with declined serum ferritin level. However, it is unexpected to find serum ferritin level raised in anaemic patients. A study was conducted by Harper et al. (2007) on a cohort of 405 patients diagnosed with coeliac disease. Iron deficiency was observed in around one-third of the total male patients and one-fifth of the total female patients. IDA was the feature in one-fifth of the total patients. Majority of anaemic patients showed low serum ferritin level, except one-tenth of the total anaemic patients who showed high serum ferritin level. This observation further asserts that iron deficiency anemia is associated with anaemic of chronic disease in coeliac disease patients. Ferritin is an acute phase reactant protein (IDI 2009).

In coeliac disease, autoimmunity is directed towards the intestinal mucosa. So inflammation of the mucosa of the intestine elevates the serum ferritin level.

References

Abdalla SH (1988) Peripheral blood and bone marrow leucocytes in Gambian children with malaria: numerical changes and evaluation of phagocytosis. Ann Trop Paediatr 8(4):250–258

Abdalla SH (1990) Hematopoiesis in human malaria. Blood Cells 16:401–416

Abdalla S, Weatherall DJ, Wickramasinghe SN, Hughes M (1980) The anemia of P. falciparum malaria. Br J Haematol 46:171–183

Abrahams P (2003) Human geophagy: a review of its distribution, causes, and implications. In: Catherine H, Skinner W, Berger AR (eds) Geology and health: closing the gap. Oxford University Press, Oxford

Acharya U (2007) Serum levels of iron, zinc and lead in patients of Pica: a prospective study. Rajasthan Med J: 11–14

Adam I, Khamis AH, Elbashir MI (2005) Prevalence and risk factors for anemia in pregnant women of eastern Sudan. Trans R Soc Trop Med Hyg 99:739–743

Adedoja A, Tijani BD, Akanbi AA, Ojurongbe TA, Adeyeba OA, Ojurongbe O (2015) Co-endemicity of plasmodium falciparum and intestinal Helminths infection in school age children in rural communities of Kwara State Nigeria. PLoS Negl Trop Dis 9(7):e0003940

Adegnika AA, Kremsner PG (2012) Epidemiology of malaria and helminth interaction: a review from 2011 to 2011. Curr Opin HIV AIDS 7(3):221–224

Advani S, Kochhar G, Chachra S, Dhawan P (2014) Eating everything except food (PICA): a rare case report and review. J Int Soc Prev Commun Dent 4(1):1–4

Agarwal N, Prchal JT (2009) Anemia of chronic disease (anemia of inflammation). Acta Haematol 122(2–3):103–108

Akcam M, Ozdem S, Yilmaz A, Gultekin M, Artan R (2007) Serum ferritin, vitamin B(12), folate, and zinc levels in children infected with Helicobacter pylori. Dig Dis Sci 52:405–410

Albonico M, Stoltzfus RJ, Savioli L (1998) Epidemiological evidence for a differential effect of hookworm species, Ancylostoma duodenale or Necator americanus, on iron status of children. Int J Epidemiol 27(3):530–537

American Academy of Pediatrics (1997) Work group on breastfeeding breastfeeding and the use of human milk. Pediatrics 100:1035–1039

American Psychiatric Association (1994) Diagnostic and statistical manual of mental disorders: DSM-IV. APA, Washington, DC, p 866. Available at: http://www.psychiatryonline.com/DSMPDF/dsm-iv.pdf

Amodu OK, Adeyemo AA, Olumese PE, Gbadegesin RA (1998) Intraleucocytic malaria pigment and clinical severity of malaria in children. Trans R Soc Trop Med Hyg 92:54–56

Anell B, Lagercrantz S (1958) Geophagical customs. Almquist and Wiksells Boktryckeri, Uppsala

Annibale B, Marignani M, Monarca B, Antonelli G, Marcheggiano A, Martino G, Mandelli F, Caprilli R, Delle Fave G (1999) Reversal of iron deficiency anemia after Helicobacter pylori eradication in patients with asymptomatic gastritis. Ann Intern Med 131:668–672

Annibale B, Capurso G, Lahner E, Passi S, Ricci R, Maggio F, Delle Fave G (2003) Concomitant alterations in intragastric pH and ascorbic acid concentration in patients with Helicobacter pylori gastritis and associated iron deficiency anemia. Gut 52(4):496–501

Antelman G, Msamanga GI, Spiegelman D, Urassa EJ, Narh R (2000) Nutritional factors and infectious disease contribute to anemia among pregnant women with human immunodeficiency virus in Tanzania. J Nutr 130:1950–1957

Anude CJ, Eze E, Onyegbutulem HC (2013) Immuno-virologic outcomes and immuno-virologic discordance among adults alive and on anti-retroviral therapy at 12 months in Nigeria. BMC Infect Dis 13:113

Anyona SB, Schrier SL, Gichuki CW, Waitumbi JN (2006) Pitting of malaria parasites and spherocytes formation. Malar J 5:64–65

Arese P, Schwarzwer E (1997) Malarial pigment (haemozoin): a very active “inert” substance. Ann Trop Med Parasitol 91:501–516

Artavanis-Tsakonas K, Riley EM (2002) Innate immune response to malaria: rapid induction of IFN-gamma from human NK cells by live Plasmodium falciparum-infected erythrocytes. Immunol J 169(6):2956–2963

Arvas A, Elgormus Y, Gur E, Alikasifoglu M, Celebi A (2000) Iron status in breast-fed full-term infants. Turk J Pediatr 42(1):22–26

Ashorn M, Ruuska T, Makipernaa A (2001) Helicobacter pylori and iron deficiency anemia in children. Scand J Gastroenterol 36:701–705

Asmuth DM, DeGirolami PC, Federman M, Ezratty CR, Pleskow DK, Desai G, Wanke CA (1994) Clinical features of microsporidiosis in patients with AIDS. Clin Infect Dis 18(5):819–825

Atoda H, Hyuga M, Morita T (1991) The primary structure of coagulation factor IX/factor X-binding protein isolated from the venom of Trimeresurus flavoviridis. Homology with asialoglycoprotein receptors, proteoglycan core protein, tetranectin, and lymphocyte Fc epsilon receptor for immunoglobulin E. J Biol Chem 266:14903–14911

Aufreiter S, Hancock RGV, Mahaney WC, Stambolic-Robb A, Sanmugadas K (1997) Geochemistry and mineralogy of soils eaten by humans. Int J Food Sci Nutr 48:293–305

Awasthi YC, Yang Y, Tiwari NK, Patrick B, Sharma A, Li J, Awasthi S (2004) Regulation of 4-hydroxynonenal-mediated signaling by glutathione S-transferases. Free Radic Biol Med 37(5):607–619

Axelsson II, Gebre-Medhin M, Hernell O, Jakobson I, Michaelsen KF, Samuelson G (1999) Recommendations for prevention of iron deficiency. Delay cow’s milk intake as a beverage to infants until 10-12 months of age. Lakartidningen 96(18):2206–2208

Badamosi EJ, Ibrahim LM, Temple VJ (1995) Nutritional evaluation of a locally formulated weaning food, JUTH-PAP. West Afr J Biol Sci 3:85–93

Bahu Mda G, da Silveira TR, Maguilnick I, Ulbrich-Kulczynski J (2003) Endoscopic nodular gastritis: an endoscopic indicator of high-grade bacterial colonization and severe gastritis in children with Helicobacter pylori. J Pediatr Gastroenterol Nutr 36:217–222

Bain BJ (1999) Pathogenesis and pathophysiology of anemia in HIV infection. Curr Opin Hematol 6(2):89–93

Baker SJ (1981) Nutritional anemias. Pan 2. Tropical Asia. Clin Haematol 3:843–871

Barabino A, Dufour C, Marino CE, Claudiani F, De Alessandri A (1999) Unexplained refractory iron-deficiency anemia associated with Helicobacter pylori gastric infection in children: further clinical evidence. J Pediatr Gastroenterol Nutr 28:116–119

Barrera V, Skorokhod OA, Baci D, Gremo G, Arese P, Schwarzer E (2011) Host fibrinogen stably bound to hemozoin rapidly activates monocytes via TLR-4 and CD 11b/CD18 integrins: a new paradigm of hemozoin action. Blood 117(21):5674–5682

Bates I, Bedu Addo G, Bevan DH (1991) Use of immunoglobin gene rearrangement to show clonal lymphoproliferative is hypereactive malarial splenomegaly. Lancet 337:505–507

Beaver PC, Jung RC, Cupp EW (1984) Clinical parasitology. Lea and Febiger, Philadelphia, pp 284–285

Bella J, Kolatkar PR, Marlor CW, Greve JM, Rossmann MG (1998) The structure of two amino terminal domains of human ICAM-1 suggests how it functions as a rhinovirus receptor and as an LFA-1 integrin ligand. Proc Natl Acad Sci U S A 95(8):4140–4145

Belperio PS, Rhew DC (2004) Prevalence and outcomes of anemia in individuals with human immunodeficiency virus: a systematic review of the literature. Am J Med 116(Suppl 7A):27S–43S

Benchley JM, Price DA, Douek DC (2006) HIV disease: fallout from a mucosal catastrophe? Nat Immunol 7:235–239

Bergamaschi G, Markopoulos K, Albertini R, Di Sabatino A, Biagi F, Ciccocioppo R, Arbustini E, Corazza GR (2008) Anemia of chronic disease and defective erythropoietin production in patients with celiac disease. Haematologica 93:1785–1791

Berkley J, Mwarumba S, Bramham K, Lowe B, Marsh K (1999) Bacteraemia complicating severe malaria in children. Trans R Soc Med Hyg 93:283–286

Bernstein CN, Leslie WD, Leboff MS (2003) AGA technical review on osteoporosis in gastrointestinal diseases. Gastroenterology 124(3):795–841

Bethony J, Chen J, Lin S, Xiao S, Zhan B, Li S, Xue H, Xing F, Humphries D, Yan W, Chen G, Foster V, Hawdon JM, Hotez PJ (2002) Emerging patterns of hookworm infection: infuence of aging on the intensity of Necator infection in Hainan Province, Peoples Republic of China. Clin Infect Dis 35:1336–1344

Bethony J, Brooker S, Albonico M, Geiger SM, Loukas A, Diemert D, Hotez PJ (2006) Soil-transmitted helminth infections: ascariasis, trichuriasis, and hookworm. Lancet 367:1521–1532

Bhatia J (2007) Human milk and the premature infant. J Perinatol 27:S71–S74

Bhuiyan TR, Quadri F, Saha A, Svennerholm AM (2009) Infection by Helicobacter pylori in Bangladeshi children from birth to two years: relation to blood group, nutritional status, and seasonality. Pediatr Infect Dis J 28(2):79–85

Blanc B, Finch CA, Hallberg L (1968) Nutritional anemias: report of a WHO Scientific Group. WHO Tech Rep Ser 405:1–40

Blanche P, Robert F, Dupouy-Camet J, Sicard D (1994) Toxoplasmosis-associated hemophagocytic syndrome in a patient with AIDS: diagnosis by the polymerase chain reaction. Clin Infect Dis 19:989–990

Bleakley H (2003) Disease and development: evidence from hookworm eradication in American South. J Eur Econ Assoc 1:376–386

Booth M (2006) The role of residential location in apparent helminth and malaria associations. Trends Parasitol 22:359–362

Borigato EV, Martinez FE (1998) Iron nutritional status is improved in Brazilian preterm infants fed food cooked in iron pots. J Nutr 128(5):855–859

Bothwell TH (1972) Iron deficiency. Med J Aust 2:433–438
Bousema T, Okell L, Felger I, Drakeley C (2014) Asymptomatic malarial infections:detectability, transmission and public health relevance. Nat Rev Microbiol 12:833–840
Brabin BJ, Premji Z, Verhoeff F (2001a) An analysis of anemia and child mortality. Nutr J 131(2S-2):636S–645S
Brabin JB, Premji Z, Verhoeff F (2001b) Iron deficiency anemia: reexamining the nature and magnitude of the public health problem. J Nutr 131:636S–648S
Bradley JE, Jackson JA (2004) Immunity, immunoregulation and the ecology of trichuriasis and ascariasis. Parasite Immunol 26:429–441
Brooker S, Clements ACA (2009) Spatial heterogeneity of parasite co-infection: determinants and geostatistical prediction at regional scales. Int J Parasitol 39:591–597
Brooker S, Michael E (2000) The potential of geographical information systems and remote sensing in the epidemiology and control of human helminth infections. Adv Parasitol 47:245–288
Brooker S, Peshu N, Warn PA, Mosobo M, Guyatt HL, Marsh K, Snow RW (1999) The epidemiology of hookworm infection and its contribution to anemia among pre-school children on the Kenyan coast. Trans R Soc Trop Med Hyg 93(3):240–246
Brooker S, Clements A, Bundy DAP (2006) Global epidemiology, ecology and control of soil-transmitted helminth infections. Adv Parasitol 62:223–265
Brooker SJ, Pullan RL, Gitonga CW, Ashton RA, Kolaczinski JH, Kabatereine NB, Snow RW (2012) Plasmodium-helminth coinfection and its sources of heterogeneity across East Africa. J Infect Dis 205:841–852
Brown LM (2000) Helicobacter pylori:epidemiology and routes of transmission. Epidemiol Rev 22(2):283–297
Brown KN, Brown IN (1965) Immunity to malaria antigenic variation in chronic infections in P.knowlesi. Nature 208:1286–1288
Brown AE, Webster HK, Teja-Isavadharm P, Keeratithakul D (1990) Macrophage activation in falciparum malaria as measured by neopterin and interferon-gamma. Clin Exp Immunol 82:97–101
Bruce MC, Alano P, Duthie S, Carter R (1990) Commitment of the malaria parasite Plasmodium falciparum to sexual and asexual development. Parasitology 100:191–200
Burgmann H, Looareesuwan S, Kapiotis S, Viravan C, Vanijanonta S, Hollenstein U, Weisinger E, Presteri E, Winkler S, Graninger W (1996) Serum levels of erythropoietin in acute Plasmodium falciparum malaria. Am J Trop Med Hyg 54:280–283
Cairncross S (2003) Sanitation in the developing world: current status and future solutions. Int J Environ Health Res 13(S1):S123–S131
Calandra T, Bucala R (1995) Macrophage migration inhibitory factor: a counter-regulator of glucocorticoid action and critical mediator of septic shock. Inflamm J 47(1–2):39–51
Calandra T, Roger T (2003) Macrophage migration inhibitory factor: a regulator of innate immunity. Nat Rev Immunol 3(10):791–800
Calenda V, Chermann JC (1992) Severe in vitro inhibition of erythropoiesis and transient stimulation of granulopoiesis after bone-marrow infection with eight different HIV-2 isolates. AIDS 6(9):943–948
Calis JC, Phiri KS, Vet RJ (2010) Erythropoiesis in HIV-infected and uninfected Malawian children with severe anemia. AIDS 24(18):2883–2887
Cappello M (2004) Global health impact of soil-transmitted nematodes. Pediatr Infect Dis J 23:663–664
Cappello M, Clyne LP, McPhedran P, Hotez PJ (1993) J Infect Dis 167:1474–1477
Cappello M, Vlasuk GP, Bergum PW, Huang S, Hotez PJ (1995) Ancylostoma caninum anticoagulant peptide: a hookworm-derived inhibitor of human coagulation factor Xa. Proc Natl Acad Sci U S A 92:6152–6156
Capurso G, Marignani M, Delle Fave G, Annibale B (1999) Iron-deficiency anemia in premenopausal women: why not consider atrophic body gastritis and Helicobacter pylori role? Am J Gastroenterol 94:3084–3085

Capurso G, Lahner E, Marcheggiano A, Caruana P, Carnuccio A, Bordi C, Delle Fave G, Annibale B (2001) Involvement of the corporal mucosa and related changes in gastric acid secretion characterize patients with iron deficiency anemia associated with Helicobacter pylori infection. Aliment Pharmacol Ther 15:1753–1761

Carlson M, Nakamura Y, Payson RO, Connell P, Leppert M, Lathrop GM, Lalouel JM, White R (1988) Isolation and mapping of a polymorphic DNA sequence on chromosome 18 (D18S24). Acids Research 16(9):4188

Cartwright GE, Lee GR (1971) The anemia of chronic disorders. Br J Haematol 21(2):147–152

Casals-Pascual C, Kai O, Cheung JO, Williams S, Lowe B, Nyanoti M, Williams TN, Maitland K, Molyneux M, Newton CR, Peshu N, Watt SM, Roberts DJ (2006) Suppression of erythropoiesis in malarial anemia is associated with hemozoin in vitro and in vivo. Blood 108(8):2569–2577

Cashman KD (2006) Milk minerals (including trace elements) and bone health. Int Dairy J 16:1389–1398

Catassi C, Fasano A (2008) Celiac disease. Curr Opin Gastroenterol 24:687–691

Centers for Disease Control and Prevention (2011a) Hygiene related diseases. Available at: http://www.cdc.gov/healthywater/hygiene/disease/chronic_diarrhea.html

Centers for Disease Control and Prevention (2011b) Neglected tropical diseases. Available at:http: www.cdc.gov/globalhealth/ntd/.

Centers for Disease Control and Prevention (2014) Parasites. Available at: http://www.cdc.gov/parasites/children.html

Centers for Disease Control and Prevention (2015) Anopheles mosquitos. Available at: https://www.cdc.gov/malaria/about/biology/mosquitoes/

Centers for Disease Control and Prevention (Reviewed 2016) Malarial parasite-biology. Available at: http://www.cdc.gov/malaria/about/biology/index.html

Centers for Diseases Control and Prevention (2014) Neglected tropical diseases. Available at: http://www.cdc.gov/globalhealth/ntd/diseases/index.html

Cesur S (2005) Neopterin: a marker used for monitoring infections. Mikrobiyol Bul 39(2):251–260

Chan MS (1997) The global burden of intestinal nematode infections: fifty years on. Parasitol Today 13:438–443

Chan SH, Perussia B, Gupta JW, Kobayashi M, Pospisil M, Young HA, Wolf SF, Young D, Clark SC, Trinchieri G (1991) Induction of IFN-g production by NK cell stimulatory factor (NKSF): characterization of the responder cells and synergy with other inducers. J Exp Med 173:869–879

Chan SH, Kobayashi M, Santoli D, Perussia B, Trinchieri G (1992) Mechanisms of IFN-g induction by natural killer cell stimulatory factor (NKSF/IL-12): role of transcription and mRNA stability in the synergistic interaction between NKSF and IL-2. J Immunol 148:92–98

Charlton RW, Bothwell TH (1983) Iron absorption. Annu Rev Med 34:55–68

Chatterjee P, Gettman JH (1972) Lead poisoning: subcultural as a facilitating agent? Am J Clin Nutr 25:324–330

Cheifetz AS, Brown A, Curry M, Moss AC (2011a) Oxford American hand book of gastroenterology and hepatology. Oxford University Press, Oxford, p 234

Cheifetz AS, Brown A, Curry M, Moss AC (2011b) American handbook of gastroenterology and hepatology. Oxford University Press, Oxford, p 234

Chiron D, Bekeredjian-Ding I, Pellat-Deceunynck C, Bataille R, Jego G (2008) Toll-like receptors: lessons to learn from normal and malignant human B cells. Blood 112:2205–2213

Choe YH, Kim SK, Son BK, Lee DH, Hong YC, Pai SH (1999) Randomized placebo-controlled trial of Helicobacter pylori eradication for iron-deficiency anemia in preadolescent children and adolescents. Helicobacter 4:135–139

Christian P, Khatry SK, West KP Jr (2004) Antenatal anthelmintic treatment, birthweight, and infant survival in rural Nepal. Lancet 364:981–983

Clark IA (1978) Does endotoxin cause both the disease and parasite death in acute malaria and babesiosis? Lancet 2(8080):75–77

Clark IA, Cowden WB (2003) The pathophysiology of falciparum malaria. Pharmacol Ther 99:221–260
Clark IA, Coeden WB, Butcher GA, Hunt NH (1987) Possible roles of tumor necrosis factor in the pathology of malaria. Am Pathol J 129(1):192–199
Clark TD, Mmiro F, Ndugwa C (2002) Risk factors and cumulative incidence of anemia among human immunodeficiency virus-infected children in Uganda. Ann Trop Paediatr 22:11–17
Coban C, Ishii KJ, Kawai T, Hemmi H, Sato S, Uematsu S, Yamamoto M, Takeuchi O, Itagaki S, Kumar N, Horii T, Akira S (2005) Toll-like receptor 9 mediates innate immune activation by the malaria. JEM 201(1):19–25. pigment hemozoin
Cohen HA, Shapiro RP, Frydman M, Varsano I (1991) Childhood protein-losing enteropathy associated with Helicobacter pylori infection. J Pediatr Gastroenterol Nutr 13:201–203
Cohen H, Weinstein WM, Carmel R (2000) Heterogeneity of gastric histology and function in food cobalamin malabsorption: absence of atrophic gastritis and achlorhydria in some patients with severe malabsorption. Gut 47:638–645
Collier AC, Bozzette S, Coombs RW, Causey DM, Schoenfeld DA, Spector SA, Pettinelli CB, Davies G, Richman DD, Leedom JM (1990) A pilot study of low-dose zidovudine in human immunodeficiency virus infection. N Engl J Med 323(15):1015–1021
Collins WE, Jeffery GM (2007) Plasmodium malaria: parasite and disease. Clin Microbiol Rev 20(4):579–592
Coltman CA (1969) Pagophagia and iron lack. J Am Med Assoc 207:513
Cook JD, Skikne BS, Baynes RD (1993) Serum transferrin receptor. Annu Rev Med 44:63–74
Costa FT, Avril M, Nogueira PA, Gysin J (2006) Cytoadhesion of Plasmodium falciparum-infected erythrocytes and the infected placenta: a two-way pathway. Braz J Med Biol Res 39(12):1525–1536
Coyle TE (1997) Hematologic complications of human immunodeficiency virus infection and the acquired immunodeficiency syndrome. Med Clin N Am 81(2):449–470
Crompton DW (2001) Ascaris and ascariasis. Adv Parasitol 48:285–375
Cynthia RE (2014) Practice essentials. In: Pica. Pataki C, Giardino A P(eds) Available at: http://emedicine.medscape.com/article/914765-overview
D'Andrea A, Rengaraju M, Valiante NM, Chehimi J, Kubin M, Aste M, Chan SH, Kobayashi M, Young D, Nickbarg E (1992) Production of natural killer cell stimulatory factor (interleukin 12) by peripheral blood mononuclear cells. J Exp Med 176(5):1387–1398
D'Ombrain MC, Robinson LJ, Stanisic DI, Taraika J, Bernard N, Michon P, Miller I, Schofield L (2008) Association of early interferon-gamma production with immunity to clinical malaria: a longitudinal study among Papua New Guinean children. Clin Infect Dis 47(11):1380–1387
Dainak N, Worthington M, Riordan MA, Kreczko S, Goldman L (1988) 3-Azido-3deoxythymidine (AZT) inhibits proliferation in vitro of human haematopoietic progenitor cells. Br J Haematol 69:299–304
Dale A, Thomas JF, Darboe MK, Coward WA, Harding M, Weaver LT (1998) Helicobacter pylori infection, gastric acid secretion, and infant growth. J Pediatr Gastroenterol Nutr 26:393–397
Dauncey MJ, Davies CG, Shaw JC, Urman J (1978) The effect of iron supplements and blood transfusion on iron absorption by low birthweight infants fed pasteurized human breast milk. Pediatr Res 12(9):899–904
De Francesco MA, Baronio M, Fiorentini S, Signorini C, Bonfanti C, Poiesi C, Popovic M, Grassi M, Garrafa E, Bozzo L (2002) HIV-1 matrix protein p17 increases the production of proinflammatory cytokines and counteracts IL-4 activity by binding to a cellular receptor. Proc Natl Acad Sci U S A 99(15):9972–9977
de Silva NR, Brooker S, Hotez PJ, Montresor A, Engels D, Savioli L (2003) Soil-transmitted helminth infections: updating the global picture. Trends Parasitol 19(12):547
Deehan MR, Goodridge HS, Blair D, Lochnit G, Dennis RD, Geyer R, Harnett MM, Harnett W (2002) Immunomodulatory properties of Ascaris suum glycosphingolipids: phosphorylcholine and non-phosphorylcholine-dependent effects. Parasite Immunol 24:463–469

del Valle A, Jones BF, Harrison LM, Chadderdon RC, Cappello M (2003) Isolation and molecular cloning of a secreted hookworm platelet inhibitor from adult Ancylostoma caninum. Mol Biochem Parasitol 129:167–177

DeMacyer E, Adiels-Tegman M (1985) The prevalence of anemia in the world. World Health Stat Q 38:302–316

Diemart DJ, Bethony JM, Hotez PJ (2008) Hook worm vaccines. Clin Infect Dis 46(2):282–288

Dillon JC (2000) Prevention of iron deficiency and iron deficiency anemia in tropical areas. Med Trop (Mars) 60(1):83–91

Diouf S, Camara B, Sall MG, Diagne I, Ndiaye O, Diallo A, Tall A, Signate HS, Moreire C, Sarr M, Fall M (2000) Protein-energy malnutrition in children less than five years old in a rural zone in Senegal. Dakar Med 45(1):48–50

Documbo OK, Sztein MB (2004) Serum levels of the proinflammatory cytokines interleukin-1 beta (IL-1beta), IL-6, IL-8, IL-10, tumor necrosis factor alpha, and IL-12(p70) in Malian children with severe Plasmodium falciparum malaria and matched uncomplicated malaria or healthy controls. Infect Immun 72:5630–5637

Dondorp AM, Omodeo-Sale F, Chotivanich K, Taramelli D, White NJ (2003) Oxidative stress and rheology in severe malaria. Redox Rep 8:292–294

Douki JB, Sterkers Y, Lepolard C, Traore B, Costa FT, Scherf A, Gysin J (2003) Adhesion of normal and Plasmodium falciparum ring-infected erythrocytes to endothelial cells and the placenta involves the rhoptry-derived ring surface protein-2. Blood 101:5025–5032

Drake L, Korchev Y, Bashford L, Diamgoz M, Wakelin D, Ashall F, Bundy D (1994) The major secreted product of the whipworm, Trichuris, is a pore-forming protein. Proc Biol Sci 257:255–261

Edington GM, Gilles HM (1976) Malaria. In: Pathology in tropics. Edward Arnold Ltd, London, pp 10–34

Ehrenpreis ED, Carlson SJ, Boorstein HL, Craig RM (1994) Malabsorption and deficiency of vitamin B12 in HIV-infected patients with chronic diarrhea. Dig Dis Sci 39(10):2159–2162

Eibach D, Ceron N, Krishnalall K, Carter K, Bonnot G, Bienvenu AL, Picot S (2012) Therapeutic efficacy of artemether-lumefantrine for Plasmodium vivax infections in a prospective study in Guyana. Malar J 11:347

el Hassan AM, Saeed AM, Fandrey J, Jelkmann W (1997) Decreased erythropoietin response in Plasmodium falciparum malaria-associated anemia. Eur J Haematol 59:299–304

Elizabeta N, Seth R, Victoria G, Charlotte K, Sarah T, Bente K, Tomas G (2004) IL-6 mediates hypoferremia of inflammation by inducing the synthesis of the iron regulatory hormone hepcidin. Clin Invest 113(9):1271–1276

Elliott S, Sinclair AM (2012) The effect of erythropoietin on normal and neoplastic cells. Biologics 6:163–189

El-Omar EM, Oien K, El-Nujumi A, Gillen D, Wirz A, Dahil S, Williams C, Ardill JE, McColl KE (1997) Helicobacter pylori infection and chronic gastric acid hyposecretion. Gastroenterology 113:15–24

Enawgaw B, Alem M, Melku M, Addis Z, Terefe B, Yitayew G (2015) Prevalence and associated risk factors of anemia among HIV infected children attending Gondar university hospital, Northwest Ethiopia: a cross sectional study. BMC Hematol 15:12

English M, Esami F, Wasunna A, Were F, Oqutu B, Warnae A, Snow RW, Peshu N (2004) Assessment of inpatient paediatric care in first referral level hospitals in 13 districts in Kenya. Lancet 363(9425):1948–1953

Eron JJ, Benoit SL, Jemsek J, MacArthur RD, Santana J, Quinn JB, Kuritzkes DR, Fallon MA, Rubin M (1995) Treatment with lamivudine, zidovudine, or both in HIV-positive patients with 200 to 500 CD4+ cells per cubic millimeter. North American HIV Working Party. N Engl J Med 333(25):1662–1669

Fagundes Neto U, Martin MC, Lima FL, Patricio FR, Toledo MR (1994) Asymptomatic environmental enteropathy among slum-dwelling infants. J Am Coll Nutr 13(1):51–56

Fagundes-Neto U, Viaro T, Wehba J, da Silva PFR, Machado NL (1984) Tropical enteropathy in early childhood: a syndrome caused by contaminated environment. Trop Pediatr 4(30):204–209
Falade C, Mokuolu O, Okafor H, Orogade A, Falade A, Adedoyin O (2007) Epidemiology of congenital malaria in Nigeria: a multi-centre study. Trop Med Int Health 12:1279–1287
Falguera M, Perez-Mur J, Puig T, Cao G (1995) Study of the role of vitamin B12 and folinic acid supplementation in preventing hematologic toxicity of zidovudine. Eur J Haematol 55:97–102
FAO (2005) Protecting and promoting good nutrition in crisis and recovery. Resource guide. FAO, United Nations, Rome
Farrell RJ, Kelly CP (2002) Celiac sprue. N Engl J Med 346:180–188
Fasano A (2005) Clinical presentation of celiac disease in the pediatric population. Gastroenterology 128:S68–S73
Fasano A (2011) Zonulin and its regulation of intestinal barrier function: the biological door to inflammation, autoimmunity, and cancer. Physiol Rev 91(1):151–175
Fearon DT, Locksley RM (1996) The instructive role of innate immunity in the acquired immune response. Science 272:50–54
Ferguson BJ, Skikne BS, Simpson KM, Baynes RD, Cook JD (1992) Serum transferrin receptor distinguishes the anemia of chronic disease from iron deficiency anemia. J Lab Clin Med 19:385–390
Ferrucci L, Corsi A, Lauretani F (2005) The origins of age-related proinflammatory state. Blood 105(6):2294–2299
Fincham JE, Markus MB, Adams VJ (2003) Could control of soiltransmitted helminthic infection influence the HIV/AIDS pandemic? Acta Trop 86:315–333
Fine KD (1996) The prevalence of occult gastrointestinal bleeding in celiac sprue. N Engl J Med 334:1163–1167
Fischl M, Galpin JE, Levine JD, Groopman JE, Henry DH, Kennedy P, Miles S, Robbins W, Miles S, Robbin W, Starrett B, Zalusky R (1990) Recombinant human erythropoietin for patients with AIDS treated with zidovudine. N Engl J Med 322:1488–1493
Fitch CD, Chevli R, Kanjananggulpan P, Dutta P, Chevli K, Chou AC (1983) Intracellular ferriprotoporphyrin IX is a lytic agent. Blood 62(6):1165–1168
Fleckenstein B, Molberg Q, Qiao SW, Schmid DG, von der Mulbe F, Elgstoen K, Jung G, Sollid LM (2002) Gliadin T cell epitope selection by tissue transglutaminase in coelic disease. Role of enzyme specificity and pH influence on the transamidation versus deamidation process. J Biol Chem 277(37):34109–34116
Fleming AF (1982) Iron deficiency in the tropics. Clin Haematol 2:365–388
Florentino RF, Guirriec RM (1984) Prevalence of nutritional anemia in infancy and childhood with emphasis on developing countries. In: Stekel A (ed) Iron nutrition in infancy and childhood. Nestle/Raven Press, Vevey, pp 61–74
Fomon SJ (2001) Infant feeding in the 20th century: formula and beikost. J Nutr 131:S409–S420
Foy H, Kondi A (1960) Hookworms in the aetiology of tropical iron deficiency anemia. Trans R Soc Trop Med Hyg 54(5):419–433
Franceschi F, Annalisa T, Teresa DR, Giovanna DA, Ianiro G, Franco S, Viviana G, Valentina T, Riccardo LL, Antonio G (2014) Role of Helicobacter pylori infection on nutrition and metabolism. World J Gastroenterol 20(36):12809–11281
Francis SE, Sullivan DJ Jr, Goldberg DE (1997) Hemoglobin metabolism in the malaria parasite Plasmodium falciparum. Annu Rev Microbiol 51:97–123
Francisco-Cruz A, Aguilar-Santelises M, Ramos-Espinosa O, Mata-Espinosa D, Marquina-Castillo B, Barrios-Payan J, Hernandez-Pando R (2014) Granulocyte-macrophage colony-stimulating factor: not just another haematopoietic growth factor. Med Oncol 31(1):774
Franz AR, Mihatsch WA, Sander S, Kron M, Pohlandt F (2000) Prospective randomized trial of early versus late enteral iron supplementation in infants with a birth weight of less than 1301 grams. Pediatrics 106(4):700–706

Freireich EJ, Ross JF, Bayles TB, Emerson CP, Finch SC (1957) Radioactive iron metabolism and erythrocyte survival studies of the mechanism of the anemia associated with rheumatoid arthritis. J Clin Invest 36(7):1043–1058

Fuchs D, Weiss G, Reibnegger G, Wachter H (1992) The role of neopterin as a monitor of cellular immune activation in transplantation, inflammatory, infectious, and malignant diseases. Crit Rev Clin Lab Sci 29(3–4):307–341

Galvani AP (2005) Age-dependent epidemiological patterns and strain diversity in helminth parasites. J Parasitol 91:24–30

Gamain B, Smith JD, Miller LH, Baruch OI (2001) Modification of CD 36 binding domain of P. falciparum variant antigen are responsible for the inability of chondrointin sulphate adherent parasites to bond CD 36. Blood 97:3268–3274

Gately MK, Warrier RR, Honosoge S, Carvajal DM, Faherty DA, Connaughton SE, Anderson TD, Sarmiento U, Hubbard BR, Murphy M (1994) Administration of recombinant IL-12 to normal mice enhances cytolytic lymphocyte activity and induces production of IFN-g in vivo. Int Immunol 6:157–167

Gause WC, Wynn TA, Allen JE (2013) Type 2 immunity and wound healing: evolutionary refinement of adaptive immunity by helminths. Nat Rev Immunol 13:607–614

Geissler AW, Mwaniki DL, Thiongo F, Friis H (1998) Geophagy as a risk factor for geohelminth infections: a longitudinal study of Kenyan primary school children. Trans R Soc Trop Med Hyg 92:7–11

Genton B, D'Acremont V, Rare L, Baea K, Reeder JC, Alpers MP, Muller I (2008) Plasmodium vivax and mixed infections are associated with severe malaria in children: a prospective cohort study from Papua New Guinea. PLoS Med 5(6):e127

George CM, Oldja L, Biswas S, Perin J, Lee G, Kosek M, Sack RB, Ahmed S, Haque R, Parvin T, Azmi IJ, Bhuyian SI, Talukder KA, Mohammed S, Faruque AG (2015) Geophagy is associated with environmental enteropathy and stunting in children in rural Bangladesh. Am J Trop Med Hyg 92(6):1117–1124

Ghosh K (2007) Pathogenesis of anemia in malaria, a concise review. Parasitol Res 101:1463–1469

Gill PS, Rarick MU, Brynes RK, Causey D, Levine AM (1987) Azidothymidine and bone marrow failure in AIDS. Ann Intern Med 107:502–505

Gleason AW (1993) Protein losing enteropathy. In: Wyllie R, Hyams JS (eds) Pediatric gastrointestinal disease. Saunders WB, Philadelphia, pp 536–541

Godschalk PC, Heikema AP, Gilbert M (2004) The crucial role of Campylobacter jejunigenes in anti-ganglioside antibody induction in Guillain-Barré syndrome. J Clin Invest 114:1659–1665

Gold BD, van Doorn L, Guarner J, Owens M, Pierce-Smith D, Qunsheng S, Hutwagner L, Sherman PM, de Mola OL, Czinn SJ (2001) Genotype, clinical, and demographic characteristics of children infected with Helicobacter pylori. J Clin Microbiol 39:1348–1352

Grau GE, Taylor TE, Molyneux ME, Wirima JJ, Vassalli P, Hommel M, Lambert PH (1989) Tumor necrosis factor and disease severity in children with falciparum malaria. N Engl Med J 320(24):1586–1591

Green PH, Cellier C (2007) Celiac disease. N Engl J Med 357:1731–1743

Green PHR, Stavropoulos SN, Panagi SG (2001) Characteristics of adult celiac disease in the USA: results of a national survey. Am J Gastroenterol 96:126–131

Griffin GE (1990) Malabsorption, malnutrition and HIV disease. Baillieres Clin Gastroenterol 4(2):361–373

Gupta A (2015a) Detrimental consequences of the habit of geophagy in children under the age of five years. Rev Res 4(4):1–9

Gupta A (2015b) To analyse effect of pallor (anemia) on nutritional status of children under age of five years. Golden Res Thoughts 4(7):1–9

Gupta RK, Gupta R (2005) Clinical profile of Pica in childhood. J K Sci 7(2):61–63

Gyan B, Kurtzhals JA, Akanmori BD, Ofori M, Goka BQ, Hviid L, Behr C (2002) Elevated levels of nitric oxide and low levels of haptoglobin are associated with severe malarial anemia in African children. Acta Trop 83(2):133–140

Haburchak DR, Dhawan VK, Christopher M (2014) Hookworm disease treatment & management Watson. In: Chandrasekar PH (ed). Available at: http://emedicine.medscape.com/article/218805-treatment#showall
Haburchak DR, Dhawan VK, Watson CM (2016) Hookworm disease. Available at: http://emedicine.medscape.com/article/218805-overview
Haghighi P, Wolf PL, Durie P (1997) Tropical Sprue and Subclinical Enteropathy: a vision for the nineties. Crit Rev Clin Lab Sci 34:313–341
Hagopian LP, Rooker GW, Rolider NU (2011) Identifying empirically supported treatments for pica in individuals with intellectual disabilities. Res Dev Disabil 32(6):2114–2120
Halfdanarson TR, Litzow MR, Murray JA (2007) Hematologic manifestations of celiac disease. Blood 109:412–421
Halliday HL, Lappin TR, McClure G (1984) Iron status of the preterm infant during the first year of life. Biol Neonate 45(5):228–235
Halsted JA (1968) Geophagia in man: its nature and nutritional effects. Am J Clin Nutr 21:1384–1393
Hambleton J (1996) Hematologic complications of HIV infection. Oncology 10:671–680
Hammond (2008) Assessment: dietary and clinical data. In: Mahan KL, Escott-Stump S (eds) Krause's food, nutrition, and diet therapy. Saunders WB, Philadelphia, pp 383–406
Harhay MO, Horton J, Olliaro PL (2010) Epidemiology and control of human gastrointestinal parasites in children. Expert Rev Anti-Infect Ther 8(2):219–234
Harper JW, Holleran SF, Ramakrishnan R, Bhagat G, Green PH (2007) Anemia in celiac disease is multifactorial in etiology. Am J Hematol 82(11):996–1000
Harriman GR, Smith PD, Horne MK (1989) Vitamin B12 malabsorption in patients with acquired immunodeficiency syndrome. Arch Intern Med 149:2039–2041
Harrison LM, Nerlinger A, Bungiro RD, Cordova JL, Kuzmic P, Cappello M (2002) Molecular characterization of ancylostoma inhibitors of coagulation factor Xa: Hookworm anticoagulant activity in vitro predicts parasite bloodfeeding in vivo. J Biol Chem 277:6223–6229
Harvey BM, Langford JE, Harthorn LF, Gillman SA, Green TD, Schartz RH, Burks AW (2014) Effects on growth and tolerance and hypoallergenicity of an amino acid–based formula with synbiotics. Pediatr Res 75:343–351
Hempelmann E (2007) hemozoin biocrystallization in Plasmodium falciparum and antimalarial activity of crystallization inhibitors. Parasitol Res 100(4):671–676
Hensmann M, Kwiatkowski D (2001) Cellular basis of early cytokine response to Plasmodium falciparum. Infect Immun 69(4):2364–2371
Herguner S, Ozyildirim I, Tanidir C (2008) Is Pica an eating disorder or an obsessive-compulsive spectrum disorder. Prog Neuro-Psychopharmacol Biol Psychiatry 32(8):2010–2011
Hill DJ, Firer MA, Shelton MJ, Hosking CS (1986a) Manifestations of milk allergy in infancy: clinical and immunologic findings. J Pediatr 109:270–276
Hill R, Pearman J, Worthy P, Caruso V, Goodwin S, Blincow E (1986b) Campylobacter pyloridis and gastritis in children. Lancet 1:387
Hill ID, Sinclair-Smith C, Lastovica AJ, Bowie MD, Emms M (1987) Transient protein losing enteropathy associated with acute gastritis and campylobacter pylori. Arch Dis Child 62:1215–1219
Ho M, Webster HK, Looareesuwan S, Supanaranond W, Phillips RE, Chanthavanich P, Warrel DA (1986) Antigen-specific immunosuppression in human malaria due to Plasmodium falciparum. J Infect Dis 153:763–771
Ho M, Schollaardt T, Snape S, Looareeuwan S, Suntharasamai P, White NJ (1998) Endogenous interleukin-10 modulates proinflammatory response in Plasmodium falciparum malaria. J Infect Dis 178(2):520–525
Hook EB (1978) Dietary cravings and aversions during pregnancy. Am J Clin Nutr 31:1355–1362
Horton J (2003) Global anthelmintic chemotherapy programs: learning from history. Trends Parasitol 19:405–499

Host A (2002) Frequency of cow's milk allergy in childhood. Ann Allergy Immunol 89(Suppl 1):33–37
Hotez PJ, Kamath A (2009) Neglected tropical diseases in Sub-Saharan Africa: review of their prevalence, distribution, and disease burden. PLoS Negl Trop Dis 3:e412
Hotez PJ, Molyneux DH (2008) One of Africa's great killers and a rationale for linking malaria and neglected tropical diseases control to achieve a common goal. PLoS Negl Trop Dis 2:270–274
Hotez PJ, Brooker S, Bethony JM, Bottazzi ME, Loukas A, Xiao S (2004) Hookworm infection. N Engl J Med 351(8):799–707
Hotez PJ, Bethony J, Bottazzi ME, Brooker S, Buss P (2005) Hookworm-'the great infection of mankind. Public Libr Sci Med 2:e67
Howden CW (2000) Vitamin B12 levels during prolonged treatment with proton pump inhibitors. J Clin Gastroenterol 30:29–33
Hsieh CS, Macatonia SE, Tripp CS, Wolf SF, O'Garra A, Murphy KM (1993) Development of TH1 CD4+ T cells through IL-12 produced by Listeria-induced macrophages. Science 260(5107):547–549
Hsieh GC, Loukas A, Wahl AM, Bhatia M, Wang Y, Williamson AL, Kehn KW, Maruyama H, Hotez PZ, Leitenberg D, Bethony J, Constant SL (2004) A secreted protein from the human hookworm Necator americanus binds selectively to NK cells and induces IFN-γ production. J Immunol 173:2699–2704
Huang FC, Chang MH, Hsu HY, Lee PI, Shun CT (1999) Long-term follow-up of duodenal ulcer in children before and after eradication of Helicobacter pylori. J Pediatr Gastroenterol Nutr 28:76–80
Humphrey JH (2009) Child undernutrition, tropical enteropathy, toilets and handwashing. Lancet 374(9694):1032–1035
Hviid L (2005) Naturally acquired immunity to Plasmodium falciparum malaria in Africa. Acta Trop 95:270–275
Inoue S, Truman JT, Lee MT (2016) Pediatric chronic anemia. Available at: http://emedicine.medscape.com/article/954598-overview#showall
Iron disorders institute (2009) Anemia of chronic disease. Available at: http://www.irondisorders.org/anemia-of-chronic-disease
Ismail S, Hill H (1996) Draft final report of a survey on anemia in three districts of Kzyl-ord, Kazakhstan. London School of Hygiene and Tropical Medicine
Issifou S, Mavoungou E, Borrmann S, Bouyou-Akotet MK, Matsiegui PI, Kremsner PG (2003) Severe malarial anemia associated with increased soluble Fas ligand concentrations in Gabonese children. Eur Cytokine Netw 14:238–241
Italian Paediatric Intestinal/HIV Study Group (1993) Intestinal malabsorption of HIV-infected children: relationship to diarrhoea, failure to thrive, enteric micro-organisms and immune impairment. AIDS 7(11):1435–1440
Iwai Y, Takanashi T, Nakao Y, Mikawa H (1986) Iron status in low birth weight infants on breast and formula feeding. Eur J Pediatr 145(1–2):63–65
Jacobson MA, Peiperl L, Volberding PA, Porteous D, Toy PT, Feigal D (1990) Red cell transfusion therapy for anemia in patients with AIDS and ARC: incidence, associated factors, and outcome. Transfusion 30(2):133–137
Jakeman GN, Saul A, Hogarth WL, Collins WE (1999) Anemia of acute malaria infections in non-immune patients primarily results from destruction of uninfected erythrocytes [In Process Citation]. Parasitology 119:127–133
Jansonn LT, Kling S, Dallman PR (1986) Anemia in children with acute infections seen in a primary care out patient pediatric clinic. Pediatr Infect Dis 5:424–427
Jenkins NE, Chakravorty SJ, Urban BC, Kai OK, Marsh K, Craig AG (2006) The effect of Plasmodium falciparum infection on expression of monocyte surface molecules. Trans R Soc Trop Med Hyg 100:1007–1012
Jensen JE (2007) Malabsorption syndrome. Colorado center for digestive disorders

Jiang T, Jeter JM, Nelsen SE, Ziegler EE (2000) Intestinal blood loss during cow milk feeding in older infants: quantitative measurements. Arch Pediatr Adolesc Med 154(7):673–678

Jones S, D'Souza C, Haboubi NY (2006) Patterns of clinical presentation of adult coeliac disease in a rural setting. Nutr J 5:24

Joosten E, Van Loon R, Billen J, Blanckaert N, Fabri R, Pelemans W (2002) Serum transferrin receptor in the evaluation of the iron status in elderly hospitalized patients with anemia. Am J Hematol 69(1):1–6

Josefsson L, Eriksson S, Sinclair E (2012) Hematopoietic precursor cells isolated from patients on long-term suppressive HIV therapy did not contain HIV-1 DNA. J Infect Dis 206(1):28–34

Jubault V, Penfornis A, Schillo F (2000) Sequential occurrence of thyroid autoantibodies and Grave's disease after immune restoration in severely immunocompromised human immunodeficiency virus-1-infected patients. J Clin Endocrinol Metab 85:4254–4257

Kachroo AH (2004) Order in stress-lessons from inanimate world. J Biosci 29(4):369–372

Kaminsky RG, Castillo RV, Flores CA (2015) Growth retardation and severe anemia in children with Trichurisdysenteric syndrome. APJTB 5(7):591–597

Kaptan K, Beyan C, Ural AU, Cetin T, Avcu F, Gülşen M, Finci R, Yalçín A (2000) Helicobacter pylori – is it a novel causative agent in Vitamin B12 deficiency? Arch Intern Med 160:1349–1353

Karapelou JW, Despommier DD (1987) Parasite life cycles. Springer Verlag, New York, pp 28–32

Karoui A, Karoui H (1993) Pica in Tunisian children. Results of a survey performed in a polyclinic of the Tunisian social security national administeration. Pediatrie 48(7):565–569

Karunaweera ND, Carter R, Grau GE, Mendis KN (1998) Demonstration of anti-disease immunity to Plasmodium vivax malaria in Sri Lanka using a quantitative method to assess clinical disease. Am J Trop Med Hyg 58:204–210

Kassebaum NJ, Jasrasaria R, Naghavi M, Wulf SK, Johns N, Lozano R, Regan M, Weatherall D, Chou DP, Eisele TP (2014) A systematic analysis of global anemia burden from 1990 to 2010. Blood 123:615–624

Katz FE, Parkar M, Stanley K, Murray LJ, Clark EA, Greaves MF (1985) Chromosome mapping of cell membrane antigens expressed on activated B cells. Eur J Immunol 15(1):103–106

Kern P, Hemmer CV, Van Damme J, Gruss HJ, Dietrich M (1989) Elevated tumor necrosis factor alpha and interleukin-6 serum levels as markers for complicated Plasmodium falciparum malaria. Am Med J 87(2):139–143

Ketema T, Bacha K (2013) Plasmodium vivax associated severe malaria complications among children in some malaria endemic areas of Ethiopia. BMC Public Health 13(637)

Khan Y, Tisman G (2010) Pica in iron deficiency: a case series. J Med Case Rep 4:86

Kleinman RE (2004) AAP pediatric nutrition handbook. American Academy of Pediatrics, Illinois

Koning F, Schuppan D, Cerf-Bensussan N, Sollid LM (2005) Pathomechanisms in celiac disease. Best practice & research. Clin Gastroenterol 19(3):373–387

Kordossis T, Piha L, Guialis A, Kosmopoulou O, Tzioufas AG, Moutsopoulos HM (1997) Non-organ specific autoantibodies against cellular components in HIV-1 infected individuals. Clin Exp Rheumatol 15:185–188

Korpe PS, Petri WA Jr (2012) Environmental enteropathy: critical implications of a poorly understood condition. Trends Mol Med 18(6):328–336

Kostaki M, Fessatou S, Karpathios T (2003) Refractory iron deficiency anemia due to silent Helicobacter pylori gastritis in children. Eur J Paediatr 162:177–179

Kotler DP (1995) HIV-associated malnutrition. J Phys Assoc AIDS Care 2(1):12–15

Kotler DP, Orenstein JM (1998) Clinical syndromes associated with microsporidiosis. Adv Parasitol 40:321–349

Kremsner PG, Winkler S, Brandts C, Wildling E, Jennel L, Graninger W, Prada J, Bienzle U, Juillard Juillard P, Grau GE (1995) Prediction of accelerated cure in Plasmodium falciparum malaria by the elevated capacity of tumor necrosis factor production. Am J Trop Med Hyg 53(5):532–538

Kreuzer KA, Rockstroh JK (1997) Pathogenesis and pathophysiology of anemia in HIV infection. Ann Hematol 75(5–6):179–187

Kreuzer KA, Rockstroh JK, Jelkmann W, Theisen A, Spengler U, Sauerbruch T (1997) Inadequate erythropoietin response to anemia in HIV patients: relationship to serum levels of tumor necrosis factor-alpha, interleukin 6 and their soluble receptors. Br J Haematol 96:235–239

Kuller LH, Tracy R, Belloso W, De Wit S, Drummond F, Lane HC, Ledergerber B, Lundgren J, Neuhaus J, Nixon D, Paton NI, Neaton JD (2008) Inflammatory and coagulation biomarkers and mortality in patients with HIV infection. PLoS Med 5(10):e203

Kurtzhals JA, Rodrigues O, Addae M, Commey JO, Nkrumah FK, Hviid L (1997) Reversible Suppression of bone marrow response to erythropoietin in Plasmodium falciparum malaria. Brit J Haematol 97:169–174

Kwiatkowski D, Cannon JG, Manoque KR, Cerami A, Dinarello CA, Greenwoog BM (1989) Tumour necrosis factor production in Falciparum malaria and its association with schizont rupture. Clin Exp Immunol 77(3):361–366

Kwiatkowski D, Hill AV, Sambou I, Twumasi P, Castracane J, Manoque KR, Cerami A, Brewster DR, Greenwood BM (1990) TNF concentration in fatal cerebral, non-fatal cerebral, and uncomplicated Plasmodium falciparum malaria. Lancet 336(8725):1201–1204

Lackey C J (1982) Prevalence of pica in pregnancy. In Alternative dietary practices and nutritional abuses in pregnancy: proceedings of a workshop. National Academy Press, Washington, DC

Ladhani S, Lowe B, Cole AO, Kowuondo K, Newton CR (2002) Changes in white blood cells and platelets in children with falciparum malaria: relationship to disease outcome. Br J Haematol 119:839–847

Lambi B, Federman M, Pleskow D, Wanke CA (1996) Malabsorption and wasting in AIDS patients with microsporidia and pathogen-negative diarrhea. AIDS 10(7):739–744

Lamikanra AA, Brown D, Potocnik A, Casala-Pascual C, Langhorne J, Roberts DJ (2007) Malarial anemia: of mice and men. Blood 10:18–28

Landzberg BR, Pochapin MB (2001) Protein-losing enteropathy and gastropathy. Curr Treat Options Gastroenterol 4:39–49

Lanzkowsky P (1959) Investigation into the etiology and treatment of pica. Arch Dis Child 34:140–148

Laufer B (1930) Geophagy: field museum of natural history and anthropology series. Field Museum of Natural History, Chicago

Le Hesran JY, Akiana J, Ndiaye el HM, Dia M, Senghor P, Konate L (2004) Severe malaria attack is associated with high prevalence of Ascaris lumbricoides infection among children in rural Senegal. Trans R Soc Trop Med Hyg 98:397–399

Le HT, Brouwer ID, Verhoef H, Nguyen KC, Kok FJ (2007) Anemia and intestinal parasite infection in school children in rural Vietnam. APJ Clin Nutr 16(4):716–723

Lechner K, Raja M (1984) Deficiency anemias in tropical countries. Wien Klin Wochenschr 96(14):539–541

Ledizet M, Harrison LM, Koskia RA, Cappello M (2005) Discovery and pre-clinical development of antithrombotics from hematophagous invertebrates. Curr Med Chem Cardiovasc Hematol Agents 3:1–10

Lee AY, Vlasuk GP (2003) Recombinant nematode anticoagulant protein c2 and other inhibitors targeting blood coagulation factor VIIa/tissue factor. J Intern Med 254:313–321

Lekana Douki JB, Traore B, Costa FT, Fusai T, Pouvelle B, Sterkers Y, Scherf A, Gysin J (2002) Sequestration of Plasmodium falciparum-infected erythrocytes to chondroitin sulfate A, a receptor for maternal malaria: monoclonal antibodies against the native parasite ligand reveal pan-reactive epitopes in placental isolates. Blood 100(4):1478–1483

Leung AKC, Sauve RS (2003) Whole cow's milk in infancy. Paediatr Child Health 8(7):419–421

Levine AM (1999) Anemia, neutropenia, and thrombocytopenia: pathogenesis and evolving treatment options in HIV-infected patients. Medscape HIV Clin Manag Ser 10:1–27

Libregts SF, Gutierrez L, de Bruin AM, Wensveen FM, Papadopoulos P, van Ijcken W, Ozgur Z, Philipsen S, Nolte MA (2011) Chronic IFN-γ production in mice induces anemia by reducing erythrocyte life span and inhibiting erythropoiesis through an IRF-1/PU.1 axis. Blood 118(9):2578–2588

Lichtin AE (2016) Iron deficiency anemia. Merck manuals professional edition. Available at: http://www.merckmanuals.com/professional/hematology-and-oncology/anemias-caused-by-deficient-erythropoiesis/iron-deficiency-anemia
Lin A, Arnold BF, Afreen S, Goto R, Huda TM, Haque R, Raqib R, Unicomb L, Ahmed T, Colford JM Jr, Luby SP (2013) Household environmental conditions are associated with enteropathy and impaired growth in rural Bangladesh. Am J Trop Med Hyg 89(1):130–137
Lochnit G, Dennis RD, Geyer R (2000) Phosphorylcholine substituents in nematodes: structures, occurrence and biological implications. Biol Chem 381:839–847
Locksley RM (1993) Interleukin 12 in host defense against microbial pathogens. Proc Natl Acad Sci U S A 90:5879–5880
Looareesuwan S, Ho M, Wattanagoon Y, White NJ, Warrel DA, Bunnag D, Harinasuta T, Wyler DJ (1987) Dynamic alteration in splenic function during acute falciparum malaria. N Engl J Med 317:675–679
Lopez LB, Ortega Soler CR, de Portela ML (2004) Pica during pregnancy: a frequently underestimated problem. Arch Latinoam Nutr 54(1):17–24
Loukas A, Prociv P (2001) Immune responses in Hookworm infections. Clin Microbiol Rev 14(4):689–703
Loukas A, Constant SL, Bethony JM (2005) Immunobiology of hookworm infection. FEMS Immunol Med Microbiol 43(2):115–124
Loukas A, Bethony J, Brooker S, Hotez P (2006) Hookworm vaccines: past, present, and future. Lancet Infect Dis 6:733–741
Lumsden WHR, Muller R, Baker JR (eds) (1979) Advances in parasitology. Academic, Cambridge, MA, p 337
Lundstrom U, Siimes MA, Dallman PR (1977) At what age does iron supplementation become necessary in low-birth-weight infants? J Pediatr 91(6):878–883
Lwambo NJ, Bundy DA, Medley GF (1992) A new approach to morbidity risk assessment in hookworm endemic communities. Epidemiol Infect 108:469–481
Lyke KE, Burges R, Cissoko Y, Sangare L, Dao M, Diarra I, Kone A, Harley R, Plowe CV, Doumbo OK, Sztein MB (2004) Serum levels of the proinflammatory cytokines interleukin-1 beta (IL-1beta), IL-6, IL-8, IL-10, tumor necrosis factor alpha, and IL-12(p70) in Malian children with severe Plasmodium falciparum malaria and matched uncomplicated malaria or healthy controls. Infect Immun 72(10):5630–5637
Madabushi A, Chakraborty S, Fisher SZ, Clemente JC, Yowell C, Agbandje-Mckenna M, Dame JB, Dunn BM, McKenna R (2005) Crystallization and preliminary X-rays analysis of aspartic protease plasmepsin 4 from the malarial parasite P. malariae. Acta Crystallogr Sect F 61(2):228–231
Maiuri L, Ciacci C, Ricciardelli I, Vacca L, Raja V, Auricchio S, Picard J, Osman M, Quaratino S, Londei M (2003) Association between innate response to gliadin and activation of pathogenic T cells in coeliac disease. Lancet 362:30–37
Maizels RM, Yazdanbakhsh M (2003) Immune regulation by helminth parasites: cellular and molecular mechanism. Nat Rev Immunol 3:733–744
Maizels RM, Balic A, Gomez-Escobar N, Nair M, Taylor MD, Allen JE (2014) Helminth parasites: masters of regulation. Immunol Rev 201:89–116
Malaguarnera L, Imbesi RM, Pignatelli S, Simpore J, Malaguarnera M, Musumeci S (2002) Increased levels of interleukin-12 in Plasmodium falciparum malaria: correlation with the severity of disease. Parasite Immunol 24(7):387–389
Malamba S, Hladik W, Reingold A, Banage F, McFarland W, Rutherford G, Mimbe D, Nzaro E, Downing R, Mermin J (2007) The effect of HIV on morbidity and mortality in children with severe malarial anemia. Malar J 6:143
Maravilla AM, Berk RN (1978) The radiographic diagnosis of pica. Am J Gastroenterol 70:94–99
Marignani M, Angeletti S, Bordi C (1997) Reversal of long standing iron deficiency anemia after eradication of Helicobacter pylori infection. Scand J Gastroenterol 32:617–622
Masawe AEJ (1981) Nutritional anemias. Part 1. Tropical Africa. Clin Haematol 3:815–842

Mata-Marín JA, Gaytán-Martínez JE, Martínez-Martínez RE, Arroyo-Anduiza CI, Fuentes-Allen JL, Casarrubias-Ramirez M (2010) Risk factors and correlates for anemia in HIV treatment-naïve infected patients: a cross-sectional analytical study. BMC Res Notes 3:230
Mayo Clinic (2016) Iron deficiency anemia causes. Available at: http://www.mayoclinic.org/diseases.../iron-deficiency-anemia/basics/causes/con-20019327
Mazigo HD, Waihenya R, Lwambo NJ, Mnyone LL, Mahande AM, Seni J, Zinga M, Kapesa A, Kweka EJ, Mshana SE, Heukelbach J, Mkoji GM (2010) Co-infections with Plasmodium falciparum, Schistosoma mansoni and intestinal helminths among schoolchildren in endemic areas of northwestern Tanzania. Parasite Vectors 3:44
McColl K, El-Omar E, Gillen D (1998) Interactions between H. pylori infection, gastric acid secretion and anti-secretory therapy. Br Med Bull 54:121–138
McGinniss MH, Macher AM, Rook AH, Alter HJ (1986) Red cell autoantibodies in patients with acquired immune deficiency syndrome. Transfusion 26:405–409
McNamara LA, Collins KL (2011) Hematopoietic stem/precursor cells as HIV reservoirs. Curr Opin HIV AIDS 6(1):43–48
McNamara LA, Onafuwa-Nuga A, Sebastian NT, Riddell JT, Bixby D, Collins KL (2013) CD133+ HPCs harbor HIV genomes in a subset of optimally treated people with long-term viral suppression. J Infect Dis 207(12):1807–1816
McSorley HJ, Maizels RM (2012) Helminth infections and host immune regulation. Clin Microbial Rev 25:585–608
Means RT, Krantz SB (1992) Progress in understanding the pathogenesis of the anemia of chronic disease. Blood 80:1639–1647
Medzhitov R, Preston-Hurlburt P, Janeway C (1997) A human homologue of the Drosophila Toll protein signals activation of adaptive immunity. Nature 388(6640):394–397
Megraud F, Brassens RMP, Denis F, Belbouri A, Hoa DQ (1989) Seroepidemiology of Campylobacter pylori infection in various populations. J Clin Microbiol 27:1870–1873
Mehta SK, Younoszai A, Pietz J, Achanti BP (2003) Pharmacological closure of the patent ductus arteriosus. Images Paediatr Cardiol 5(1):1–15
Menendez C, Kahigwa E, Hirt R, Vounatsou P, Aponte JJ, Font F, Acosta CJ, Schellenberg DM, Galindo CM, Kimario J, Urassa H, Brabin B, Smith TA, Kitua AY, Tanner M, Alonso PL (1997) Randomised placebo-controlled trial of iron supplementation and malaria chemoprophylaxis for prevention of severe anemia and malaria in Tanzanian infants. Lancet 350:844–850
Miao D, Young SL, Golden CD (2015) A meta-analysis of pica and micronutrient status. Am J Hum Biol 27(i):84–93
Miguel EA, Kremer M (2003) Worms: identifying impacts on education and health in the presence of treatment externalities. Econometrica 72:159–217
Miller TA (1966) Blood loss during Hookworm infection, determined by erythrocyte labeling with radioactive 51Chromium. II. Pathogenesis of ancylostoma braziliense infection in dogs and cats. J Parasitol 52(5):856–865
Milman N (2011) Anemia – still a major health problem in many parts of the world. Ann Hematol 90(4):369–377
Minnich V, Okcuoglu A, Tarkon Y, Arkasoy A, Cin S, Yorukoglu O, Renda F, Demirag B (1968) Pica in Turkey. II. Effect of clay upon iron absorption. Am J Clin Nutr 21(1):78–86
Mocroft A, Kirk O, Barton SE (1999a) Anemia is an independent predictive marker for clinical prognosis in HIV-infected patients from across Europe. EuroSIDA study group. AIDS 13(8):943–950
Mocroft A, Ole K, Barton SE (1999b) Anemia is an independent predictive marker for clinical prognosis in HIV-infected patients from across Europe. AIDS 13:943–950
Modiano D, Sirima BS, Konate A, Sanou I, Sawadogo A (2001) Leucocytosis in severe malaria. Trans R Soc Med Hyg 95:175–176
Mohan K, Dubey ML, Ganguly NK, Mahajan RC (1995) Plasmodium falciparum: role of activated blood monocytes in erythrocyte membrane damage and red cell loss during malaria. Exp Parasitol 80:54–63

Molberg O, McAdam S, Lundin KE (2001) T cells from celiac disease lesions recognize gliadin epitopes deamidated in situ by endogenous tissue transglutaminase. Eur J Immunol 31:1317–1323
Moore RD (1999) Human immunodeficiency virus infection, anemia and survival. Clin Infect Dis 29:44–49
Moore RD (2000) Anemia and human immunodeficiency virus disease in the era of highly active antiretroviral therapy. Semin Hematol 37(4 Suppl 6):18–23
Moore RD, Forney D (2002) Anemia in HIV-infected patients receiving highly active antiretroviral therapy. J Acquir Immune Defic Syndr 29(1):54–57
Moore DF, Sears DA (1994) Pica: iron deficiency, and the medical history. Am J Med 97:390–393
Moore RD, Keruly J, Richman DD, Creagh-Kirk T, Chaisson RE (1992) Natural history of advanced HIV disease in patients treated with zidovudine. The Zidovudine Epidemiology Study Group. AIDS 6(7):671–677
Mor SM, Tumwine JK, Naumova EN, Ndeezi G, Tzipori S (2009) Microsporidiosis and malnutrition in children with persistent diarrhea, Uganda. Emerg Infect Dis 15(1):49–52
Moyle G (2002) Anemia in persons with HIV infection: prognostic marker and contributor to morbidity. AIDS Rev 4(1):13–20
Murr C, Widner B, Wirleitner B, Fuchs D (2002) Neopterin as a marker for immune system activation. Curr Drug Metab 3(2):175–187
Naing C, Whittaker MA, Nyunt-Wai V, Reid SA, Wong SF, Mak JW (2013) Malaria and soil-transmitted intestinal helminth co-infection and its effect on anemia: a meta-analysis. Trans R Soc Trop Med Hyg 107:672–683
Nathan C, Xie QW (1994) Nitric oxide synthases – roles, tolls, and controls. Cell 78(6):915–918
National Health Service (England) (2015) Hookworm infection. Available at: http://nhs.uk/conditions/hookworm/Pages/Introduction.aspx
National Institute of Allergy and Infectious Diseases (2016) Life cycle of malarial parasite. Available at: https://www.niaid.nih.gov/topics/Malaria/Pages/lifecycle.aspx
National Institute of Health-Consensus Development Conference on Celiac Disease (2014) NIH Consens State Sci Statements 21:1–23
Navitsky RC, Dreyfuss ML, Shrestha J, Khatry SK, Stoltzfus RJ, Albonico M (1998) Ancylostoma duodenale is responsible for hookworm infections among pregnant women in the rural plains of Nepal. J Parasitol 84:647–651
Nazir G, Naz S, Ali S, Aziz S, Malik SA, Qari IH, Irum S (2011) Anemia: the neglected female health problem in developing countries. J Ayub Med Coll Abbottabad 23(2):8–11
Nemeth E, Tuttle MS, Powelson J, Vaughn MB, Donovan A, Ward DM, Ganz T, Kaplan J (2004) Hepcidin regulates cellular iron efflux by binding to ferroportin and inducing its internalization. Science 306:2090–2093
Nguyen PH, Day N, Pram TD, Ferguson DJ, White NJ (1995) Intraleucocytic malaria pigment and prognosis in severe malaria. Trans R Soc Trop Med Hyg 89:200–204
Nishihira J (2000) Macrophage migration inhibitory factor (MIF): its essential role in the immune system and cell growth. J Interf Cytokine Res 20(9):751–762
Njiru H, Elchalai U, Paltiel O (2011) Geophagy during pregnancy in Africa: a literature review. Obstet Gynecol Surv 66(7):452–459
Nyesigire RE, Bajunirwe F, Kiwanuka J (2012) Anemia in HIV-infected children: severity, types and effect on response to HAART. BMC Pediatr 12:170
O'Donnell A, Premawardhena A, Arambepola M, Allen SJ, Peto EA, Fisher CA (2007) Age related changes in adaptation to severe anemia in childhood in developing countries. Proc Natl Acad Sci 104:9440–9444
Odebunmi JF, Adefioye OA, Adeyeba A (2007) Hookworm infection among school children in Vom, Plateau state Nigeria. Am Eurasian J Sci Res 1:39–42
Office of The Registrar General & Census Commissioner, India (2011). Census data 2001/report on post enumeration survey. Available at: http://censusindia.gov.in/POST_ENUMERATION/Post_Enumaration.html

Olivares M, Walter T, Osono M, Chadud P, Schlesinger L (1989) Anemia of a mild viral infection: the measles vaccine as a model. Pediatrics 84:851–855

Olivares M, Hertrampf E, Pizarro F (1993) Effect of iron stores on heme iron absorption. Nutr Res 13:633–638

Olivares M, Walter T, Cook JD, Llaguno S (1995) Effect of acute infection on measurement of iron status: usefulness of the serum transferrin receptor. Int Pediatr Hematol Oncol 2:31–33

Olivares M, Walter T, Hertrampf E, Pizarro F (1999) Anemia and iron deficiency disease in children. Br Med Bull 55(3):534–543

Oliver M, Van Den Ham K, Shio MT, Kassa FA, Fougeray S (2014) Malarial pigment hemozoin and the innate inflammatory response. Front Immunol 5:25

Omodeo-Sale F, Motti A, Dondorp A, White NJ, Taramelli D (2005) Destabilisation and subsequent lysis of human erythrocytes induced by Plasmodium falciparum haem products. Eur J Haematol 74:324–332

Oppenheimer SJ (1989) Iron and malaria. Parasitol Today 5:77–82

Oquendo P, Hundt E, Lawler J, Seed B (1989) CD36 directly mediates cytoadherence of Plasmodium falciparum parasitized erythrocytes. Cell 58(1):95–101

Osazuwa F, Ayo OM, Imade P (2011) A significant association between intestinal helminth infection and anemia burden in children in rural communities of Edo state, Nigeria. N Am J Med Sci 3(1):30–34

Osterbauer B, Kapisi J, Bigira V, Mwangwa F, Kinara S, Kamya MR, Dorsey G (2012) Factors associated with malaria parasitaemia, malnutrition, and anemia among HIV-exposed and unexposed Ugandan infants: a cross-sectional survey. Malar J 11:432

Ouf EA, Ojurongbe O, Akindele AA, Sina-Agbaje OR, Van Tong H, Adeyeba AO, Kremsner PG, Kun JF, Veavan T (2012) Ficolin-2 levels and FCN2 genetic polymorphisms as a susceptibility factor in schistosomiasis. J Infect Dis 206:562–570

Ouyang W, Rutz S, Crellin NK, Valdez PA, Hymowitz SG (2011) Regulation and functions of the IL-10 family of cytokines in inflammation and disease. Annu Rev Immunol 29:71–109

Pagola S, Stephens PW, Bohle DS, Kosar AD, Madsen SK (2000) the structure of malarial pigment beta-hematin. Nature 404(6775):307–310

Pandey AV, Tekwani BL (1996) Formation of haemozoin/beta-haematin under physiological conditions is not spontaneous. FEBS Lett 393:189–193

Papier K, Williams GM, Luceres-Catubig R, Ahmed F, Olveda RM, McManus DP (2014) Childhood malnutrition and parasitic helminth interactions. Clin Infect Dis 59(2):234–243

Parente JML, DA Silva BB, Palha-Dias MPS, Zaterka S, Nishimura NF, Zeitune JM (2006) Helicobacter pylori infection in children of low and high socioeconomic status in Northeastern Brazil. Am J Trop Med Hyg 75(3):509–512

Parsonnet J, Blaser MJ, Perez PG, Hargrett-Bean N, Tauxe RV (1992) Symptoms and risk factors of Helicobacter pylori infection in a cohort of epidemiologists. Gastroenterology 102:41–46

Passaro DJ, Taylor DN, Meza R, Cabera L, Gilman RH, Parsonet J (2001) AcuteHelicobacter pylori infection is followed by an increase in diarrhoeal disease among Peruvian children. Pediatrics 108:e87

Pearson G, Robinson F, Beers Gibson T, Xu BE, Karandikar M, Berman K, Cobb MH (2001) Mitogen-activated protein (MAP) kinase pathways: regulation and physiological functions. Endocr Rev 22(2):153–183

Pelletier M, Girard D (2007) Biological functions of interleukin-21 and its role in inflammation. Sci World J 22(7):715–735

Perkins DJ, Kniss DA (1997) Tumor necrosis factor-alpha promotes sustained cyclooxygenase-2 expression: attenuation by dexamethasone and NSAIDs. Prostaglandins 54(4):727–743

Perkins DJ, Kremsner PG, Schmid D, Misukinis MA, Kelly MA, Weinberg JB (1999) Blood mononuclear cell nitric oxide production and plasma cytokine levels in healthy Gabonese children with prior mild or severe malaria. Infect Immun 67:4977–4981

Perkins DJ, Weinberg JB, Kremsner PG (2000) Reduced interleukin-12 and transforming growth factor-beta1 in severe childhood malaria: relationship of cytokine balance with disease severity. J Infect Dis 182(3):988–992
Pietzak MM, Thomas DW (2003) Childhood malabsorption. Pediatr Rev 24(6)
Podlaski FJ, Nanduri VB, Hulmes JD, Pan Y-C E, Levin W, Danho W, Chizzonite R, Gately MK, Stern AS (1992) Molecular characterization of interleukin 12. Arch Biochem Biophys 294:230–237
Polimeni M, Valente E, Aldieri E, Khadjavi A, Giribaldi G, Prato M (2013) Role of 15-hydroxyeicosatetraenoic acid in hemozoin-induced lysozyme release from human adherent monocytes. Biofactors 39(3):304–314
Pouvelle B, Buffet PA, Lepolard C, Scherf A, Gysin J (2000) Cytoadhesion of Plasmodium falciparum ring-stage-infected erythrocytes. Nat Med 6:1264–1268
Prasad A, Halsted J, Nadimi M (1983) Syndrome of iron deficiency anemia, hepatosplenomegaly, hypogonadism, dwarfism, andgeophagia. Nutr Rev 41:220–223
Prendergast A, Kelly P (2012) Review: enteropathy in the developing world:neglected effects on global health. Am J Trop Med Hyg 5(85):756–763
Price RN, Simpson JA, Nosten F, Luxemburger C, Hkirjaroen L, ter Kuile F, Chongsuphaiaisiddhi T, White NJ (2001) Factors contributing to anemia after uncomplicated falciparum malaria. Am J Trop Med Hyg 65:614–622
Prual A, Daouda M, Develoux B, Sellin P, Galan S, Hercberg C (1992) Consequences of Schistosoma haematobium infection on the iron status of school children in Niger. Am J Trop Med Hyg 47:291–297
Prudencio M, Rodriguez A, Mota MM (2006) The silent path to thousands of merozoites: the Plasmodium liver stage. Nat Rev Microbiol 4:849–856
Pullan RL, Kabatereine NB, Bukirwa H, Staedke SG, Brooker S (2011) Heterogeneities and consequences of *Plasmodium* species and hookworm coinfection: a population based study in Uganda. J Infect Dis 203:406–417
Punnonen K, Iejala K, Rajamaki A (1997) Serum transferrin receptor and its ratio to serum ferritin in the diagnosis of iron deficiency. Blood 89:1052–1057
Raj S, Faradi MMA, Rusia U, Singh O (2008) A prospective study of iron status in exclusively breastfed term infants up to 6 months of age. Int Breastfeed J 3:3
Ramakrishna BS, Venkataraman S, Mukhopadhya A (2006) Tropical malabsorption. Postgrad Med J 82(974)
Ramdath DD, Simeon DT, Wong MS, Grantham-McGregor SM (1995) Iron status of schoolchildren with varying intensities of Trichuris trichiura infection. Parasitology 110(3):347–351
Rao R, Georgieff MK (2010) Iron therapy for preterm infants. Clin Perinatal 36(1):27–42
Rarick MU, Espina B, Mocharnuk R, Trilling Y, Levine AM (1992) Thrombotic thrombocytopenic purpura in patients with human immunodeficiency virus infection: a report of three cases and review of the literature. Am J Hematol 40:103–109
Ravikumara M, Tuthill DP, Jenkin HR (2006) The changing clinical presentation of coeliac disease. Arch Dis Child 91(12):969–971
Rawat R, McCoy SI, Kadiyala S (2013) Poor diet quality is associated with low CD4 count and anemia and predicts mortality among antiretroviral therapy-naive HIV-positive adults in Uganda. J Acquir Immune Defic Syndr 62(2):246–253
Rawlins SC, Campbell M, Fox K, Benett F, Gibbs WN, Greene M (1991) Parasitic infections in young Jamaicans in different ecological zones of the island. Trop Geogr Med 43(1–2):136–141
Razak A, Ragab MS (2014) Helicobacter pylori infections in children of a rural community. J Bacteriol Parasitol 5:185
Reddy SC, Wirima JJ, Steketee RW, Breman JG, Heymann DL (1996) Transplacental transmission of Plasmodium falciparum in rural Malawi. Am J Trop Med Hyg 55(1 Suppl):57–60
Redig AJ, Berliner N (2013) Pathogenesis and clinical implications of HIV-related anemia in 2013. Hematology Am Soc 2013:377–381

Rees JM, Monsen ER, Merrill JE (1985) Iron fortification of infant foods: a decade of change. Clin Pediatr 24:707–710

Reyburn H, Mbatia R, Drakeley C, Bruce J, Carneiro I, Olomi R, Cox J, Nkya WM, Lemnge M, Greenwood BM, Riley EM (2005) Association of transmission intensity and age with clinical manifestations and case fatality of severe Plasmodium falciparum malaria. JAMA 293(12):1461–1470

Richman DD, Fischl MA, Grieco MH, Gottlieb MS, Volberding PA, Laskin OL, Leedom JM, Groopman JE, Mildvan D, Hirsch MS (1987) The toxicity of azidothymidine (AZT) in the treatment of patients with AIDS and AIDS-related complex. A double-blind, placebo-controlled trial. N Engl J Med 317(4):192–197

Robischon P (1971) Pica practice and other hand-mouth behavior and children's developmental level. Nurs Res 20:4–16

Roche M, Layrisse M (1966) The nature and causes of hookworm anemia. Am J Trop Med Hyg 15:1029–1102

Rockett K, Awburn MM, Cowden WB, Clark IA (1991) Killing of Plasmodium falciparum in vitro by nitric oxide derivatives. Infect Immun 59:3280

Rodriguez-Morales AJ, Sanchez E, Vargas M, Piccolo C, Colina R, Arria M (2006) Anemia and thrombocytopenia in children with Plasmodium vivax malaria. J Trop Pediatr 52:49–51

Romand S, Bouree P, GelezJ B-MB, Bisaro F, Dommergues JP (1994) Congenital malaria. Infected twins born to an asymptomatic mother. Presse Med 23(17):797–800

Romani L, Puccetti P, Bistoni F (1997) Interleukin-12 in infectious diseases. Clin Microbiol Rev 10(4):611–636

Roselle HA (1970) Association of laundry starch and clay ingestion with anemia in New York City. Arch Intern Med 125:57–61

Rosenberg I, Bowman B (1982) Intestinal physiology and parasitic diseases. Rev Infect Dis 4(4):763–767

Rosenthal PJ, Meshnick SR (1996) Hemoglobin catabolism and iron utilization by malarial parasites. Mol Biochem Parasitol 83(2):131–139

Rossi M, Young JW (2005) Human dendritic cells: potent antigen-presenting cells at the crossroads of innate and adaptive immunity. J Immunol 175:1373–1381

Rothenbacher D, Bode G, Berg G, Gommel R, Gonser T, Alder G, Brenner H (1998) Prevalence and determinants of H. pylori infection in preschool children: a population based study from Germany. Int J Epidemiol 27:135–141

Rowland M, Bourke B, Drumm B (2004) In: Walker WA, Goulet O, Kleinman RE, Sherman PM, Shneider BL, Sanderson IR (eds) Helicobacter pylori and peptic ulcer disease. BC Decker, Ontario, pp 491–512

Rule S, Reed C, Costello C (1991) Fatal haemophagocytic syndromes in HIV-antibody positive patient. Br J Haematol 79:127

Russell EC, Charalambous S, Pemba L, Churchyard GJ, Grant AD, Fielding K (2010) Low haemoglobin predicts early mortality among adults starting antiretroviral therapy in an HIV care programme in South Africa: a cohort study. BMC Public Health 10:433

Sabbatani S, Fiorino S, Manfredi R (2010) The emerging of the fifth malaria parasite (Plasmodium knowlesi). A public health concern? Braz J Infect Dis 14(3):299–309

Sabbi T, De Angelis P, Colistro F, Dall Oglio L, di Abriola GF, Castro M (2005) Efficacy of noninvasive tests in the diagnosis of Helicobacter pylori infection in pediatric patients. Arch Pediatr Adolesc Med 159(3):238–241

Salomon J, De TP, Melchior JC (2002) Nutrition and HIV infection. Br J Nutr 87(Suppl 1):S111–S119

Sandau KB, Zhou J, Kietzmann T, Brune B (2001) Regulation of hypoxia-inducible factoralpha by the inflammatory mediators nitric oxide and tumor necrosis, factor alpha in contrast to desferroxamine and phenylarsine oxide. J Biol Chem 276:39805–39811

Sarker SA, Mahalanabis D, Hildebrand P, Rahaman MM, Bardhan PK, Fuchs G, Beglinger C, Gyr K (1997) Helicobacter pylori: prevalence, transmission, and serum pepsinogen II concentrations in children of a poor peri-urban community in Bangladesh. Clin Infect Dis 25:990–995
Sasadeusz J, Buchanan M, Speed B (1990) Reactive haemophagocytic syndrome in human immunodeficiency virus infection. J Infect 20:65–68
Savilahti E (2000) Food-induced malabsorption syndromes. J Pediatr Gastroenterol Nutr 30(Suppl):S61–SS6
Say B, Ozsoylu S, Berkel I (1969) Geophagia associated with iron-deficiency anemia, hepatosplenomegaly, hypogonadism and dwarfism. A syndrome probably associated with zinc deficiency. Clin Pediatr (Phila) 8:661–668
Sayers G, Lipschitz DA, Sayers M, Seftel HC, Bothwell TH, Charlton RW (1974a) Relationship between pica and iron nutrition in Johannesburg black adults. S Afr Med J 68:1655–1660
Sayers G, Lipschitz DA, Sayers M, Seftel HC, Bothwell TH, Charlton RW (1974b) Relationship between pica and iron nutrition in Johannesburg Black adults. Afr Med J 48:1655–1660
Scadden DT (1997) Cytokine use in the management of HIV disease. J Acquir Immune Defic Syndr Hum Retrovirol 16(Suppl 1):S23–S29
Schellenberg D, Menendez C, Kahigwa E, Aponte J, Vidaj J, Tanner M, Mshinda H, Alonso P (2001) Intermittent treatment for malaria and anemia control at time of routine vaccinations in Tanzanian infants: a randomised, placebo-controlled trial. Lancet 357:1471–1477
Scherf A, Pouvelle B, Buffet PA, Gysin J (2001) Molecular mechanisms of Plasmodium falciparum placental adhesion. Cell Microbiol 3(3):125–131
Schwartz JE, Scuderi P, Wiggins C, Rudoph A, Hersh EM (1989) A phase I trial of recombinant tumor necrosis factor (rTNF) administered by continuous intravenous infusion in patients with disseminated malignancy. Biotherapy 1(3):207–214
Schwarzer E, Alessio M, Ulliers D, Arese P (1998) Phagocytosis of the malarial pigment, hemozoin, impairs expression of major histocompatibility complex class II antigen, CD54, and CD11c in human monocytes. Infect Immun 66:1601–1606
Schwenk A, Büger B, Wessel D, Stützer H, Ziegenhagen D, Diehl V, Schrappe M (1993) Clinical risk factors for malnutrition in HIV-1-infected patients. AIDS 7(9):1213–1219
Scott SP, Chen-Edinboro LP, Caulfield LE, Muaay-Kolg LE (2014) The impact of anemia on child mortality: an updated review. Forum Nutr 6(12):5915–5932
Seidman E (1995) Protein losing gastroenteropathy. In: Roy C, Silverman A, Alagille D (eds) Pediatric clinical gastroenterology. Mosby Year Book Co, Totonto, pp 362–366
Semba RD (2003) Iron deficiency anemia and the cycle of poverty among human immunodeficiency virus infected women in the inner city. Clin Infect Dis 37:S105–S111
Senczuk AM, Reeder JC, Kosmala MM, Ho M (2001) Plasmodium falciparum erythrocyte membrane protein 1 functions as a ligand for P-selectin. Blood 98(10):3132–3135
Seo JK, Ko JS, Choi KD (2002) Serum ferritin and Helicobacter pylori infection in children: a seroepidemiologic study in Korea. J Gastroenterol Hepatol 17:754–757
Shaman R, Niranga MD, de Silva HJ (2009) Helicobacter Pylori infection in children. Saudi J Gastroenterol 15(2):86–94
Shan L, Molberg Q, Parrot I, Hausch F, Filiz F, Gray GM, Sollid LM, Khosla C (2002) Structural basis for gluten intolerance in celiac sprue. Science 297:2275–2279
Sheldon W (1959) Celiac disease. Pediatrics 23(1 Pt 1):132–145
Shen H, Cheng T, Preffer FI (1999) Intrinsic human immunodeficiency virus type 1 resistance of hematopoietic stem cells despite coreceptor expression. J Virol 73(1):728–737
Sheth TN, Choudhry NK, Bowes M, Detsky AS (1997) The relation of conjunctival pallor to the presence of anemia. J Gen Intern Med 12:102–106
Shio MT, Kassa FA, Bellemare MJ, Olivier M (2010) Innate inflammatory response to malarial pigment hemozoin. Microbes Infect 12(12–13):889–899
Silvestrini F, Alano P, Williams JL (2000) Commitment to the production of male and female gametocytes in the human malaria parasite Plasmodium falciparum. Parasitology 121:465–471

Sinden RE (2002) Molecular interactions between Plasmodium and its insect vectors. Cell Microbiol 4:713–724
Singhi S, Ravishanker R, Singhi P, That R (2003) Low plasma zinc and iron in pica. Indian J Paediatr 70(2):139–143
Sinha N, Deshmukh PR, Garg BS (2008) Evaluation of WHO haemoglobin colour scale & palmar pallor for screening of anemia among children (6–35 months) in rural Wardha, India. Indian J Med Res 128(3):278–281
Sipsas NV, Kokori SI, Ioannidis JP, Kyriaki D, Tzioufas AG, Kordossis T (1999) Circulating autoantibodies to erythropoietin are associated with HIV-1 related anemia. J Infect Dis 180:2044–2047
Skorokhod OA, Caione L, Marrocco T, Migliardi G, Barrera V, Arese P, Piacibello W, Schwarzer E (2010) Inhibition of erythropoiesis in malarial anemia: role of hemozoin generated 4 hydroxynonemal. Blood 116(20):4328–4337
Slater AF, Swiggard WJ, Orton BR, Flitter WD, Goldberg DE, Cerami A, Henderson GB (1991) An iron-carboxylate bond links the heme units of malaria pigment. Proc Natl Acad Sci U S A 88:325–329
Sleisenger MH (1969) Malabsorption syndrome. N Engl J Med 281:1111–1117
Smith TG, Lourenço P, Carter R, Walliker D, Ranford-Cartwright LC (2000) Commitment to sexual differentiation in the human malaria parasite, Plasmodium falciparum. Parasitology 121:127–133
Smith TG, Robbins PA, Ratcliffe PJ (2008) The human side of hypoxia inducible factor. Br J Haematol 141(3):325–334
Snow RW, Guerra CA, Noor AM, Myint HY, Hay SI (2005) The global distribution of clinical episodes of Plasmodium falciparum malaria. Nature 434:214–217
Sorensen WC, Cappello M, Bell D, DiFedele LM, Brown MA (2011) Polyhelminthic infection in East Guatemalan school children. J Glob Infect Dis 3(1):25–31
Spivak JL, Barnes DC, Fuchs E, Quinn TC (1989) Serum immunoreactive erythropoietin in HIV-infected patients. JAMA 261:3104–3107
Stephenson LS, Holland CV, Cooper ES (2000) The public health significance of Trichuris trichiura. Parasitology 121:S73–S95
Stoltzfus RJ, Dreyfuss ML, Chwaya HM, Albonico M (1997a) Hookworm control as a strategy to prevent iron deficiency. Nutr Rev 55:223–232
Stoltzfus RJ, Dreyfuss M, Hababuu MPH, Chwaya HM, Albonico M (1997b) Hookworm control as a strategy to prevent iron deficiency. Nutr Rev 55:223–232
Stoltzfus RJ, Chwaya HM, Montresor A, Tielsch JM, Jape JK, Albonico M, Savioli L (2004a) Low dose daily iron supplementation improves iron status and appetite but not anemia, whereas quarterly anthelmintic treatment improves growth, appetite, and anemia in Zanzibari preschool children. J Nutr 134:348–356
Stoltzfus RJ, Mullany L, Black RE (2004b) Comparative quantification of health risks: global and regional burden of disease attributable to selected major risk factors. World Health Organization, Geneva, pp 163–209
Stoltzfus RJ, Chwaya HM, Montresor A, Albonico M, Savioli L, Tielsch JM (2000) Malaria, hookworms and recent fever are related to anemia and iron status indicators in 0- to 5-y old Zanzibari children and these relationships change with age. J Nutr 130:1724–1733
Stopeck A (2000) Links between Helicobacter pylori infection, cobalamin deficiency, and pernicious anemia. Arch Intern Med 160:1229–1230
Stratton R, Slapak G, Mahungu T, Kinloch-de Loes S (2009) Autoimmunity and HIV. Curr Opin Infect Dis 22:49–56
Suarez J (1931) PerniciPernicious anemia and sprue.pdf.-Libraria. Available at:http://libraria.rcm.upr.edu:8080/.../Pernicious%20Anemia%20and%20Sprue.p...
Suerbaum S, Michetti P (2002) Helicobacter pylori infection. N Engl Med 347:1175–1186
Suharno D, West CE, Muhilal KD, Hautvast JGAJ (1993) Supplementation with Vitamin A and iron for nutritional anemia in pregnant women in West Java, Indonesia. Lancet 342:1325

Sullivan DJ (2002) Theories on malarial pigment formation and quinoline action. Int J Parasitol 32(13):1645–1653

Sullivan PS, Hanson DL, Chu SY, Jones JL, Ward JW (1988) Epidemiology of anemia in human immunodeficiency virus (HIV)-infected persons: results from the multistate adult and adolescent spectrum of HIV disease surveillance project. Blood 91(1):301–308

Sullivan PB, Thomas JE, Wight DG, Neale G, Easthan EJ, Corrah T, Lloyd-Evans N, Greenwood BM (1990) Helicobacter pylori in Gambian children with chronic diarrhoea and malnutrition. Arch Dis Child 65:189–191

Takahashi M, Ozaki N, Kawakita S, Nozaki M, Saeki Y (1985) Vascular effects of 15-hydroperoxyeicosatetraenoic acid and 15-hydroxyeicosatetraenoic acid on canine arteries. Jpn J Pharmacol 37(4):325–334

Takenaka H, Maruo S, Yamamoto N, Wysocka M, Ono S, Kobayashi M, Yagita H, Okumura K, Hamoka T, Trinchieri G, Fujiwara H (1997) Regulation of T cell dependent and independent IL-12 production by the Th2 type cytokines IL-10, IL-6 and IL-4. Leukoc Biol 61:80–87

Tasker PWG (1961) Blood loss from hook worm infection. Trans Royal Soc Trop Med Hyg 55(1):36

Taylor MD, van der Werf N, Maizels RM (2012) T cells in helminth infection: the regulators and the regulated. Trends Immunol 33:181–189

Teach SJ, Fleisher GR (1994) Rectal bleeding on the pediatric emergency department. Ann Emerg Med 23:1252

Thomas KE, Sapone A, Fasano A, Vogel SN (2006) Gliadin stimulation of murine macrophage inflammatory gene expression and intestinal permeability are MyD88-dependent: role of the innate immune response in celiac disease. J Immunol 176:2512–2521

Tracy N (2014) Eating disorder: Pica. Available at: http://www.healthyplace.com/eating-disorders/other-eating-disorders/eating-disorders-pica/

Trinchieri G (1998) Interleukin 12; a cytokine at the interface of inflammation and immunity. Adv Immunol 70:83–243

Tsiakalos A, Kordossis T, Ziakas PD, Kontos AN, Kyriaki D, Sipsas NV (2010) Circulating antibodies to endogenous erythropoietin and risk for HIV-1-related anemia. J Infect 60:238–243

Tsiakalos A, Routsias JG, Kordossis T, Moutsopoulos HM, Tzioufas AG, Sipsas NV (2011) Fine epitope specificity of anti-erythropoietin antibodies reveals molecular mimicry with HIV-1 p17 protein: a pathogenetic mechanism for HIV-1-related anemia. J Infect Dis 204(6):902–911

Tsuyuoka R, Bailey JW, Nery Guimaraes AM, Gurgel RQ, Cuevas LE (1999) Anemia and intestinal parasitic infection in primary school students in Aracaju, Sergipe, Brazil. Caderno de Saude Publica 15(2):413–421

Ubino JR, Ijaz MK (2010) The global war against intestinal parasites – should we use a holistic approach? IJSD 14. Issue 9:e732–e738

Uc A, Chong SK (2002) Treatment of Helicobacter pylori gastritis improves dyspeptic symptoms in children. J Pediatr Gastroenterol Nutr 34:281–285

United Nations Children' Emergency Fund (1995) Nutrition: the progress of nations. Available at: http://www.unicef.org/pon95/nutr0014.html

United Nations Children' Emergency Fund (2014) Water, sanitation and hygiene (WASH). Available at: http://www.unicef.org/media/media_45481.html]

United Nations Millenium Development Goals Report (2012) United Nations, New York

United States National Library of Medicine (2016) Chondroitin sulphate. Available at: https://pubchem.ncbi.nlm.nih.gov/compound/Chondroitin_sulfate#section=Top

Urganci N, Arapoglu M, Telhan L, Kayaalp N (2002) A rare case of protein losing enteropathy in childhood: Helicobacter pylori gastritis – case report. Turk Arch Ped 37:238–240

Urquhart AD (1994) Putative pathophysiological interactions of cytokines and phagocytic cells in severe human falciparum malaria. Clin Infect Dis 9:117–3110

van der Windt DA, Jellema P, Mulder CJ, Kneepkens CM, van der Horst HE (2010) Diagnostic testing for celiac disease among patients with abdominal symptoms: a systematic review. JAMA 303(17):1738–1746

Vanasse GJ, Berliner N (2010) Anemia in elderly patients: an emerging problem for the 21st century. Hematology Am Soc Hematol Educ Program 2010:271–275
Vandenplas Y, Brueton M, Dupont C, Hill D, Isolauri E, Koletzko S, Oranje AP, Staiano A (2007) Guidelines for the diagnosis and management of cow's milk protein allergy in infants. Arch Dis Child 92:902–908
Vanderberg J, Rhodin J (1967) Differentiation of nuclear and cytoplasmic fine structure during sporogonic development of Plasmodium berghei. J Cell Biol 32:C7–C10
Verhoef H, West CE, Kraaijenhagen R, Nzyuko SM, King R, Mbandi MM, van Laatum S, Hogervorst R, Schep C, Kok FJ (2002) Malarial anemia leads to adequately increased erythropoiesis in asymptomatic Kenyan children. Blood 100:3489–3494
Vermer DE, Frate DA (1979) Geophagia in rural Mississippi: environmental and cultural contexts and nutritional implications. Am J Clin Nutr 32(10):2129–2135
Volberding P (2000) Consensus statement: anemia in HIV infection – current trends, treatment options, and practice strategies. Anemia in HIV Working Group. Clin Ther 22(9):1004–1020
Volberding P (2002) The impact of anemia on quality of life in human immunodeficiency virus-infected patients. J Infect Dis 185(Suppl 2):S110–S114
Volberding PA, Lagakos SW, Koch MA, Pettinelli C, Myers MW, Booth DK, Balfour HH, Reichman RC, Bartlett JA, Hirsch MS (1990) Zidovudine in asymptomatic human immunodeficiency virus infection. A controlled trial in persons with fewer than 500 CD4-positive cells per cubic millimeter. The AIDS Clinical Trials Group of the National Institute of Allergy and Infectious Diseases N Engl J Med 322(14):941–949
Volberding PA, Levin AM, Dieterich D (2004) Anemia in HIV infection: clinical impact and evidence-based management strategies. Clin Infect Dis 38:1454–1463
von Bonsdorff B (1977) Pica: a hypothesis. Br J Haematol 35:476–477
Vyas D, Chandra RK (1984) Functional implications of iron deficiency. In: Nestle AS (ed) Iron nutrition in infancy and childhood. Veney/Raven Press, New York, pp 45–59
Walker M (2003) What is tropical sprue? J Gastroenterol Hepatol 18(8):887–890
Walker RE, Parker RI, Kovacs JA, Masur H, Lane HC, Carleton S, Kirk LE, Gralnick HR, Fauci AS (1988) Anemia and erythropoiesis in patients with the acquired immunodeficiency syndrome (AIDS) and Kaposi sarcoma treated with zidovudine. Ann Intern Med 108:372–376
Walter T, Hertrampf E, Arredondo M (1990) Gastrointestinal iron losses in infancy: effect of the diet. In: Hercberg S, Galan P, Dupin H (eds) Colloque INSERM. INSERM, Paris, pp 283–290
Wang Q, Fujioka H, Nussenzweig V (2005) Exit of plasmodium sporozoites from oocysts is an active process that involves the circumsporozoite protein. PLoS Pathog 1(1):e9
Warren JR, Marshall B (1983) Unidentified curved bacilli on gastric epithelium in active chronic gastritis. Lancet 321:1273–1275
Weichold FF, Zella D, Barabitskaja O (1998) Neither human immunodeficiency virus-1 (HIV-1) nor HIV-2 infects most-primitive human hematopoietic stem cells as assessed in long-term bone marrow cultures. Blood 91(3):907–915
Weidanz WP (1982) Malaria and alterations in immune reactivity. Br Med Bull 38:167–172
White NJ (2008) Plasmodium knowlesi: the fifth human malaria parasite. Clin Infect Dis 46(2):172–173
WHO, UNICEF (2008) World malaria report 2008. WHO, Geneva
WHO, World Health Statistics (2013) Available at: http://www.who.int/gho/publications/world_health_statistics/2013/en/
Wickramasinghe SN, Abdalla SH (2000) Blood and bone marrow changes in malaria. Clin Haematol 13:277–299
Williamson AL, Brindley PJ, Abbenante G, Prociv P, Berry C, Girdwood K, Pritcahrd DI, Fairlie DP, Hotez PJ, Dalton JP, Loukas A (2003) Cleavage of hemoglobin by hookworm cathepsin D aspartic proteases and its potential to host specificity. FASEB J 16:1458–1460
Windle HJ, Kelleher D, Crabtree JE (2007) Childhood Helicobacter pylori infection and growth impairment in developing countries: a vicious cycle? Pediatrics 119(3):e754–e759

Wirleitner B, Schroecksnadel K, Winkler C, Fuchs D (2005) Neopterin in HIV-1 infection. Mol Immunol 42(2):183–194
Wiwanitkit V (2008) Malarial parasite infection and increase of erythrocyte sedimentation rate. Irn J Med Hypotheses Ideas 2:20–23
Wolfram W (2016) Pediatric gastrointestinal bleeding. In: Minkes RK (ed) Available at: http://emedicine.medscape.com/article/1955984-overview#showall
World Health Organization (1995) Physical status: the use and interpretation of anthropometry, WHO technical report series No. 854. WHO, Geneva
World Health Organization (2000) Severe falciparum malaria; World Health Organization, communicable diseases cluster. Trans R Soc Trop Med Hyg 94(Suppl 1):S1–90
World Health Organization (2001) The optimal duration of exclusive breastfeeding. Available at: http://www.who.int/inf-pr-2001/en/note2001-07.html
World Health Organization (2004) Severe falciparum malaria. Trans R Soc Trop Med Hyg 94:1–90
World Health Organization (2005) Deworming for health and development. Report of the third global meeting of the partners for parasite control. WHO, Geneva
World Health Organization (2008) World malaria report 2008. WHO, Geneva
Xie YW, Wolin MS (1996) Role of nitric oxide and its interaction with superoxide in the suppression of cardiac muscle mitochondrial respiration. Involvement in response to hypoxia/reoxygenation. Circulation 94:2580–2586
Xu J, Zhong NS (1997) The interaction of tumour necrosis factor alpha and endothelin-1 in pathogenetic models of asthma. Clin Exp Allergy 27(5):568–573
Yamaoka Y (2008) Helicobacter pylori: molecular genetics and cellular biology. Caister Academic Press, Norfolk
Yazdanbakhsh M, van den Biggelaar A, Maizels R (2001) Th2 responses without atopy: immunoregulation in chronic helminth infections and reduced allergic disease. Trends Immunol 22:372–377
Yehuda S, Mostofsky DI (2010) Iron deficiency and overload: from basic bilogy to clinical medicine. Springer Science & Business Media, Berlin, p 302
Yeung DL (Retrieved 2016) Iron and micronutrients: complementary food fortification. Available at: http://archive.unu.edu/unupress/food/V192e/ch11.htm
Yip R, Dallman PR (1988) The roles of inflammation and iron deficiency as causes of anemia. Am J Clin Nutr 48:1295–1300
Yip R, Limburg PJ, Ahlquist DA, Carpenter HA, O'Neill A, Kruse D, Stitham S, Gold BD, Gunter EW, Looker AC, Parkinson AJ, Nobmann ED, Petersen KM, Ellefson M, Schartz S (1997) Pervasive occult gastrointestinal bleeding in an Alaska native population with prevalent iron deficiency. JAMA 77:1135–1139
Youdim MB (1977) Pica hypothesis. Br J Haematol 36
Young SL (2010) Pica in pregnancy: new ideas about an old condition. Annu Rev Nutr 30:403–422
Yu SH, Shen WX (1990) In: Schad GA, Warren KS (eds) Hookworm disease: current status and new directions. Taylor & Francis, London
Zandman-Goddard G, Shoenfeld Y (2002) HIV and autoimmunity. Autoimmun Rev 1:329–337
Ziegelbauer K, Speich B, Mausezahl D, Bos R, Keiser J, Utzinger J (2012) Effect of sanitation on soil-transmitted helminth infection: systematic review and meta-analysis. PLoS Med 9(1):e1001162
Ziegler EE, Fomon SJ, Nelson SE, Rebouche CJ, Edward BB, Rogers RR, Lehman LJ (1990) Cow milk feeding in infancy: further observations on blood loss from the gastrointestinal tract. Pediatrics 116:11–18
Zuckerman A (1977) Current status of the immunology of blood and tissue protozoa: II. Plasmodium. Exp Parasitol 42:374–446

Chapter 6
Laboratory Estimation of Iron in Body

Abstract Haemoglobinometry is the effective method of assessing haemoglobin concentration in blood. Cyanmethaemoglobin method is based on the conversion of haemoglobin into cyanmethaemoglobin. Its absorbance is recorded at 540 nm in a colorimeter, and its intensity of colour is proportional to concentration of Hb. HemoCue method is an easy, portable and rapid measurement system. Colour scale method by the World Health Organization is another inexpensive, easy and rapid method for estimation of haemoglobin. This method is highly suited to field study where limited resources are the main hindrance to the access of medical facilities. Serum ferritin level is an important biochemical indicator of body iron store. Ferritin level in body correlates with the iron store of body. Serum iron test measures iron content available in plasma. Serum iron, TIBC and transferrin saturation, collectively, can be a valuable tool for diagnosis of iron deficiency anemia and anemia of inflammation in children.

6.1 Haemoglobin Estimation

Haemoglobinometry is an effective method of assessing haemoglobin concentration in blood which portrays anemia. Hb can be estimated by methods like haemoglobin colour scale, Sahli technique, Lovibond–Drabkin technique, Tallquist technique, HemoCue, cyanmethaemoglobin method and automated haematology analysers. Each method is based on its unique principle. Some methods are obsolete, whereas a few methods are expensive and need skilled staff and equipments.

6.1.1 Haldane Method

Haldane method is the oldest technique of Hb assessment. This method assesses the carbon monoxide carrying capacity of blood. But method is obsolete today due to availability of newer methods which are reliable and easier to perform. This method needs a source of CO as a standard for reference. Moreover, commercial reference for CO is not reliable (Clegg and king 1942). So Haldane method of Hb estimation

A. Gupta, *Nutritional Anemia in Preschool Children*,
DOI 10.1007/978-981-10-5178-4_6

is not practised by laboratories. However, every method has its unique feature and is associated with advantages and disadvantages.

6.1.2 Dare Method

In the Dare method, whole blood fills the glass chamber by capillary action. The chamber is provided with an electric bulb as a source of illumination. The colour of blood in the chamber is compared with a standard by observing through an eyepiece (Stone et al. 1984; Elwood and Jacob 1966). Colour comparison is the basis of Hb estimation in Dare method.

6.1.3 Lovibond–Drabkin Method

Lovibond–Drabkin method estimates cyanmethaemoglobin. Blood sample is mixed with a solution and kept undisturbed for the next 3 min. Thereafter, the colour of blood is matched with a colour standard on discs, which gives the value of haemoglobin. According to a study by van Lerberghe et al. (1983), this method is better in comparison to others which employ comparator for Hb estimation.

However, the cost of discs is high, and the method has subjective bias due to interpretation of colour.

6.1.4 Tallquist Method

In the Tallquist method, a blood drop is placed on a strip of blotting paper. Again, the colour of blood sample is compared with a colour standard, and Hb concentration is estimated. In this method, colour matching is difficult (Elwood and Jacob 1966).

6.1.5 Spencer Method

Spencer method makes use of a drop of blood which is placed over one half of glass chamber. A stick smeared with saponin is used to stir the blood drop. Glass chamber is covered by cover glass, and it is introduced into a slot provided in the instrument. An electric bulb transmits green light through the blood drop and is compared with a standard on a split screen. Haemoglobin concentration is interpreted in (g/100 ml) unit (Stone et al. 1984; Elwood and Jacob 1966).

6.1.6 Copper Sulphate Method

Copper sulphate method is based on gravity principle to assess Hb level. Copper sulphate solution of known specific gravity is used. Its gravity is equivalent to the gravity of blood which contains 100 g/L of haemoglobin. Blood droplet is allowed to move in copper sulphate solution. Next time, test is performed with a copper sulphate solution of specific gravity equivalent to blood containing 80 g/L of Hb (Stone et al. 1984). This method is suitable for Hb estimation in blood donor selection. Blood samples are grouped into less than 80 g/l, 80–100 g/l or above 100 g/l value, and according to a study by Sawant et al. (2007), this method gives the value of Hb in a range and beset with subjective bias. The traditional method being used for donor screening at many blood centres. Though a traditional and cheap method of donor screening, it does not provide accurate value of Hb (Philipps et al. 1950; Ross et al. 1986; Radtke et al. 2005). Precision by copper sulphate method is not acceptable (Ugwu et al. 1986; Morse et al. 1987; Sawant et al. 2007).

6.1.7 Direct Cyanmethaemoglobin Method

Cyanmethaemoglobin method is based on the conversion of haemoglobin into cyanmethaemoglobin. Its absorbance is recorded at 540 nm in a colorimeter (Nkrumah et al. 2011). Intensity of the colour produced is proportional to the concentration of Hb. A fixed amount of blood is added to 5 ml of Drabkin's solution. It is allowed to stand for 2–4 h. The colour of cyanmethaemoglobin is read in a spectrophotometer or colorimeter at a wavelength of 540 nm against a standard blank. This method was regarded as gold standard for Hb estimation (DeMaeyer et al. 1989). Cyanmethaemoglobin method is advocated by the International Council for Standardization in Haematology, according to Zwart et al. (1996) and ICSH (1987). However, its main drawback is the necessity of venipuncture. This method requires a laboratory setup along with the need of a technical person to handle spectrophotometer. Its use in remote areas and in field survey screening of anemia is difficult.

According to Shah et al. (2011), direct cyanmethaemoglobin method (HiCN) is employed by most of the laboratories in rural areas. Cyanide method has associated disadvantages. First, the reagent is a potential threat to the environment. Second, turbidity in cyanide method causes inaccuracy while recording absorbance at 540 nm wavelength.

6.1.8 Indirect Cyanmethaemoglobin Method

Indirect cyanmethaemoglobin method finds its suitability in an area which has limited laboratory setup. It can be used in field surveys for screening of anemia. In this method, 20 μL of blood is taken, and it is poured over the Whatman No. 1 filter paper. After the filter paper has dried up, it is transferred into a plastic bag which is properly labelled. The bag is packaged for laboratory analysis of Hb. Dried filter paper is inserted into a test tube containing 5 ml of Drabkin's solution. Dried blood spot and Drabkin's solution are thoroughly mixed over a vortex for 5 min, and thereafter, it is allowed to stand for 2 h. The colour of the solution is read in a spectrophotometer at a wavelength of 540 nm. Total time between procurement of blood drop and lab analysis is recommended to be 6 days as per ICMR (2001).

6.1.9 HemoCue Method

It is an easy, portable and less time-consuming Hb measurement system. It is made up of a specially designed battery-operated photometer and a set of microcuvettes. A precise amount of blood, 10 μL, is taken and transferred in a microcuvette. Undiluted blood is used for Hb estimation. Blood in microcuvette interacts with sodium deoxycholate. Erythrocytes undergo haemolysis and release haemoglobin. Sodium nitrite in the procedure changes haemoglobin to methaemoglobin, which in turn reacts with sodium azide giving rise to azidemethaemoglobin. Its absorbance is recorded at two wavelengths of 565 nm and 880 nm (Adam et al. 2012; von Schenck et al. 1986). The entire process takes around 40–60 s (Ingram and Lewis 2000; van den Broek et al. 1999; Lewis et al. 1998; Lewis and Emmanuel 2001; Critchley and Bates 2005; Akhtar et al. 2008).

HemoCue system is an easy, reliable and quantitative method of Hb estimation. It is best suited in field surveys (Van Schenck et al. 1986). It is a single-step procedure which does not require any reagents. Paramedical staff without any specific training can work on this apparatus. HemoCue system is a satisfactory, accurate and precise laboratory method (John and Lewis 1989).

HemoCue method has a sensitivity of 94.1% and specificity of 95.2% when compared with 90.1% and 94.2% for direct cyanmethaemoglobin method (Akhtar et al. 2009).

Long-term field experience has also shown the instrument to be stable and durable. These features make it possible to include haemoglobin determinations in multipurpose health and nutrition surveys (Dacie and Lewis 2001).

6.1.10 Sahli's Method

In Sahli's method, a fixed amount of blood is taken in a Hb test tube. It is mixed with HCl of 1/10 normality. Hb is converted into dark brown-coloured hematinic acid. Colour intensity is matched with a standard. Sahli's method is routinely practised in labs situated in remote areas and in hospitals. It is an easy, cost-effective and simple method for Hb assessment. However, its disadvantage is the subjective bias and interobserver variability when the colour of acid haematin is matched with the colour of standard. Further, the colour of standard is liable to fade with time. Hence, despite its simplicity, its sensitivity and reliability are poor in comparison with cyanmethaemoglobin method for screening of anemia (Balasubramaniam and Malathi 1992).

6.1.11 Automated Haematology Analyser

Automated haematology analyser is an expensive and accurate equipment for Hb estimation. The system requires trained staff and in-house establishment. Precision of HemoCue system is lower in comparison to automated analyser.

6.1.12 NBM-200

NBM-200 utilizes a sensor and is a non-invasive instrument for Hb estimation. Sensor is kept on the thumb, and optical reading is taken after temporary blood flow occlusion (Kim et al. 2013).

In rural areas and remote places where lab facility is unavailable, anemia is diagnosed by clinical sign like pallor over the conjunctiva, palm, nail beds and lingual mucosa. This method is easy and rapid and needs a little knowledge to diagnose moderate to severe anemia. In rural areas, haemoglobin colour scale is a lab method; as per WHO, it has a sensitivity and specificity of 95% and 99.5%, respectively (WHO 2001).

However, a study by Anand et al. (2009) and Aldridge et al. (2012) has asserted that haemoglobin colour scale does not furnish clinically acceptable accuracy. Contrarily, in another study by Montresor et al. (2000) and Timan et al. (2004), HCS has been confirmed for field study. In a latest study by Darshana and Uluwaduge (2014), HCS does hold necessary accuracy for screening anemia in blood donation. However, it can very well replace older copper sulphate method employed in blood donation. HCS can also be used for estimation of Hb levels in infants up to 4 months of age (van Rheenen and de Moor 2007).

Moreover, haemoglobin colour scale is subject to variable reading for the same sample if observed by different operators. It cannot detect a change in Hb concentration of less than 1 g/100 ml (Bates et al. 2007).

Sahli's method is used in primary health-care centres (Barduagni et al. 2003; Lindblade et al. 2006) for Hb estimation. It is still used in hospitals, but it is subject to subjective bias in colour comparison, fading of comparator easily and poor sensitivity. However, cyanmethaemoglobin method has better sensitivity than Sahli's method for use in hospitals (Balasubramaniam and Malathi 1992). Cyanmethaemoglobin method is gold standard, but its use is limited to hospitals and well-established labs.

Automated haematology analyser has high accuracy and precision than HemoCue system (Adam et al. 2012). NBM-200 method has a lower sensitivity and specificity in comparison to HemoCue (Kim et al. 2013).

6.1.13 WHO Haemoglobin Colour Scale Method

Children and pregnant women have the greater vulnerability towards anemia, especially more in developing countries than developed ones. Haemoglobin content of blood is the reliable indicator of anemia as far as clinical findings are concerned (WHO 2001; Ingram and Lewis 2000).

Colour scale method by the World Health Organization is the most inexpensive, easy and rapid method for estimation of haemoglobin. This method is highly suited to field study where limited resources are the main hindrance to access to medical facility. Colour scale uses a small card which has six shades of red colour. Shades represent haemoglobin concentration at 4, 6, 8, 10, 12 and 14 g/dL (WHO 2001).

Shades of red represent standards with which the colour of a drop of blood is matched. Shades follow each other in a row incessantly. Series of shades are mounted on a white-coloured rigid sheet made up of polyvinyl chloride or polypropylene. Each standard blood drop is captured on a Whatman 31ET filter. Blood drop forms a regular spot. The colour of blood sample on filter paper is matched with the colour standards consecutively (Stott et al. 1998).

WHO colour scale was initially prepared for anemia screening in obstetrical clinics and paediatric wards, to assess anemia associated with malaria and hookworm infections, in donor selection before blood transfusion and epidemiological surveys.

According to Gies et al. (2003), sensitivity for the haemoglobin differed at diverse Hb concentration. It was 42.9%, 33.3% and 43.5% for Hb values <9 g/dL, <10 g/dL and <11 g/dL, accordingly. Nonetheless, haemoglobin specificity was high for all three categories of Hb.

Colour scale by WHO is a semiqualitative method of Hb estimation and has been adopted and practised successfully for screening anemia in remote areas, primary health centres and field studies. Its reading can be correlated with the clinical finding; hence, colour scale aids in the confirmation of mild to moderate anemia.

However, its usefulness is unpredictable in severe anemia. According to Crichtley and Bates (2005), its sensitivity and specificity were very high in laboratory-based studies however reduced significantly in field studies.

Authors (Montresor et al. 2000) designed a study to assess the performance of WHO colour scale developed by Stott and Lewis, in the detection of anemia in a primary health centre.

The study covered preschool children. A significant correlation was noticed between the observations of colour scale and HemoCue system. Colour scale reading detected mild anemia in 353 (85.2%) of the children out of the total 415 anaemic children as determined by HemoCue. Further in the same study, colour scale assessed severe anemia in 13 (73.6%) children out of the total 19 children. Colour scale method is a beneficial tool in Hb estimation. It is a cost-effective, simple, time-sparring and user-friendly method in remote settings with inadequate laboratory facility.

6.2 Serum Ferritin

Serum ferritin level is an important biochemical indicator of body iron store. Ferritin level in body correlates with the iron store of body. Another advantage of ferritin level over Hb in anemia detection is that ferritin level does not fluctuate in altitude. A drop in serum ferritin level always reflects a fall in body iron store, hence a prerequisite for iron deficiency excluding the presence of any infection or inflammation in body. Presently, it is the most important parameter for the iron status measurement. Because ferritin is an acute-phase response protein, hence, in infection, the presence of alcoholic liver disease and occasional blood transfusion, its concentration increases during the above conditions, thereby no more reflecting the status of the iron store.

Therefore, serum ferritin in the normal range reflects iron sufficiency in the absence of any infection or inflammation. Utility of ferritin value in the screening of anemia in field study is questionable in areas where there is a high prevalence of malaria, helminth infection and environment enteropathy.

According to Gibson (2005), normal ferritin level changes with age. Except during the first 2 months of postnatal life, its level decreases in later part of infancy. Later on, by the end of the first year of life, ferritin level starts rising and keeps on increasing until adulthood.

A cut-off value for serum ferritin has been generally considered as 15 ng/mL for children $\geq$5 years of age and 12 ng/mL for children <5 years of age. These values are indicative of low body iron store (WHO 2001).

6.3 Serum Iron/Total Iron-Binding Capacity (TIBC)

Serum iron test measures the iron content available in plasma. About 65% of the body iron is present in the form of haemoglobin in blood circulation. Myoglobin contains around 45% of iron. Another 30% of the iron remains distributed in the bone marrow, spleen, liver and other tissues. Serum iron test cannot estimate the iron in these tissues. Plasma contains a fraction (1%) of the total body iron. Serum iron test depicts the presence of iron in circulation that is available for erythropoiesis. Its normal value in children ranges between 50 and 150 μg/100 ml. A value lower than the 50 μg/100 ml is indicative of iron deficiency in the body. Nonetheless, serum iron value undergoes diurnal variation. It becomes a flaw in its utility in screening for anemia in population survey and field study. Iron molecules remain bound to the binding sites on transferrin molecule.

TIBC test describes the amount of total protein that is associated with iron molecules. However, transferrin is the chief protein that remains attached to iron. Therefore, TIBC is indicative of the amount of transferrin present in circulation. Its normal value is between 240 and 450 μg/dL.

Normal value may vary slightly among different laboratories. TIBC is inversely proportional to iron level in the body. Its value rises in iron deficiency in the body. But its value lower than normal conceives anemia of inflammation, pernicious anemia and haemolytic anemia (Yee et al. 2008).

6.4 Transferrin Saturation Test

Transferring saturation is expressed in percentage, and it is the ratio of the serum iron to the total iron-binding capacity. Its normal value is between 20 and 50%. Its value is indicative of the percentage of saturation of transferrin. At a time, nearly 30% of the sites on transferrin remain occupied by iron molecules. A value of $<20\%$ is descriptive of iron deficiency in the body. Its utility is limited by physiological and diurnal variability. Furthermore, transferrin saturation remains low iron deficiency and the anemia of inflammation. Its assay requires availability of sophisticated laboratory and skilled manpower. Serum iron, TIBC and transferrin saturation, collectively, can be a valuable tool for diagnosis of iron deficiency anemia and anemia of inflammation in children. Furthermore, it is dubious whether these tests can be performed in slum areas, in village settings and in remote areas with inadequate lab equipments and paucity of skilled technician (Williamson et al. 2011).

6.5 Red Blood Cell Zinc Protoporphyrin

Iron is associated with the heme in haemoglobin molecule. In condition of iron deficiency in body, zinc gets the opportunity to associate with protoporphyrin to form zinc protoporphyrin. So estimation of red blood cell zinc protoporphyrin content is indicative of iron deficiency state (Labbe et al. 1999).

According to Cook (2005), a rise in the concentration of zinc protoporphyrin posits iron deficiency in the body. However, the presence of acute inflammation and/or infection limits the release of iron from body store into circulation despite the adequacy of iron store. Therefore, an increase in ZPP level is truly indicative of iron store of body in the absence of inflammation. Another factor that restricts the widespread use of ZPP in screening of anemia is the high cost and the need of sophisticated laboratory and skilled technician. So the test cannot be employed in anemia detection in areas which have high prevalence of malaria, helminth infection, diarhoea and/or poor environmental conditions (Cook 2005).

6.6 Serum Transferrin Receptor Concentration

Serum transferrin receptor (sTfR) concentration is not influenced by the presence of inflammation or infection in the body (Weiss 1999). According to Feelders et al. (1999), sTfR concentrations are correlated with the number of receptors present on the surfaces of RBCs. As in iron-deficient state, serum transferrin receptor concentration rises due to raised expression of transferrin receptors on iron-deficient erythroblasts. Therefore, sTfR concentration is an indicator of erythropoietic activity and functional iron deficiency in erythroblasts.

Unlike ferritin, transferrin saturation and ZPP tests, sTfR concentration is not affected by the inflammation in the body.

References

Adam I, Ahmed S, Mahmoud MH, Yassin MI (2012) Comparison of HemoCue® hemoglobinmeter and automated hematology analyzer in measurement of hemoglobin levels in pregnant women at Khartoum hospital, Sudan. Diagn Pathol 7:30

Akhtar K, Sherwani RK, Rahman K, Hasan J, Shahid M (2008) HemoCue photometer: a better alternative of hemoglobin estimation in blood donors? Indian J Hematol Blood Transfus 24:170–172

Aldridge C, Foster HM, Albonico M, Ame SM, Montresor A (2012) Evaluation of the diagnostic accuracy of the Haemoglobin Colour Scale to detect anemia in young children attending primary healthcare clinics in Zanzibar. Tropical Med Int Health 17:423–429

Anand H, Mir R, Saxena R (2009) Hemoglobin color scale a diagnostic dilemma. Indian J Pathol Microbiol 52:360–362

Balasubramaniam P, Malathi A (1992) Comparative study of hemoglobin estimated by Drabkin's and Sahli's methods. J Postgrad Med 38:8–9
Barduagni P, Ahmed AS, Curtale F, Raafat M, Soliman L (2003) Performance of Sahli and colour scale methods in diagnosing anemia among school children in low prevalence areas. Tropical Med Int Health 8:615–618
Bates I, McKew S, Sarkinfada F (2007) Anemia: a useful indicator of neglected disease burden and control. PLoS Med 4:e231
Clegg JW, King EJ (1942) Estimation of haemoglobin by the alkaline haematin method. Br Med J 2:329–333
Cook JD (2005) Diagnosis and management of iron deficiency anemia. Best Pract Res Clin Haematol 18(2):319–332
Critchley J, Bates I (2005) Haemoglobin colour scale for anemia diagnosis where there is no laboratory: a systematic review. Int J Epidemiol 34(6):1425–1434
Dacie JV, Lewis SM (2001) Practical hematology. Churchill Livingstone, Edinburgh, p 23
Darshana LG, Uluwaduge DI (2014) Validation of the WHO Hemoglobin Color Scale method. Anemia 2014:531670
De Maeyer EM, Dallman P, Gurney JM, Hallberg L, Sood SK, Srikantia SG (1989) Preventing controlling iron deficiency anemia through primary health care:a guide for health administrators and programme managers. World Health Organization, Geneva
Elwood PC, Jacobs A (1966) Haemoglobin estimation: a comparison of different techniques. Br Med J 1:20–24
Feelders RA, Kuiper-Kramer EP, van Eijk HG (1999) Structure, function and clinical significance of transferrin receptors. Clin Chem Lab Med 37:1–10
Gibson R (2005) Principles of nutritional assessment. Oxford University Press, Oxford
Gies S, Brabin BJ, Yassin MA, Cuevas LE (2003) Comparison of screening method for anemia in pregnant women in Awassa, Ethiopia. Tropical Med Int Health 8(4):301–309
Indian Council of Medical Research (2001) Micronutrient deficiency disorder in 16 districts of India. Part 1 report of an ICMR task force study-district nutrition project. ICMR, New Delhi
Ingram CF, Lewis SM (2000) Clinical use of WHO hemoglobin colour scale: validation and critique. J Clin Pathol 53:933–937
International Committee for Standardization in Haematology (1987) Recommendations for reference method for haemoglobinometry in human blood (ICSH standard 1986) and specifications for international haemiglobincyanide reference preparation. Clin Lab Haematol 9(1):73–79
John WL, Lewis SM (1989) Primary health screening haemoglobinometry in a tropical community. Bull World health Organ 67:627–633
Kim MJ, Park Q, Kim MH, Shin JW, Kim HO (2013) Comparison of the accuracy of noninvasive hemoglobin sensor (NBM-200) and portable hemoglobinometer (HemoCue) with an automated hematology analyzer (LH500) in blood donor screening. Ann Lab Med 33:261–267
Labbe RF, Vreman HJ, Stevenson DK (1999) Zinc protoporphyrin: a metabolite with a mission. Clin Chem 45(12):2060–2072
Lewis SM, Emmanuel J (2001) Validity of the haemoglobin colour scale in blood donor screening. Vox Sang 80(1):28–33
Lewis SM, Stott GJ, Wynn KJ (1998) An inexpensive and reliable new haemoglobin colour scale for assessing anemia. J Clin Pathol 51(1):21–24
Lindblade KA, Mwololo K, van Eijk AM, Peterson E, Odhiambo F, Williamson J, Slutsker L (2006) Evaluation of the WHO Haemoglobin Colour Scale for diagnosis of anemia in children and pregnant women as used by primary health care nurses and community health workers in western Kenya. Tropical Med Int Health 11:1679–1687
Montresor A, Albonico M, Khalfan N, Stoltzfus RJ, Tielsch JM, Chwaya HM, Savioli L (2000) Field trial of a haemoglobin colour scale: an effective tool to detect anemia in preschool children. Tropical Med Int Health 5:129–133
Morse EE, Cable R, Pisciotto P, Kakaiya R, Kiraly T (1987) Evaluation of iron status in women identified by copper sulfate screening as ineligible to donate blood. Transfusion 27(3):238–241

Nkrumah B, Nguah SB, Sarpong N, Dekker D, Idriss A, May J, Adu Sarkodie Y (2011) Hemoglobin estimation by the HemoCue® portable hemoglobin photometer in a resource poor setting. BMC Clin Pathol 11:5

Philipps RA, Van Slyke DD, Hamilton PB, Dole VP (1950) Measurement of specific gravities of whole blood and plasma by standard copper sulphate solutions. J Biol Chem 183:305–330

Radtke H, Polat G, Kalus U, Salama A, Kiesewetter H (2005) Hemoglobin screening in prospective blood donors: comparison of different blood samples and different quantitative methods. Transfus Apher Sci 33:31–35

Ross DG, Gilfillan AC, Houston DE, Heaton WA (1986) Evaluation of hemoglobin screening methods in prospective blood donors. Vox Sang 50:78–80

Sawant RB, Bharucha ZS, Rajadhyaksha SB (2007) Evaluation of hemoglobin of blood donors deferred by the copper sulphate method for hemoglobin estimation. Transfus Apher Sci 36:143–148

Shah VB, Shah BS, Puranik GV (2011) Evaluation of non cyanide methods for hemoglobin estimation. Indian J Pathol Microbiol 54:764–768

Stone JE, Simmons WK, Jutsum PJ, Gurney JM (1984) An evaluation of methods of screening for anemia. Bull World Health Organ 62:115–120

Stott GJ, Lewis SM, Wynn KJ (1998) An inexpensive and relaible new haemoglobin color scale for assessing anemia. J Clin Pathol 5(1):21–24

Timan IS, Tatsumi N, Aulia D, Wangsasaputra E (2004) Comparison of haemoglobinometry by WHO Haemoglobin Colour Scale and copper sulphate against haemiglobincyanide reference method. Clin Lab Haematol 26:253–258

Ugwu AC, Reid HL, Famodu AA (1986) The copper sulphate screening test for haemoglobin levels in blood donors: a re-assessment. Med Lab Sci 43(2):174–176

van den Broek NR, Ntonya C, Mhango E, White SA (1999) Diagnosing anemia in pregnancy in rural clinics: assessing the potential of the Haemoglobin Colour Scale. Bull World Health Organ 77(1):15–21

van Lerberghe W, Keegels G, Cornelis G, Ancona C, Mangelschots E, van Balen H (1983) Haemoglobin measurement: the reliability of some simple techniques for use in a primary health care setting. Bull World Health Organ 61:957–965

van Rheenen PF, de Moor LT (2007) Diagnostic accuracy of the haemoglobin colour scale in neonates and young infants in resource-poor countries. Trop Dr 37:158–161

von Schenck H, Falkensson M, Lundberg B (1986) Evaluation of "HemoCue," a new device for determining hemoglobin. Clin Chem 32:526–529

Weiss G (1999) Iron and anemia of chronic disease. Kidney Int suppl 69:S12–S17

Williamson MA, Synder LM, Wallach JB (2011) Wallach's interpretation of diagnostic tests. Wolters Kluwer/Lippincott Williams and wilkins Health, Philadelphia

World Health Organization, Department of blood safety and Clinical Technology (2001) Haemoglobin color scale practical answer to a vital need. World Health Organization, Geneva

Yee DL, Bollard CM, Geaghan SM (2008) Normal blood values: selected reference values for neonatal pediatric and adult populations. In: Hoffman R, Benz EJ, Shattil SS (eds) Haematology: basic principles and practice. Elsevier Churchill Livingstone, Philadelphia. Chapter 164

Zwart A, van Assendelft OW, Bull BS, England JM, Lewis SM, Zijlstra WG (1996) ICSH recommendations for reference method for hemoglobinometry in human blood (ICSH standards 1995) and specifications for international hemoglobincyanamyde standard. J Clin Pathol 49:271–274

Chapter 7
Consequences of Iron Deficiency Anemia

Abstract Iron deficiency anemia impedes the normal growth and development of children. Severity of its effect is noticeable even in infancy. Children lacked proper motor functioning and cognitive development as memory and recall. These children score poorer in arithmetic and written work.

Early childhood embodies an active period of growth and development and is remarkable for maximum growth of neural tissues in the central nervous system. Myelination of the neurons in the inner region of the brain occurs during this period. The process continues for the first 4 years of life, and adequate iron store with micronutrients is highly essential in this period. Iron deficiency and associated anemia, perpetually, deviate the normal structure and functioning of the brain, especially, in children during the period of rapid growth spanning around 0–4 years. Iron deficiency limits the oxygen-carrying capacity of blood and constrains the supply of oxygen to body tissues, visceral organs and skeletal muscles. Iron is crucial in the differentiation and proliferation of macrophages, monocytes and lymphocytes. Iron deficiency results into attenuated cytotoxic activity of macrophages. Lymphocyte proliferation was reduced significantly in the spleen and in purified T-cell population in iron deficiency state.

7.1 Overview

At birth, iron store of the body is sufficient to meet the body's demand in the coming 4–6 months of infancy. However, after the sixth month of infancy, the body growth increases rapidly. Blood volume and cell mass expand. Inherent iron store of body is consumed. These factors, collectively, impose a higher demand for iron in the body. According to (IOM 2000), it has been estimated that daily dietary iron demand rises by 0.78 mg in children during the second half of infancy (IOM 2000).

Leung and Chan (2001) asserted the need for higher daily intake of dietary iron during the period between 6 and 24 months of life owing to rapid growth of body tissues. Therefore, children during the abovesaid periods are prone to iron deficiency and iron deficiency anemia.

Adolescence is another period of active growth in the life of human. Onset of puberty creates high demand for micronutrients, especially iron and semi-essential amino acids in the body. Hence, children during puberty stage are again predisposed

A. Gupta, *Nutritional Anemia in Preschool Children*,
DOI 10.1007/978-981-10-5178-4_7

to ID and IDA. Adolescence is associated with the menarche in girls and has higher prevalence to ID and IDA due to blood loss in comparison to adolescent boys (Thane et al. 2003).

According to Ferrara et al. (2006), multiple factors determine the prevalence of ID and IDA in infants, preschool children and adolescents. Inadequacy of dietary iron is one of the important predisposing factors for ID. Additionally, low socioeconomic status, food habits, sanitation and hygiene, malabsorption syndrome and environmental enteropathy, definitely, influence the iron store of the body and, hence, are determinants of ID and IDA in children.

According to WHO (2002), about 20% of the children below the age of 4 years are suffering from iron deficiency in industrialized countries, whereas the prevalence of iron deficiency in children in the same age group in developing countries is 39%. Non-industrialized developing countries are facing a high burden of iron deficiency and iron deficiency anemia in children during preschool age group.

According to WHO and FAO (2006), the highest (30–55%) prevalence of IDA is noted in adolescents. Truly, the fact synchronizes with heavy iron requirement due to rapid growth in puberty, and additionally, onset of menarche in girls results into blood loss. Preschool children and school children suffer from 25 to 40% prevalence of IDA, respectively.

With the lack of a globally acclaimed cut-off value for iron level in serum, it becomes difficult to assess the iron deficiency in infants and children. IDA affects the health of infants and children negatively. IDA impairs cognitive development, delays psychomotor skill acquisition, retards physical growth, modulates immunity and adversely affects the social and academic traits (Idjradinata and Pollitt 1993).

Infancy and preschool age represent the periods of rapid growth which requires additional dose of iron. Research substantiates the need of iron for proper growth and development. On the contrary, physical growth, cognitive development and motor skills are severely affected in preschool children (Lozoff 2003).

According to Lozoff et al. (2000), iron deficiency anemia impedes the normal growth and development of children. The severity of its effects is noticeable even in infancy. Children lacked proper motor functioning and cognitive development as memory and recall. These children scored poorer in arithmetic and written work. Anaemic children exhibited attention deficiency, anxiety and problems in social interaction. These children had problematic behaviour. Another important fact was the continuation of ill effects of iron deficiency anemia on children at developmental level and behaviour which persisted for the next 10 years after the institution of iron therapy. Hence, iron deficiency anemia leaves indelible marks on the physique, psyche and intellectuality of preschool children.

Early childhood embodies an active period of growth and development and is remarkable for maximum growth of neural tissues in CNS. Myelination of the neurons in the inner region of the brain occurs during this period. The process continues for the first 4 years of life, and adequate iron store with micronutrients is highly essential in this period. Circuit is an interconnection of components which serves a particular function. Neural circuit is the interlinking and interplay among neurons in the brain. Neural circuits are the chief processors of vast information which is received, modified and relayed to body tissues. These are developed in the brain during the first 3–4 years of life. The role of iron in the neurodevelopment is not yet elucidated fully. Iron is necessary for myelination of neurons and neural circuit formation (Lozoff and Black 2004).

Iron is necessary for biosynthesis of neurotransmitters like dopamine, serotonin and adrenaline. Iron is a cofactor in the catalysis of tyrosine to L-DOPA and tryptophan to serotonin by enzymes tyrosine hydroxylase and tryptophan hydroxylase. ID is responsible for inactivity of these enzymes and, hence, a depletion of neurotransmitters which are necessary for cognitive and motor development in children (Beard et al. 2003).

Iron deficiency and associated anemia, perpetually, deviate the normal structure and functioning of the brain, especially in children during periods of rapid growth spanning around 0–4 years. Consequence of iron deficiency anemia has a wide spectrum during the period of rapid growth. Ill effects of iron deficiency anemia on children persist throughout life. Iron is the basic and necessary element of the human body that is essential for extensive physiological activities. Manifestations of iron deficiency range from cellular dysfunctioning to organ failure. Abnormality in the electron transport system is the innocuous effect of iron deficiency. Nonetheless, iron deficit may have profound injury on the organ. The oxygen-carrying capacity of blood is decreased. The brain is the highly sensitive organ to hypoxic hypoxia. IDA severely cripples the psychomotor development, cognitive development and growth of children. Children become susceptible to infections.

Iron deficiency anemia is the major health hazard affecting nearly two billion population of the world, and iron deficiency is one of the chief nutritional deficiencies as projected by Stoltzfus and Dreyfuss (1998).

Children in growing age group, pregnancy and lactation demand high dose of iron due to raised anabolic processes during the former physiological periods in human life. Iron deficiency exacerbates, particularly, in those regions where daily food consumption comprises strictly the vegetarian diet. Moreover, children facing pitiable environmental conditions and those exposed to worm infection require comparatively more iron to supplement its daily loss due to cumulative effect. According to WHO/International Nutritional Anemia Consultative Group/United Nations Children's Fund (WHO/INACG/UNICEF 2001) guidelines, iron and folic acid should be given as supplements in addition to dietary intake to children and pregnant women throughout the world in those areas with high prevalence of nutritional anemia.

Iron deficiency anemia has a causal relationship with mortality in children especially in regions of the world where there is high prevalence of malaria and associated severe anemia in children. This cause and effect relation holds true in sub-Saharan Africa. The effect is further compounded in the presence of malnutrition. These factors exhaust the sources for calorie intake, increase the demand for energy and aggravate the energy loss and derange the calorie homeostasis of the body. So iron deficiency and malnutrition, collectively, increase the mortality in children.

Preschool-aged children in sub-Saharan Africa survive in a precarious state of body iron store and have higher susceptibility to iron deficiency and IDA. Endogenous iron store of the body depletes during the age of 4–6 months. Demand for dietary iron is raised during this period due to increase in blood volume and body tissues. Rapid erythropoiesis further marginalizes the iron store of the body. Infants with low birth weight and born to mothers with iron deficiency become prone to higher prevalence of iron deficit and associated anemia. Inappropriate periods of breast feeding coupled with cereal-based supplementary foods exaggerate the iron store of the body. Plasmodium-induced malaria and helminth infections cause blood loss. Overall, malnutrition, diarrhoea and iron deficiency anemia predominantly affect the health of children (Miller et al. 2003).

Iron deficiency anemia decreases the capacity to do physical work. It is attributed to diminished oxygen supply to muscles and brain tissues. Iron deficiency and oxygen-carrying capacity are closely related to each other (Horton and Levin 2001). This consequence of iron deficiency is estimated as a direct sequela of iron deficiency and is not presented here as a risk factor.

7.2 Effect on Intelligence or Cognitive Development

Iron is the important mineral for the growth and development of body organs; especially, its role is unique and principal in brain development. A number of metabolic processes are dependent on the adequate iron store of the body. It is needed in oxidation-reduction reaction, biosynthesis of neurotransmitters and myelination of nerve fibres. Intrauterine life and toddler age are the stages of human life where iron is essentially needed; defiantly, iron deficiency during either of the abovesaid stages is responsible for impaired brain development and its consequences in the body (Thompson and Nelson 2001).

Deficiency of iron and/or any other nutrient negatively affects the brain development, and its severity is dependent on the amount of deficient nutrient, its duration, area of brain affected and need of nutrient in that area. During intrauterine life, iron status and pregnancy-induced hypertension and diabetes mellitus of the mother predominantly determine the foetal iron status. The foetus undergoes hypoxic hypoxia especially in the body of diabetic mothers. Body iron is diverted to the bone marrow to meet the raised demand for iron for exaggerated erythropoiesis. These infants have polycythaemia at birth. Serum iron is deficit at birth. The growth of the brain is affected (Georgieff et al. 1990).

According to Petry et al. (1992), iron store in the liver is reduced in infants born to diabetic mothers. Concomitantly, brain iron level is also reduced affecting neurodevelopment.

Infants born to diabetic mothers have higher risk for perinatal iron deficiency in the brain which might affect the hippocampus development in infants. It was studied by Siddappa et al. (2004) in infants of diabetic mothers who were categorized into two groups. One group of infants were with potential risk for brain iron deficiency (cord ferritin ≤34 μg/L), and the second group of infants belonged to adequate brain iron store (cord ferritin >34 μg/L). It was observed that motor development was slow in infants with brain iron deficit in comparison to infants with sufficient brain iron store at the age of 1 year. Auditory recognition memory was impaired in infants with cord serum ferritin concentrations <35 mcg/L. These infants did not recognize a familiar voice, say a mother's voice, and the ability to discriminate a mother's voice from a stranger's voice was lost. However, auditory recognition memory was intact in infants with cord serum ferritin concentrations >35 mcg/L. Former infants born to diabetic mothers had lower psychomotor developmental score.

The above study reflects the abnormal functioning of the hippocampus due to deficiency of iron in neonates. Maternal anemia limits the accessibility of iron to foetus. The growth of the brain and foetus is restricted (Siddappa et al. 2007).

Effect of iron deficiency on the neurodevelopment has been studied thoroughly on animal models. These studies indicate abnormality of the dendritic structure in the hippocampal lobe. Another important impairment has been observed in monoamine oxidase metabolism and myelination (Lozoff and Georgieff 2006).

Iron is an element that is needed for biosynthesis of neurotransmitters like serotonin and dopamine in the brain tissues. Iron deficiency, especially, during the period of rapid growth alters secretory activity of dopaminergic and serotonergic neurons in a particular part of the brain which in turn affects the functioning of body organs. The effect is more pronounced in preterm infants born to anaemic mothers modifying the release of iron-dependent neurochemicals. These chemicals affect the temperament of iron-deficient infants.

7.3 Effect on Psychomotor Development

Most of the studies in the past focused on the role of iron on the cognitive development during and after age of infancy. But proper psychomotor development is necessary for cognitive development in infancy and preschool age. A child who can control and coordinate activities during walking can approach new areas and see and touch objects; therefore, fine locomotor ability is necessary for development of exploration, visual perception and concept formation. Iron has an important role in normal cognitive and motor skills of a child.

Motor ability is necessary for the interaction of infant with new object, its manipulation and development of new perception. It was Walka and Pollitt (2000) who developed a conceptual model depicting the poorer development of intellectual and cognitive processes in undernourished infants. Impaired motor development is the important variable that delays the cognitive outcome. Motor development also affects positively the development of normal behaviour in children. Fine walking ability in a child is necessary for gaining self-confidence, self-esteem and assertiveness (Biringen et al. 1995; Corapci et al. 2006).

There is a significant interplay among iron deficiency, foetal growth, preterm infant and low birth weight. Iron deficiency anemia causes reduced partial pressure of oxygen in circulation and tissues. Hypoxic hypoxia stimulates the release of angiotensin, adrenaline and aldosterone in the mother's body. These endocrine substances severely affect the haemodynamic of the mother and foetus. Maternal and foetal circulation faces the impact of high secretion of cortisol. Intense vasoconstriction causes pregnancy-induced eclampsia. Cortisol in foetal circulation impairs the growth of foetus. High predisposition towards preterm delivery is anticipated. According to Allen (2001), an association subsists between Hb level of the mother and levels of human chorionic gonadotrophin (hCG) and human placental lactogen (hPL). Low body iron and haemoglobin level inversely affect the secretion of these two hormones.

The study by Allen (2001) on the effect of maternal haemoglobin level on the health of foetus and birth weight of infant is further supplemented by the work of Rasmussen (2001). He asserted the presence of a strong association among maternal haemoglobin level, birth size of infant at birth, gestational period and neonatal and/or perinatal mortality. There exists a rise in the prevalence of low birth weight infants with a decline in haemoglobin level of mothers.

Moreover, psychomotor impedance causes delayed learning of new skills in response to external stimuli. Finally, the behaviour of children is affected constraining the manifestation of self-confidence, high self-esteem and assertiveness and prompting the appearance of apathetic, defiant and hostile behaviour. Children with iron deficiency are prone to higher risk of hindrance in the development of appropriate emotional regulation (Pollitt 2001; Adolph et al. 1993).

Authors (Grantham-McGregor and Ani 2001) reviewed the studies and concluded the presence of poor cognitive development and delayed motor skill acquisition in iron deficiency anaemic children.

Iron deficiency during the first year of life worsens the neurodevelopment of infants that may continue to preschool-aged group. These children do not show any improvement in behaviour and motor skills even after the short-term iron supplementation.

Several hypotheses have been put forward explaining the possible cause-effect relationship between iron and cognitive development. One view is that iron deficit in the body lowers the brain iron level. It unfavours the synthesis of neurohormones like serotonin and dopamine in the brain. Hypomyelination and impaired neural circuits alter the cognitive development. A hypothesis has been tested in rat model by Felt and Lozoff (1996).

Iron deficiency in infancy and/or early preschool age has profound and long-lasting effect on brain iron level. Dopamine D2 receptors were decreased. Myelination and dopamine secretion were altered.

A study was performed by Cankaya et al. (2003) on auditory brainstem response in anaemic and non-anaemic children. It is a measure of the activation of auditory pathways from the distal part of the acoustic nerve to the lateral lemniscus. Auditory brainstem response was recorded in 33 iron deficiency anaemic children and was compared with that of 31 healthy children. There was an increase in ABR latent period. The fact can be explained on the basis of delayed myelination of neurons. Possibly, iron deficit in the brain resulted in hyposecretion of iron-dependent neurohormones in the brain and trigged a complex response.

Iron deficiency in infants and preschool children provokes low response to environmental stimuli by children that in turn affect, unfavourably, the behaviour of infants. Iron-deficient children are less attentive, timid, dependent on care takers and less social. They have low psychomotor development scores (Lozoff and Jimenez 1991). Workers (Weinberg et al. 1980) studied the behaviour of iron deficiency in rat model. Rats had low brain iron, were less active and responded less to environmental stimuli when compared with the control rats.

7.4 Effect on Physiological Functions of the Body in Early Childhood

Iron deficiency limits the oxygen-carrying capacity of blood and constrains the supply of oxygen to body tissues, visceral organs and skeletal muscles. Partial hypoxia reduces the respiratory capacity and contractility of muscles of the body (Zhu and Haas 1999).

Muscles get easily fatigued; thus performance is reduced. Overall, iron deficiency and IDA lower the functional capacity of muscles. It might be the plausible cause for delayed motor development and acquisition of new motor skills in anaemic children in comparison to non-anaemic children (Harahap et al. 2000).

Iron is imperative in carrying out oxidation-reduction reactions as the oxidoreductase enzymes in the mitochondria are iron dependent in the context of their structure and catalytic activity. Iron is also the chief structural constituent of haemoglobin and myoglobin, which are the inter-tissue and intra-tissue oxygen carrier for metabolic activities. Enzymes in the electron transport chain are iron dependent. Overall, energy production in an individual is dependent on iron store of the body. Oxygen consumption of the body is related to physical fitness. Endurance is the ability of the body to withstand a given workload over a period of time. It demands efficient oxygen supply to muscles for energy production in carrying out the work. It has been observed by Haas and Brownlie (2001) that iron deficiency anemia reduces aerobic capacity, wherein an individual utilizes maximum oxygen for energy production. Therefore, physical activity and performance of child deteriorate in iron deficiency anemia.

Lassitude, laziness and indolence are the common clinical manifestations of iron deficiency anemia in children. Iron deficiency in the body is expressed in terms of reduced synthesis of haemoglobin as well as diminished production of myoglobin and cytochrome series of enzymes in mitochondria. Effects were studied by Dallman (1986) through their work on rats.

It was confirmed that the decline in Hb synthesis was proportional to the decline in myoglobin and cytochrome. Conclusively, a fall in oxygen transportation, decreased oxygen diffusion into skeletal muscles and finally a drop in oxidative capability of muscles were observed (McLane et al. 1981). Muscles become easily fatigued resulting into affected endurance and physical activity.

7.5 Effect on Immunity Development

Iron adequacy is fundamental in the normal development and functioning of the immune system of the body. Iron is crucial in the differentiation and proliferation of macrophages, monocytes and lymphocytes. Several studies have been conducted to describe the role of iron deficiency on immunity. Non-specific immunity is impaired by iron deficiency. Several possible mechanisms have been proposed to elucidate the interrelationship between iron deficiency and iron deficiency anemia and altered functioning of components of non-specific immunity.

Innate immunity is a non-specific immunity. It is dependent on the activity of tissue macrophage system. Macrophages are the giant cells of the reticuloendothelial system. Synthesis of reactive oxygen species is necessary for the innate immune response by macrophage ROS, and hydrogen peroxide production by macrophages is reduced when the macrophages are deficient of iron. Further, secretion of macrophage-inhibiting factor and TNF-α is reduced. Overall, iron deficiency results into attenuated cytotoxic activity of macrophages (Hallquist et al. 1992).

Further in iron deficiency, antibacterial activity of neutrophils is also decreased. It is due to decreased activity of iron-dependent myeloperoxidase enzyme in neutrophils. This enzyme produces hypohalous acid which confers antibacterial activity to neutrophils (Spear and Sherman 1992).

T-lymphocyte Proliferation Is Another Factor That Regulates the Non-specific Immunity in ID and IDA Kuvibidila et al. (1999) worked on the role of iron deficiency on the activity of protein kinase C in the different population of mice. Enzyme has a profound impact on the T-cell proliferation. Lymphocyte proliferation was reduced significantly in the spleen and in purified T-cell population in iron deficiency state.

Activity of PKC enzyme from mice with iron deficiency was reduced. Translocation of PKC is necessary for T-cell mitosis. In iron deficiency state, reduced translocation may play a crucial role in the reduced lymphocyte proliferation and compromised innate response to the pathogens (Kuvibidila et al. 1999).

In fact, diacylglycerol (DAG) and inositol 1,4,5-trisphosphate (IP_3) actively regulate the T-cell proliferation. These cellular signalling molecules are produced by the hydrolysis of phosphatidyl-inositol-4,5-bisphosphate (PIP_2) through the action of phospholipase C enzyme present in the cell membrane. PIP_2 is a phospholipid and a constituent of the cell membrane, whereas phospholipase C is an intracellular enzyme (Rhee and Choi 1992).

DAG remains attached to the inner surface of the cell membrane. It helps in the translocation of protein kinase C (PKC) from the cytosol to the inner surface of the membrane. IP_3 diffuses into the cytoplasm and attaches on the receptors present on the endoplasmic reticulum and helps in the release of calcium into the cytosol and results into the rise of intracellular calcium level. DAG and Ca^{2+} ion constitute the regulatory domain of the PKC enzyme (Lenz et al. 2002).

However, in iron deficiency, activity of phospholipase C is reduced; consequently, synthesis of DAG is reduced, and ultimately, translocation of PKC is impaired affecting negatively the proliferation of T lymphocytes. Both phospholipase C and PKC are not iron-dependent enzymes but are dependent on the zinc content of cytosol. It is presumed that iron deficiency affects the activity of these enzymes at the gene expression level.

7.5.1 Humoral Immunity

Humoral immunity is conferred by the B-cell activation and synthesis of immunoglobulins in the plasma. B-lymphocytes express low levels of transferrin receptors still in resting stage. So they persistently draw iron from circulation. After activation, expression of transferrin receptors is increased. Additionally, synthesis of prostaglandins is increased in infection. PGE_2 regulates the activation and proliferation of T-helper cells (TH_2) which in turn controls the immunoglobulin synthesis. Therefore, iron deficiency, largely, does not affect the humoral immunity (Bagchi et al. 1980). Necessity of iron in the process of monocyte differentiation and activation and phagocytotic activity of macrophages is noteworthy. NADPH-dependent oxidative reactions are carried out in the presence of iron.

Iron is necessary for DNA synthesis. Ribonucleotide reductase is an iron-dependent enzyme. It catalyses the conversion of ribonucleotides into deoxyribonucleotides (Elledge et al. 1992).

Iron deficiency reduces the activity of ribonucleotide reductase affecting the DNA synthesis, thereby reducing the proliferation of cells of the reticuloendothelial system and cells of the immune system.

Iron deficit state is associated with decreased level of interleukin IL-6 and lower activity of killer T cells. Hence, the delayed hypersensitivity reaction of antigen was sluggish (Keusch 2003).

7.6 Critical Review

Early childhood period is a phase of active growth and development of children which stresses on a higher demand of energy and micronutrients. Dietary insufficiency results into inadequacy of growth which is difficult to regain in later life. However, contrasting reports have been published regarding the role of iron in growth.

Results from iron supplementation trials conducted by Fischer et al. (2005) have not shown growth benefits in children who were anaemic. Reasonable mechanism is the inhibition of absorption of other micronutrients like zinc, copper, etc. in the presence of iron supplements.

The study in Honduras and Sweden was conducted by Dewey et al. (2002). Authors (Dewey et al. 2002) observed the impact of iron supplements during infancy over the growth of infants. Full-term infants were selected by authors (Dewey et al. 2002), and effect of placebo and iron supplements were noticed by authors. All the infants were those who were breastfed exclusively for 6 months. It was observed that increase in length and head circumference was more in children (Dewey et al. 2002) who received placebo in comparison to those who were given iron supplements; hence, impact of iron on growth is negated by this study.

Similarly, the study in India found an adverse effect of iron supplementation on weight gain and linear growth in iron-adequate children aged 6–24 months. Double-blind and placebo-controlled trial was conducted by Majumdar et al. (2003). The study (Majumdar et al. 2003) observed the influence of iron therapy over the growth of children who were partly iron sufficient and partly with iron deficiency out of the total study population. Children were in ages between 6 and 24 months. The study population was divided into two groups (Majumdar et al. 2003) as iron replete and iron deficient. Each group was given (Majumdar et al. 2003) iron supplements and placebo. Authors observed that there was weight increase and gain in linear height with iron supplements in the group which was iron deficient. However, to the surprise of authors (Majumdar et al. 2003), it was observed that iron supplements counteracted weight gain and increased the linear growth of children who had adequate iron store in the body (iron replete).

Contrary to regressive effect on growth by iron supplements in iron-sufficient children, another study by Idjradinata et al. (1994) was conducted in Indonesia involving iron-adequate children in age group of 12–18 months. Children were divided into (Idjradinata et al. 1994) two groups and were given iron supplements and placebo. After 4 months of treatment, authors found that increase in weight was more in children who received placebo than those who had iron therapy (Idjradinata et al. 1994). In this study, there was comparative weight gain by iron therapy unlike inhibitory effect on weight gain as found in earlier study.

Hence, reviewed literature posits that iron therapy could be of little value in the growth of children who already have adequate iron store in the body. Moreover, therapy may have deleterious effects on the growth of children with iron-replete status. So the use should be judiciously decided after thorough estimation of body iron store. Iron supplements definitely benefit the children with iron deficiency.

Authors have found a profound ill effect of iron deficiency anemia on the children's cognitive development and psychomotor skills (Sachdev et al. 2005). According to Roncagliolo et al. (1998) and Lozoff et al. (1998), IDA disturbs behaviour and impairs development of infants (Lozoff et al. 1991).

Authors (Moffatt et al. 1994) studied infants belonging to low socioeconomic status regarding the effect of iron supplements on psychomotor development. Infants (Moffatt et al. 1994) were fed on dietary supplements with iron fortification and without iron fortification. Psychomotor development was assessed by authors (Moffatt et al. 1994) on Bayley scales for infant development. Authors (Moffatt et al. 1994) found that development was better in infants fed with iron-fortified dietary supplements in comparison to the other group of infants.

Iron deficiency impairs the synthesis and functioning of neurohumorals and alters the neuro-bioenergetics. It delays the development of CNS (Beard 2008).

Generally, observational studies involving children in age range from 6 to 24 months have been conducted by researchers in divergent geographical locations, socioeconomic conditions, size of study population and last but not the least through the use of different assessment tests. Therefore, consistency in the outcomes is lacking among the work of scholars.

From literature, it is highly volatile to determine the association between iron deficiency anemia and retardation in cognitive and psychomotor development. Socioeconomic status of family is implicated in the availability of quality and quantity of food, approach to medical facility, diarrhoea, infections, care of infant and health status of the mother. These factors supervise the development of infants. Therefore, iron deficiency alone and/or in association with other factors may contribute to delayed development of infants. Iron deficiency and impaired development are the sequel of underprivileged environment.

7.7 Harmful Effects of Iron Supplements

Iron supplementation is possibly associated with some harmful consequences as has been noted by researchers. Its absolute benefits on the growth and development are dependent on its concentration, body iron store and presence of infection in the body. Earlier, its mixed response to the growth and development has been discussed.

7.7.1 Interference in Divalent Metal Ion Absorption

Competitive inhibition is responsible for either the retardation or inhibition in the absorption of dietary iron in the lumen of the small intestine. Iron supplements in infant formulae, possibly, hinder the absorption of zinc. It is due to a common transporter named as divalent metal transporter (Einstein and Blemings 1998).

The postulated hypothesis becomes strongly true in case the ratio of iron to zinc in the supplements is >2:1. Higher concentration of iron inhibits the absorption of zinc due to its higher affinity for DMT-1. Iron, copper and zinc frequently constitute the infant formulae. These divalent metals have common pathway and transporter for their absorption; hence, an undesirable interaction could result into deficiency and/or harmful effects on the body of infants (Andrews 1999).

In a study by Arredondo et al. (2006), similar negative interaction was observed. Authors showed that copper or zinc inhibited the absorption of iron. Further, iron also retards absorption of copper, whereas copper absorption remains unaffected by zinc. Authors (Arredondo et al. 2006) further insisted that simultaneous administration of three metals in molar ratio (Arredondo et al. 2006) of 1:1:1 interfered with the absorption of copper and iron. Authors suggested a practical concern in the infant formulae, which are fortified with copper, iron and zinc, according to authors, in molar ratio of 1:20:5 (Arredondo et al. 2006). It was observed by authors that copper absorption was inhibited more than 50% if the copper, iron and zinc were available together in ratio of 1:10:5 (Arredondo et al. 2006).

Other authors (Lonnerdal and Hernell 1994; Haschke et al. 1986) observed similar inhibitory effects on the absorption of copper by competitive inhibition. Authors confirmed that infant formulae contain high iron concentration (7 mg per L) that could have regressive effect on the absorption of copper in the small intestine.

Authors (Lonnerdal and Hernell 1994) confirmed inhibitory effect of iron in infant formulae on the copper absorption. Authors used infant formulation with 4 mg/l and 7 mg/l of iron content and fed two groups of infants. Authors (Lonnerdal and Hernell 1994) observed a lower serum coper concentration in infants who were nurtured with 7 mg/l of iron concentration in infant formulation indicating a thwarting impact of high concentration of iron on copper absorption.

Authors (Haschke et al. 1986) in their study regarding the effect of iron fortification on the absorption of copper and zinc confirmed a reduced absorption of copper with a rise in the content of iron in the infant formulae. Authors (Haschke et al. 1986) asserted that a marginal reduction in copper absorption is clinically irrelevant and raised a doubt over the interaction and absorption capability of one metal over the other in the body. Further, authors said that zinc absorption is not affected with rise in iron concentration in formulation.

The inhibitory effect of iron content on the absorption of zinc was studied by Solomons (1986). The author studied that iron had inhibitory effect on the uptake and absorption of zinc when iron and zinc were present in ratio of 2:1, according to the author, which was detectable on human studies. The author insisted that the study involving pregnant women who were administered with multimineral supplements containing high content of iron during prenatal periods showed the presence of impaired and delayed growth in infants in postnatal life, and these children had decreased serum zinc concentration. The study confirms a deleterious effect of high iron content in infant formulation on absorption of zinc.

7.7.2 Growth Factor for Pathogens

A few virulent pathogens require iron as growth factor. Non-judicious iron supplementation to infants and children could have harmful effect on body immunity and a predisposing factor for opportunistic infections. Further, it is hypothesized that iron supplements given to breastfed normal infants, probably, saturate lactoferrin and hinder its antibacterial property. According to Andersson et al. (2000), breastfed infants have better immune status in comparison to formula-fed or cow milk-fed infants. Human milk has known antimicrobial property. The author cultured fungi named as *Candida albicans* and *Rhodotorula rubra* in liquid culture medium and observed antifungal effect of human milk on these fungi. The author put forwarded that antifungal property of milk is attributed to the presence of lactoferrin in milk which showed potent growth inhibitory effect, i.e. fungistatic effect.

Lactoferrin is a glycoprotein of globular size and shape. It is observable in body fluids like milk, saliva, tears, sweat and nasal secretion. Neutrophil granules also contain appreciable concentration of lactoferrin. Colostrum, the first human milk after delivery, contains the highest concentration of lactoferrin (Sanchez et al. 1992).

Lactoferrin can bind with two ions of iron, zinc or copper. But its affinity to iron is much higher; hence, it is responsible for the regulation of free iron in blood and secretions. Its antimicrobial effect is mediated by its property to sequester free iron in circulation and secretions of the body, which is a favourable nutrient for the growth of bacteria (Xanthou 1998).

Authors (Andersson et al. 2000) found that an increase in iron content of culture medium resulted into diminished/abolished antifungal effect of lactoferrin. Therefore, unnecessary iron supplements in normal full-term infants predispose a higher incidence of bacterial and/or fungal infection in children.

The author (Oppenheimer 1989) posited a strong clinical evidence against the iron supplements in children and occurrence of infection. Iron overload in the body is associated with higher prevalence of infection whose pathogenesis is uncertain. The author said that parenteral administration of iron preparations to infants and children has witnessed serious outcomes in terms of incidence of malaria and respiratory diseases which have further increased the morbidity in children.

Current issue of iron supplements and infection is still unresolved and further requires more studies to reach a uniform conclusion over the dose, route, age and content of iron in preparations.

For these reasons, it has been suggested that iron supplementation, or the addition of iron to infant formulae, might increase the incidence of gastroenteritis and other infections in infants. However, several studies have failed to confirm this theory for iron-fortified formulae. One study even showed that added iron provided a small but significant protection against diarrhoea.

7.7.3 Iron-Induced Pro-oxidant Effect

A little iron is excreted daily, so its rampant use in the form like iron supplements to infants and/or to children poses potential threat towards its harmful effects in the body owing to the tendency of iron to form free radicals through its interaction with hydrogen peroxide. Iron is a transition metal having a very high tendency towards involvement in redox reactions. Iron forms hydroxyl and hydroperoxyl free radicals through Fenton reaction (Winterbourn 1995).

Authors (Kadiiska et al. 1995), with the use of electron spin resonance spectroscopic method, have demonstrated the generation of hydroxyl free radical in rats which were fed with diet rich in iron. Free radicals are the atoms which have unpaired electron in their outer energy shell. Atoms are highly reactive towards interaction with other atoms to gain electron and achieve stable state. Free radicals are strongly implicated in the lipid peroxidation, protein denaturation and damage of DNA (Aust et al. 1985).

Therefore, iron is nefarious for its pro-oxidant effect. Excess and free iron has a potential to carcinogenesis (Stevens et al. 1994) as well as a predisposing role in coronary artery disease (Ascherio et al. 1994). Tendency of dietary iron towards higher prevalence of colorectal malignancy in adults has been disposed by Deneo-Pellegrini et al. (1999) and Kato et al. (1999).

Authors (Sobotka et al. 1996) had studied the effect of dietary iron overload on the behaviour changes in rats. Authors fed male rats with diets containing different concentrations of iron in ppm. An iron-deficient group of rats was included for the comparison. Authors observed that rats fed upon diet containing iron at 2,000 ppm showed decreased response after repeated stimuli. Body weight decreased. Overall changes were accompanied by systemic iron toxicity.

References

Adolph KE, Eppler MA, Gibson EJ (1993) Crawling versus walking infants' perception of affordances for locomotion over sloping surfaces. Child Dev 64:1158–1174

Akhtar K, Sherwani RK, Rahman K, Hasan J, Shahid M (2009) HemoCue photometer: a better alternative of hemoglobin estimation in blood donors? Indian J Hematol Blood Transfus 24(4):170–172

Allen LH (2001) Biological mechanisms that might underlie iron's effect on foetal growth and preterm birth. J Nutr 131:581S–589S

Andersson Y, Lindquist S, Lagerqvist C, Hernell O (2000) Lactoferrin is responsible for the fungistatic effect of human milk. Early Hum Dev 59:95–105

Andrews NC (1999) Disorders of iron metabolism. N Engl J Med 341:1986–1995

Arredondo M, Martinez R, Nunez MT, Ruz M, Olivares M (2006) Inhibition of iron and copper uptake by iron, copper and zinc. Biol Res 39(1):95–102

Ascherio A, Willett WC, Rimm EB, Giovannucci EL, Stampfer MJ (1994) Dietary iron intake and risk of coronary disease among men. Circulation 89:969–974

Aust SD, Morehouse LA, Thomas CE (1985) Role of metals in oxygen radical reactions. Free Radic Biol Med 1:3–25

Bagchi K, Mohanram M, Reddy V (1980) Humoral immune response in children with iron-deficiency anemia. Br Med J 280:1249–1251

Beard JL (2008) Why iron deficiency is important in infant development. J Nutr 138:2534–2536

Beard J, Erikson KM, Jones BC (2003) Neonatal iron deficiency results in irreversible changes in dopamine function in rats. J Nutr 133:1174–1179

Biringen Z, Emde RN, Campos JJ, Appelbaum MI (1995) Affective reorganization in the infant, the mother, and the dyad: the role of upright locomotion and its timing. Child Dev 66:499–514

Cankaya H, Oner AF, Egeli E, Caksen H, Uner A, Akcay G (2003) Auditory brainstem response in children with iron deficiency anemia. Acta Paediatr Taiwan 44(1):21–24
Corapci F, Radan AE, Lozoff B (2006) Iron deficiency in infancy and mother-child interaction at 5 years. J Dev Behav Pediatr 27:371–378
Dallman PR (1986) Biochemical basis for the manifestations of iron deficiency. Anun Rev Nutr 6:13–40
Deneo-Pellegrini H, De Stefani E, Boffetta P, Ronco A, Mendilaharsu M (1999) Dietary iron and cancer of the rectum: a case-control study in Uruguay. Eur J Cancer Prev 8:501–508
Dewey KG, Domellof M, Cohen RJ, Rivera LL, Hernell O, Lonnerdal B (2002) Iron supplementation affects growth and morbidity of breast-fed infants: results of a randomized trial in Sweden and Honduras. J Nutr 132(11):3249–3255
Einstein RS, Blemings KP (1998) Iron responsive elements and iron homeostatis. J Nutr 128:2295–2298
Elledge SJ, Zhou Z, Allen JB (1992) Ribonucleotide reductase: regulation, regulation, regulation. Trends Biochem Sci 17(3):119–123
Felt BT, Lozoff B (1996) Brain iron and behaviour of rats are not normalized by treatment of iron deficiency anemia during early development. J Nutr 126:693–701
Ferrara M, Coppola L, Coppola A, Capozzi L (2006) Iron deficiency in childhood and adolescence: retrospective review. Hematology 11:183–186
Fischer WC, Kordas K, Stoltzfus RJ, Black RE (2005) Interactive effects of iron and zinc on biochemical and functional outcomes in supplementation trials. Am J Clin Nutr 82:5–12
Georgieff MK, Landon MB, Mills MM, Hedlund BE, Faassen AE, Schmidt RL, Ophoven JJ, Widness JA (1990) Abnormal iron distribution in infants of diabetic mothers: spectrum and maternal antecedents. J Pediatr 117:455–461
Grantham-McGregor S, Ani C (2001) A review of studies on the effect of iron deficiency on cognitive development in children. J Nutr 131:649S–668S
Haas JD, Brownlie T IV (2001) Iron deficiency and reduced work capacity: a critical review of the research to determine a causal relationship. J Nutr 131:676S–690S
Hallquist NA, McNeil LK, Lockwood JF, Sherman AR (1992) Maternal-iron-deficiency effects on peritoneal macrophage and peritoneal natural-killer-cell cytotoxicity in rat pups. Am J Clin Nutr 55:741–746
Harahap H, Jahari AB, Husaini MA, Saco-Pollitt C, Pollitt E (2000) Effects of an energy and micronutrient supplement on iron deficiency anemia, physical activity, and motor and mental development in undernourished children in Indonesia. Eur J Clin Nutr 54:S114–S119
Haschke F, Ziegler EE, Edwards BB, Fomon SJ (1986) Effect of iron fortification of infant formula on trace mineral absorption. J Pediatr Gastroenterol Nutr 5:768–773
Health Care A guide for health administrators and programme managers (1989). WHO, Geneva
Horton S, Levin C (2001) Commentary on "evidence that iron deficiency anemia causes reduced work capacity". J Nutr 131:S691–S696
Idjradinata P, Pollitt E (1993) Reversal of developmental delays in iron-deficient anaemic infants treated with iron. Lancet 341:1–4
Idjradinata P, Watkins WE, Pollitt E (1994) Adverse effect of iron supplementation weight gain of iron-replete young children. Lancet 343:1252–1254
Institute of Medicine (2000) Dietary reference intakes for vitamin A, vitamin K, arsenic, boron, chromium, copper, iodine, iron, manganese, molybdenum, nickel, silicon, vanadium, and zinc. National Academy Press, Washington, DC
Kadiiska MB, Burkitt MJ, Xiang QH, Mason RP (1995) Iron supplementation generates hydroxyl radical in vivo. An ESR spin-trapping investigation. J Clin Invest 96(3):1653–1657
Kato I, Dnistrian AM, Schwartz M (1999) Iron intake, body iron stores and colorectal cancer risk in women: a nested case-control study. Int J Cancer 80:693–698
Keusch GT (2003) The history of nutrition: malnutrition, infection and immunity. J Nutr 133:336S–340S

Kuvibidila SR, Kitchens D, Baliga BS (1999) In vivo and in vitro iron deficiency reduces protein kinase C activity and translocation in murine splenic and purified T cells. J Cell Biochem 74:468–478
Lenz JC, Reusch HP, Albrecht N, Schultz G, Schaefer M (2002) Calcium controlled competitive diacylglycerol binding of protein kinase c isoenzymes in living cells. J Cell Biol 159:291–202
Leung AK, Chan KW (2001) Iron deficiency anemia. Adv Pediatr Infect Dis 48:385–408
Lonnerdal B, Hernell O (1994) Iron, zinc, copper and selenium status of breast-fed infants and infants fed trace element fortified milk-based infant formula. Acta Paediatr 83:367–373
Lozoff B (2003) Iron deficiency in infancy: impact on behavior and development. UNSCN Annual Meeting, Chennai, India, 3–7, unpublished presentation
Lozoff B, Black MM (2004) Impact of micronutrient deficiencies on behavior and development. In: Pettifor JM, Zlotkin S (eds) Micronutrient deficiencies during the weaning period and the first years of life. Nestec Ltd, Basel, pp 119–135
Lozoff B, Georgieff MK (2006) Iron deficiency and brain development. Semin Pediatr Neurol 13:158–165
Lozoff B, Jimenez E, Wolf AW (1991) Long-term developmental outcome of infants with iron deficiency. N Engl J Med 325:687–694
Lozoff B, Klein NK, Nelson EC, McClish DK, Manuel M, Chacon ME (1998) Behavior of infants with iron-deficiency anemia. Child Dev 69:24–36
Lozoff B, Jimenez E, Hagen J, Mollen E, Wolf AW (2000) Poorer behavioral and developmental outcome more than 10 years after treatment for iron deficiency in infancy. Pediatrics 105:11
Majumdar I, Paul P, Talib VH, Ranga S (2003) The effect of iron therapy on the growth of iron-replete and iron deplete children. J Trop Pediatr 49:84–88
McLane JA, Fell RD, McKay RH, Winder WW, Brown EB, Holloszy JO (1981) Physiological and biochemical effects of iron deficiency on rat skeletal muscle function. Am J Phys 241:C47–C54
Miller MF, Stoltzfus RJ, Mbuya NV, Malaba LC, Iliff PJ, Humphrey JH (2003) Total body iron in HIV-positive and HIV-negative Zimbabwean newborns strongly predicts anemia throughout infancy and is predicted by maternal hemoglobin concentration. J Nutr 133:3461–3468
Moffatt ME, Longstaffe S, Besant J, Dureski C (1994) Prevention of iron deficiency and psychomotor decline in high-risk infants through use of iron-fortified infant formula: a randomized clinical trial. J Pediatr 125:527–534
Oppenheimer SJ (1989) Iron and infection: the clinical evidence. Acta Paediatr Scand Suppl 361:53–62
Petry CD, Eaton MA, Wobken JD, Mills MM, Johnson DE, Georgieff MK (1992) Iron deficiency of liver, heart, and brain in newborn infants of diabetic mothers. J Pediatr 121:109
Pollitt E (2001) The developmental and probabilistic nature of the functional consequences of iron-deficiency anemia in children. J Nutr 131:669S–675S
Rasmussen K (2001) Is there a causal relationship between iron deficiency or iron deficiency anemia and weight at birth, length of gestation and perinatal mortality? J Nutr 131:590S–603S
Rhee SG, Choi KD (1992) Regulation of inositol phospholipid-specific phospholipase C isozymes. J Biol Chem 267(18):12393–12396
Roncagliolo M, Garrido M, Walter T, Peirano P, Lozoff B (1998) Evidence of altered central nervous system development in infants with iron deficiency anemia at 6 months: delayed maturation of auditory brainstem responses. Am J Clin Nutr 68:683–690
Sachdev H, Gera T, Nestel P (2005) Effect of iron supplementation on mental and motor development in children: systematic review of randomised controlled trials. Public Health Nutr 8:117–132
Sanchez L, Calvo M, Brock JH (1992) Biological role of lactoferrin. Arch Dis Child 67(5):657–661
Siddappa AM, Georgieff MK, Wewerka S, Worwa C, Nelson CA, Deregnier RA (2004) Iron deficiency alters auditory recognition memory in newborn infants of diabetic mothers. Pediatr Res 55(6):1034–1041

Sobotka TJ, Whittaker P, Sobotka JM, Brodie RE, Quander DY, Robl M, Bryant M, Barton CN (1996) Neurobehavioral dysfunctions associated with dietary iron overload. Physiol Behav 59(2):213–219

Solomons NW (1986) Competitive interaction of iron and zinc in the diet: consequences for human nutrition. J Nutr 116:927–935

Spear AT, Sherman AR (1992) Iron deficiency alters DMBA-induced tumor burden and natural killer cell cytotoxicity rats. J Nutr 122:46–55

Stevens RG, Graubard BI, Micozzi MS, Neriishi K, Blumberg BS (1994) Moderate elevation of body iron level and increased risk of cancer occurrence and death. Int J Cancer 56:364–369

Stoltzfus RJ, Dreyfuss ML (1998) Guidelines for the use of iron supplements to prevent and treat iron deficiency anemia. ILSI Press, Washington, DC

Thane CW, Bates CJ, Prentice A (2003) Risk factors for low iron intake and poor iron status in a national sample of British young people aged 4–18 years. Public Health Nutr 6:485–496

Thompson RA, Nelson CA (2001) Developmental science and the media: early brain development. Am Psychol 56(1):5–15

Walka H, Pollitt E (2000) A preliminary test of a developmental model for the study of undernourished children in Indonesia. Eur J Clin Nutr 54:S21–S27

Weinberg J, Dallman P, Levine S (1980) Iron deficiency during early development in the rat: behavioural and physiological consequences. Pharmacol Biochem Behav 12:493–502

WHO, INACG, UNICEF (2001) Iron-deficiency anemia: reexamining the nature and magnitude of the public health problem. Beard JL, Stoltzfus R (eds). J Nutr 131:2SII

Winterbourn CC (1995) Toxicity of iron and hydrogen peroxide: the Fenton reaction. Toxicol Lett 82(83):969–974

World Health Organization (2002) Nutrition: global strategy for infant and young child feeding. WHO, Geneva

World Health Organization, Food and Agricultural Organization of the United Nations (2006) In: Allen L, de Benoist B, Dary O, Hurrell R (eds) Guidelines on food fortification with micronutrients. WHO, Geneva

Xanthou M (1998) Immune protection of human milk. Biol Neonate 74(2):121–133

Zhu YI, Haas JD (1999) Altered metabolic response of iron depleted nonanemic women during a 15-km time trial. J Physiol 84:1768–1775

Chapter 8
Vitamin B_{12} and Folic Acid in Nutritional Anemia in Children

Abstract Vitamin B_{12} is a water-soluble, essential, octahedral macromolecule containing cobalt. It is vitally needed for functioning of the brain and nervous system. It has an important role in DNA synthesis and erythropoiesis in bone marrow, hence, termed as anti-pernicious anemia factor. Megaloblastosis in bone marrow cells results into megaloblastic anemia. It is notably caused by a deficiency of folic acid and/or B_{12} in the body. Common cause of folate inadequacy in the body is its impaired absorption in the small intestine due to malabsorption syndrome, whereas cobalamin deficiency is attributed to lack of its absorption as a result of deficit amount of intrinsic factor in gastric secretion.

Folic acid is called as pteroylglutamic acid or vitamin B9, and its deficiency causes anorexia, loss of body weight, headache and megaloblastic anemia in children. Also irritability and a change in the behaviour of child have been observed. Preschool children face stunted growth. Folate deficiency before conception is fatal for the growth and development of foetus and is associated with serious consequences. Preconceptual folate deficiency is largely responsible for neural tube defects in intrauterine life.

8.1 Structure of Vitamin B_{12}

Vitamin B_{12} is a water-soluble, essential, octahedral cobalt containing micronutrient called as cobalamin. It is vitally needed for functioning of the brain and nervous system. It has an important role in DNA synthesis and erythropoiesis in bone marrow, hence termed as anti-pernicious anemia factor (Combs 1992; Krautler 2012; Yamada 2013; Allen et al. 1993).

Cobalamin is a stable metal–carbon bonded biomolecule. It is composed of corrin ring to which an additional DBI ring is strapped (Johnson et al. 2012; Randaccio et al. 2010; Wuerges et al. 2006).

Cobalamin is a subclass of corrinoids. Corrin ring is a tetrapyrrole structure, wherein rings are numbered from A to D and are linked together by methylene bridges except ring A and D which are interconnected directly, hence removing one methylene bridge. A heavy metal, cobalt, is located in the centre of the tetrapyrrole nucleus. Cobalt atom is coordinated to nitrogen atom in each pyrrole ring, thereby forming a corrin ring system. Benzimidazole derivative, 5-6-dimethylbenzimidazole,

A. Gupta, *Nutritional Anemia in Preschool Children*,
DOI 10.1007/978-981-10-5178-4_8

is attached to 5-C ribose sugar through nitrogen atom and is termed as DBI ring. It is attached to cobalt atom through its nitrogen atom on one side and propionic acid in IV pyrrole ring through phosphate bonding. Naturally occurring vitamin B_{12} is produced by bacteria and is called as hydroxocobalamin. Synthetically, a new form of cobalamin is produced by cyanide group substitution and is labelled as cyanocobalamin. Its action is similar to hydroxocobalamin, but it has higher stability and low cost; hence, it is extensively marketed by pharmaceutical companies as food supplements (Brown 2005).

Other variants of cobalamin are the methylcobalamin and 5-deoxyadenosylcobalamin which are actively responsible for human metabolism. Cobalamin also acts as cofactor with the enzymes, methionine synthase and L-methylmalonyl–CoA mutase. Former enzyme is involved in the conversion of homocysteine into methionine in the presence of Vit B_{12} as cofactor. Methionine in the form of S-adenosylmethionine is a universal methyl donor in methylation metabolism of the body, whereas the latter enzyme is concerned with the conversion of L-methylmalonyl–CoA to succinyl–CoA during lipid metabolism. Succinyl–CoA has its value in haemoglobin synthesis (Clarke 2008).

8.2 Dietary Sources

Cobalamins are richly available in animal food like liver, fish, eggs and meat. Milk and its products also contain trace amount of cobalamins (USDA 2011).

In the liver, vitamin B_{12} is stored in large amount, and it is the important animal source of this vitamin. Plant food sources contain small amount of vit B_{12} (Smith and Cobalt 1987). In human, vitamin b12 is synthesized by bacteria present in the colon and caecum. Vitamin B_{12} is present in human faeces (Herbert et al. 1984).

Pseudomonas and *Klebsiella* may synthesize appreciable amount of vitamin B_{12} in the large intestine (Kalser et al. 1966). Microorganisms can synthesis vitamin B_{12}. These microbes live in soil, water and the colon of animals. Synthesized vitamin B_{12} from the colon cannot be absorbed. Human is absolutely reliant on food sources for the dose of this vitamin.

8.3 Absorption of Vitamin B_{12}

Vitamin B_{12} is present in dietary sources in bound form with proteins. After consumption, its complexed form reaches the stomach. Hydrochloric acid in gastric fluid helps to hydrolyse the complex, and vitamin B_{12} is released (Cooper and Castle 1960).

Pepsin also has an additional role in the liberation of vitamin B_{12} from its complexed form in foods. Cooking also aids to protein denaturation and release of cyanocobalamin from particular foods like the liver (Kittang and Schjqnsby 1987). In the free state, vitamin B_{12} is susceptible to denaturation in the presence of hydrochloric acid in the stomach. It complexes with a glycoprotein called transcobalamin-1 or haptocorrin or R-protein in the stomach. This protein is synthesized by salivary glands in the mouth. In the stomach at low pH, affinity of vitamin B_{12} for R-factor or haptocorrin is high; hence, it forms a vitamin B_{12}–haptocorrin complex in the stomach. Haptocorrin–B_{12} complex is immune to the insult of HCl in the stomach. Vitamin b12 can safely pass through the pylorus and reaches proximal part of the duodenum (Morkbak et al. 2007).

Proximal part of the duodenum contains a proteolytic enzyme called as pancreatic protease. It is secreted by the pancreas and released into the lumen of the small intestine. This enzyme partly hydrolyses the haptocorrin–B_{12} complex. Vitamin B_{12} is again released in the duodenum. This part of the duodenum marks the rise in pH of the lumen. At high pH, vitamin B_{12} has higher affinity for intrinsic factor in comparison to R-factor (Weir and Scott 1999).

Intrinsic factor is a glycoprotein. It has a molecular weight of 45,000 dalton. It is secreted by parietal cells of the stomach in the fundus and body of the stomach. IF enters the duodenum. In the proximal part of the ileum, intrinsic factor binds with the exogenous factor, cobalamin, and forms a B_{12}–intrinsic factor complex. Earlier it was presumed that the complex is formed in the gastric lumen. However, concept has changed with more research and understanding of the affinity of vitamin B_{12} to R-factor and IF factor under different pH of the medium (Alpers and Russell-Jones 2013).

Vitamin B_{12}–intrinsic factor complex moves downstream into distal part of the ileum. Its epithelium is endorsed with the specialized giant endocytic receptors called as cubilin and megalin (Moestrup et al. 1998). These receptors were unknown in the past and have been discovered recently. These are cell surface receptors and can bind with a wide range of ligands. Megalin is a transmembrane protein. Its extracellular domain is for binding with ligands and is comprised of cysteine-rich four clusters named as I, II, III and IV as ligand binding domains (Davis et al. 1987).

Cubilin is a peripheral protein involved in multiligand endocytosis. It is expressed by epithelial cells of the renal proximal convoluted tubule, visceral yolk sac, ileum and placenta (Christensen et al. 1998). It is made up of 100 amino acid N-terminal stretch containing free cysteine of the molecule which is followed by eight epidermal growth factor and 27 CUB (complement C1ruC1s, Uegf and bone morphogenic protein-1) domains (Kozyraki et al. 1998). Each CUB domain has 110 amino acids which are linked to two disulphide bonds. CUB domains have the ability to attach with a number of ligands like proteins, carbohydrates and/or lipids. Unlike megalin, it is devoid of transmembrane domain.

Cubilin is internalized in association with megalin, which is co-localized with cubilin in the epithelium of the ileum. The two together facilitate the absorption of intrinsic factor–vitamin B_{12} complex (Verroust 2002). Vitamin B_{12}–intrinsic factor complex binds to cubilin receptors on the epithelium in the distal ileum. The linkage is very specific, and the receptors do not bind with ligand apart from cobalamin bound to IF.

Complex undergoes endocytosis and reaches enterocyte cytosol. IF is released at the cell surface for its reuse. So vitamin B_{12} undergoes cubilin-mediated endocytosis across the enterocytes (Viola-Villegas et al. 2009).

8.4 Transportation in Blood Circulation

Cobalamin is internalized by enterocytes. Peters and Hoffbrand (1970) administered a physiological dose of CoBI2 to the guinea pig. High mitochondrial radioactivity in the mucosa of the ileum was noted in the first 2 h of ingestion. Later on, radioactivity was decreased with rise in hepatic activity. It is retained within mitochondria of enterocytes before it enters the circulation. This was termed as intracellular absorption of cobalamin. Its exact cause and metabolic activity within mitochondria need explanation.

Afterwards, it attaches to transcobalamin II within enterocytes, and a complex is formed as holotranscobalamin II. It is carried by hepatic portal circulation to the liver. However, its binding with transcobalamin II, whether it occurs in enterocytes or hepatocytes, needs more research. But indirect evidence by Chanarin et al. (1978) is suggestive of its binding within the intestinal mucosal cells. Transcobalamin II is synthesized by ileal enterocytes.

Human body stores an average of 3,000 μg cobalamin especially in non-vegetarians. The liver stores around 60% of total body cobalamin, whereas 30% of it remains stored in the muscles. Cobalamin storage inside the liver is maximum to the concentration of 4 mg (Von Schenck et al. 1997).

Transcobalamin II deficiency causes failure of cobalamin absorption; hence, its synthesis is necessary for cobalamin absorption in blood circulation. It was observed by Hakami et al. (1971) that congenital deficiency of transcobalamin II in two siblings effected into megaloblastic anemia. Possibly it could be attributed to impairment in cobalamin absorption due to lack of carrier molecule.

It is excreted along with bile into the small intestine. Vitamin B_{12} in bile remains associated with R-binder. Its amount in bile is around 4 μg. Bile also contains non-cobalamin corrinoids. However, out of total corrinoids in bile, more than half of the corrinoids represent vitamin B_{12}.

Vitamin B_{12} analogues comprise a large percentage of total corrinoids in the body. But their source and significance in the body is not clear. It has been found that the level of B_{12} analogues was normal in patients with pernicious anemia, ulcerative colitis and irritative bowel syndrome.

But its level was elevated in individuals who had a pathology with the ileum, or it was resected surgically, suggesting that the ileum is the site of absorption for B_{12} analogues in the body. Its pathway of absorption is entirely different from B_{12} absorption in the body (Shaw et al. 1989).

Bile R-protein which binds to corrinoids is proteolysed by pancreatic enzymes. Vitamin B_{12} binds to intrinsic factor in the small intestine. Hence, a large portion of biliary B_{12} might be reabsorbed. Nonetheless, B_{12} analogues do not bind with intrinsic factor and, hence, are excreted from the body in stool and urine. Enterohepatic circulation serves to conserve biliary B_{12} in the body and supports to excrete vitamin B_{12} analogues (Kanazawa et al. 1983).

8.5 Excretion

It is not excreted in urine, except, in parenteral administration, wherein cyanocobalamin excretion may go up to 0.3 μg/day.

8.6 Metabolically Active Form of Vitamin B_{12}

Cobamide is the biologically active form of cobalamin. It possesses adenine nucleoside attached to central cobalt atom through C→CO bond, and it lacks cyanide group. During the activation process, cobalt atom undergoes repeated reduction catalysed by enzyme B_{12} reductase in the presence of NADH and FAD as follows (Fig. 8.1):

Fig. 8.1 Active forms of cobalamin

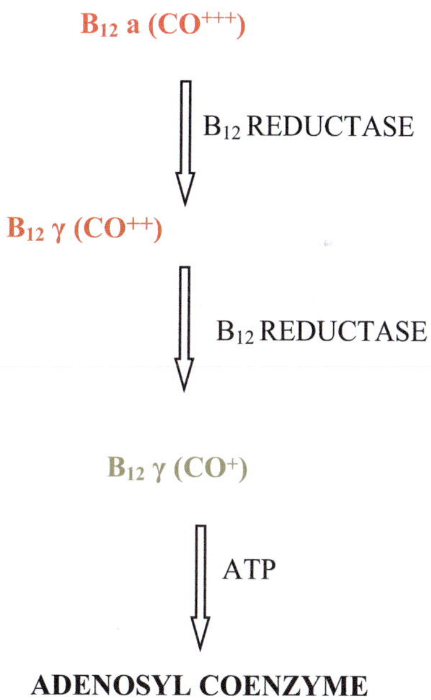

Cobamides have different varieties as:
1. Dimethyl-Benzimidazole cobamide(DBC)
2. Benzimidazole cobamide(BC)
3. Adenyl cobamide(AC)
4. Methyl cobamide(MC)

8.7 Physiological Significance of Cobalamin-Dependent Enzymes

8.7.1 Restoration of Methionine from Homocysteine

Methionine synthase is the mammalian enzyme that catalyses the conversion 5-methyltetrahydrofolate (N^5-MeTHF) into tetrahydrofolate (THF). This enzyme in mammals is cobalamin dependent requiring cobalamin as cofactor for its activity. But enzyme in plants has cobalamin-independent activity (Zydowsky 1986).

In this reaction, methyl group is transferred from 5-methyltetrahydrofolate to cobalamin forming a methylcobalamin complex. Cobalamin in free state contains cobalt(I), which is activated into cobalt(III) state in bound state of cobalamin. Methyl group in the complex is in activated form. It is further transferred to homocysteine, hence reviving methionine. Cobalt returns to (+1) oxidation state. Cobalamin-dependent form of methionine synthase helps to regenerate methionine from homocysteine and THF which is a cofactor in the synthesis of DNA (Zhang et al. 2012; Baneerjee and Matthews 1990; Chanarin et al. 1985; Gruber and Kratky 2006).

8.7.2 Conversion of Ribonucleotides into Deoxyribonucleotides

Deoxythymidine monophosphate (dTMP) is the monomeric component of deoxyribonucleotides. It is synthesized from uridine monophosphate. In this reaction,

$$\text{5,10-methylenetetrahydrofolate} + \text{dUMP} \Longleftrightarrow \text{dihydrofolate} + \text{dTMP}.$$

The reaction is catalysed by thymidylate synthetase enzyme. It is a cobamide-dependent enzyme.

Pathway provides the source of production of dTMP. This enzyme is also helpful in the oxidation of 5,10-methylenetetrahydrofolate to dihydrofolate (Mitchell et al. 1941; Stover 2009). Hence, vitamin B_{12} is necessary for reductive methylation of ribonucleotides to deoxyribonucleotides.

Its evidence was provided by Downing and Schweigert (1956) in their isotopic studies in *Lactobacillus leichmannii.*

Another evidence was provided by Williams et al. (1963) in their study in human bone marrow cells. In the deficiency of vitamin B_{12}, synthesis of DNA was affected. It showed need of cobalamin in the impaired conversion of ribonucleotides into deoxyribonucleotides.

8.8 Effects of Vitamin B_{12} Deficiency on the Body

8.8.1 Hypocobalaminaemia

8.8.1.1 Inadequate Dietary Intake of Vitamin B_{12}

Children in early childhood are at higher risk of developing B_{12} deficiency due to inadequate dietary intake. Storage capacity of B_{12} in their bodies is limited. The need for vitamins and minerals is raised during early childhood as a result of rapid growth and development. Vitamin B_{12} is found in animal sources of diet as eggs, meat, fish, liver and milk. Another non-conventional source of B_{12} is the green algae, called as *Chlorella*. It has been claimed to be a good source of macro- and micronutrients. After drying up, it is comprised of around 45% proteins, 20% lipids, another 20% carbohydrates, 5% fibre and 10% micronutrients. Its mass production is being under trial stage to yield good source of nutrients (Watanabe et al. 2002; Croft et al. 2005).

Children and women in families adopting strict vegetarian diet are susceptible to cobalamin deficiency (Emery et al. 1997; Korenke et al. 2004; Simsek et al. 2004).

Vegetarians as well as vegans have higher potential to develop B_{12} deficiency as a result of poor dietary intake. They mainly rely on plant sources of B_{12}. Poverty is a common cause of cobalamin deficiency in the body of children and pregnant women in different parts of the world, especially in developing countries (Casterline et al. 1997; Rogers et al. 2003; Guerra-Shinohara et al. 2004; Garcia-Casal et al. 2005; Pruthi and Tefferi 1994).

Unlikely, B_{12} deficiency has also been observed in people who are predominantly dependent on animal sources of food (McBride 2000).

8.8.1.2 Impaired Absorption of Vitamin B_{12}

Intrinsic factor is secreted by parietal cells of gastric mucosa and is necessary for absorption of B_{12}. Chronic atrophic gastritis causes damage to parietal cells of stomach. Autoimmunity is directed against parietal cells and autoantibodies are synthesized against intrinsic factor resulting in its deficiency in gastric secretion.

Congenital intrinsic factor deficiency is an autosomal recessive hereditary disorder that has been detected in a few cases. It generally expresses in early childhood before age of 5 years. Children show signs and symptoms of anemia. Children are stunted and underweight. Allelic mutation on the GIF gene on chromosome 11 is responsible for impaired synthesis and deficiency of intrinsic factor (Gordon et al. 2004). Deranged absorption of B_{12} as a consequence of diminished availability of intrinsic factor manifests into pernicious anemia (Anthony 2012).

Achlorhydria is characterized by the absence of hydrochloric acid secretion in the gastric lumen. It is largely caused by drugs as proton pump inhibitors and H2 receptor antagonists, infection by *Helicobacter pylori*, atrophic gastritis, gastric carcinoma and gastric surgical resection. HCl deficiency interferes with liberation of B_{12} from dietary food sources and hence decreasing its absorption (Lam et al. 2013). Achlorhydria is more common in old-aged persons. Prolonged use of ranitidine, a H2 antagonist, probably contributes to hypocobalaminaemia (Yeomans et al. 1982).

Coeliac disease is a chronic inflammatory disorder of the small intestine. It damages the architecture of intestinal mucosa and deranges its absorptive capability of B_{12}. Oral hypoglycaemic drugs like metformin have potential to decrease the level of serum B_{12} and folic acid, thus raising the concentration of homocysteine in the body. A cross-sectional cohort study on type 2 diabetic patients by Hermann et al. (2004), who were on metformin therapy, reported the prevalence of 5.8–33% deficiency of B_{12}. So metformin-induced B_{12} deficiency is uncommon in children.

Giardiasis is a protozoan and contagious disease of the alimentary canal. It is caused by pathogenic endoparasite, *Giardia lamblia*, which enters the intestine through faecal–oral route. Use of contaminated water and edibles and unhygienic practices are the main predisposing factors to giardiasis. It has been shown by Cordingley and Crawford (1986) that infection by *Giardia* could result into B_{12} deficiency and anemia in children and adults.

8.8.2 *Megaloblastic Anemia*

Megaloblastosis in bone marrow cells results into megaloblastic anemia. It is notably caused by a deficiency of folic acid and/or B_{12} in the body. Common cause of folate inadequacy in the body is its impaired absorption in the small intestine due to malabsorption syndrome, whereas cobalamin deficiency is attributed to lack of its absorption as a result of deficit amount of intrinsic factor in gastric secretion.

Megaloblasts are the giant cells with large-sized immature nucleus having three to four nucleoli and, hence, are precursors of erythrocytes. These cells proliferate rapidly. In the scarcity of folate and/or B_{12} in the body, DNA synthesis is deranged at S phase (Waxman et al. 1969). Unwinding of DNA replication fork is delayed. Ligation of Okazaki fragments is prolonged (Wickremasinghe 1980). Therefore, karyokinesis does not progress to G2 stage of cell division as a consequence S phase of mitosis is protracted (Steinberg et al. 1983).

However, continuity of transcription and translation results into asynchronization in the volume of nucleus to cytoplasm. Cells continue to grow in size with immature nucleus, called as macrocytosis.

Megaloblasts have higher volume of cytoplasm and RNA content in comparison to their normal counterparts; however, the amount of DNA is decreased due to defective synthesis (Myhre 1964).

It is a pathological process. Defective DNA synthesis is the prime cause of these abnormalities, signifying that synthesis of RNA and protein is faster in comparison to that of DNA (Steinberg et al. 1983).

Bone marrow undergoes hyperactive proliferation characterized into disturbed megaloblasts to late normoblasts population in bone marrow, called as megaloblastic hyperplasia of bone marrow. Count of megaloblasts exceed beyond its normal value (30%) to 70%, thereby depleting the count of late normoblasts to 30% only in bone marrow aspirate (Wang et al. 2009).

Megaloblastosis, further, disturbs the normal cell lines causing dyspoiesis of cell lines. It causes intramedullary haemolysis, reticulocytopenia and leucopenia. Neutrophils become hypersegmented. Large-sized RBC enters the blood circulation (Lichtin 2013).

It is indistinguishable to narrate the role of either folate or cobalamin deficiency in the causation of DNA impairment. Tetrahydrofolate is the active form of folate in the body for biosynthesis of purines (Duthie and McMillan 1997).

5,10-Methylene-THF is reduced to 5-methyltetrahydrofolate, which in turn is converted into THF in the presence of cobalamin. It transfers methyl group to homocysteine, and methionine is synthesized which is vital for methylation of proteins, nucleic acids and neurotransmitters. 5,10-Methylene act as a methyl group donor in conversion of dUMP into dTMP. Reaction is catalysed by thymidylate enzyme. Methyl group is transferred to uracil residue. Deficiency of folate or cobalamin inhibits the synthesis of dTMP, and hence its phosphorylation into dTDP and dTTP is curtailed. It is necessary for DNA synthesis (Das et al. 1986).

DNA polymerase misincorporates dUTP into replicating polynucleotide strands instead of dTTP. Uracil DNA glycosylase enzyme in DNA machinery helps to excise the uracil base from the new DNA strand. In this way, it prevents mutagenesis by removing uracil from the strand that could have been misincorporated (Longo et al. 1990).

However, a small amount of dU remains lodged in newly formed strands of DNA. It is fatal for the chromatin network and induces apoptosis. Overall, effect is the prejudiced synthesis of DNA in folate and/or cobalamin deficiency megaloblasts in bone marrow. Neonates may exhibit irritability, emaciation, anorexia, denial of solid foods, apathy and delayed developmental as a sequel of cobalamin deficiency. In later phase of infancy and early childhood, megaloblastic anemia and neurological manifestations complicate the prognosis of disorder. Cobalamin therapy for these patients might result into temporary, self-limited infantile tremor syndrome (Emery et al. 1997; Goraya 1998).

However, in few cases, neurological manifestations do not regress and may persist even after successful cobalamin therapy (Emery et al. 1997; Goraya 1998; CDC 2003, 2004; Korenke et al. 2004; Simsek et al. 2004; Weiss et al. 2004; Codazzi et al. 2005; Reghu et al. 2005; Roschitz et al. 2005). In another cases of cobalamin deficiency, severe cerebral atrophy and impaired myelination were observed by Lovblad et al. (1997) and von Schenck et al. (1997) through magnetic resonance imaging of the brain.

A study by Lowenstein et al. (1960) noticed a decrease in serum cobalamin level in almost one-fifth of the total pregnancies. Although low serum cobalamin level is normalized soon after delivery, however, it is detrimental to the health of neonates. In another study by Bjorke Monsen et al. (2001) and Michael Whitehead (2006), raised serum levels of tHcy and MMA which accompany depleted cobalamin level in women could be translated as low cobalamin level in neonates which persisted for 6 weeks, according to the authors. Hence, it was concluded by authors that maternal cobalamin status was strongly correlated with that of newborns (Bjorke Monsen et al. 2001; Michael Whitehead 2006).

Congenital pernicious anemia is a disorder of limited prevalence. It is characterized by hereditary deficiency of intrinsic factor in the neonates. This protein is necessary for cobalamin absorption in the small intestine. These infants have normal gastric acid secretion (Gordon et al. 2004; Yassin et al. 2004; Tanner et al. 2004, 2005). Consequently, cobalamin absorption, transport and its intracellular concentration are impaired.

8.9 Diagnosis of B_{12} Deficiency

Megaloblastic anemia is suspected in anaemic patients with macrocytic indices. Diagnosis is usually based on peripheral smear. When fully developed, the anemia is macrocytic, with MCV >100 fL/cell. The smear shows macro-ovalocytosis, anisocytosis and poikilocytosis. The RBC distribution width (RDW) is high. Howell–Jolly bodies (residual fragments of the nucleus) are common. Reticulocytopenia is present. Hypersegmentation of the granulocytes develops early; neutropenia develops later. Thrombocytopenia is often present in severe cases, and platelets may be bizarre in size and shape. If the diagnosis is questionable, a bone marrow examination may be needed (Lichtin 2013).

Haematological findings in cobalamin deficiency can be interpreted as:

Decreased red cell count and concentration of haemoglobin (MedlinePlus 2012). Mean corpuscular volume and mean corpuscular haemoglobin are raised. MCV is higher than 95 fL (normal value 80–100 fL). MCH value is >36 g/dL (normal value in between 32 and 36 g/dL) (Medline Plus 2014).

Thrombocytopenia due to destruction of precursors of all line (dyspoiesis). Neutrophils become hypersegmented. It is a compensatory mechanism for the reduced life span of neutrophils. These are called senile neutrophils. It is the unique and incipient manifestation of B_{12} and/or folate deficiency anemia in children (Kumar and Abbas 2007).

Generally, a neutrophil has three lobed nuclei. But in megaloblastic neutrophils, the lobes can go up to six or more in number. These lobes have a diagnostic value (McKenzie and Williams 2010). Nuclear remains persist within the erythrocytes which are called as Howell–Jolly bodies. These represent clusters of DNA which otherwise are removed by the spleen when the erythrocytes circulate through it. Failure to do so is indicative of malfunctioning of spleen or splenectomy (Harrod et al. 2007).

Homocysteine and methylmalonic acid levels are raised in vitamin B_{12} deficiency and are considered a definite diagnostic test for the screening of megaloblastic anemia in children (Savage et al. 1994). Additionally, the normal levels are confirmatory to exclude possibility of B_{12} and/or folate deficiency in the body. Gastrointestinal manifestations like diarrhoea or constipation mark a change in bowel movements (Chiara et al. 2013).

In advanced stage of anemia, neurological manifestations like the absence of reflexes, reduced touch sensation and seizures appear in children.

Normal serum B_{12} level is in the range 500–1,300 pg/mL as reported by Mitsuyama and Kogoh (1988). Serum vitamin B_{12} level <150 ng/L is considered as hypocobalaminaemia. Otherwise, individuals with serum B_{12} level in between 150 and 400 pg/ml are considered borderline cases. Overall, a serum vitamin B_{12} level <180 pg/ml may have to impetuously develop megaloblastic anemia (Mayo Clinic).

Elevated serum homocysteine level is an early indicator of B_{12} and folate deficiency. Serum homocysteine level that is between 4 and 15 μmol/L is considered as normal. Any value above 15 μmol/L is indicative of hyperhomocysteinaemia. Raised level of methylmalonic acid (MMA) is a sensitive indicator of B_{12} deficiency, whereas hyperhomocysteinaemia is diagnostic of B_{12} or folate deficiency in the body (McMullin et al. 2001).

Urine or blood level of methylmalonic acid also helps to detect B_{12} deficiency (Donalson 2000).

8.9.1 *Nieweg's Hypothesis*

Cobalamin deficiency in the body leads to neurological disorders like subacute degeneration of cord, whereas folate deficiency is devoid of any such lesions of the brain; however, both are associated with megaloblastic anemia.

Hypothesis says that cobalamin is essential for DNA and RNA synthesis, while folic acid is necessary for DNA metabolism.

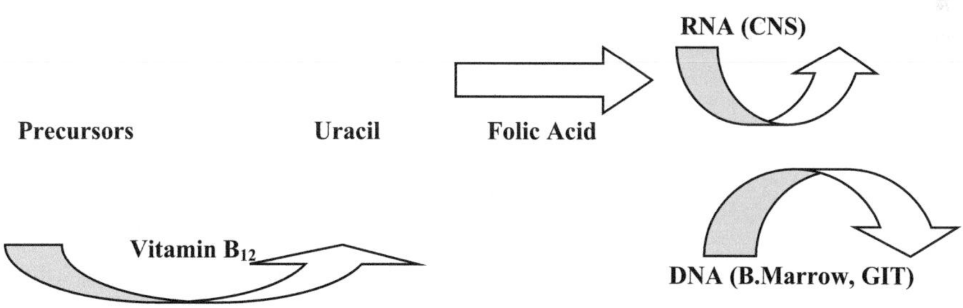

NIEWEG'S Hypothesis Flow Chart

8.10 Folic Acid Structure

Folic acid is called as pteroylglutamic acid, vitamin B_9 or folate and represents the heterogenous compound. Folate is a water-soluble vitamin predominantly found in natural foods and in supplementary foods (O'Broin et al. 1975; Berry et al. 2010; Eitenmiller and Landen 1999). It is made up of 2-amino-4-hydroxy-pteridine joined to para-aminobenzoic acid (pteroic acid), which are associated with variable number of glutamic acid moieties, generally ranging between one to nine depending on the type of food. Mammalian cells are unable to synthesize folates due to the absence of necessary enzymes. Cells are completely dependent on the exogenous source of folates which could be food or fortified foods. Folic acid derives its name from leaves in which (green leafy vegetables) it is present in abundance (Birn 2006).

Folate plays an important role in the single-carbon transfer necessary in biosynthesis of nucleic acids and metabolism of amino acids (Bailey and Gregory 2006; IOM/FNB 1998). Adversely, folate deficiency might impair the biosynthesis of purine and thymidylate, hence, interrupt de novo biosynthesis of DNA and RNA and, additionally, interfere with remethylation of homocysteine and further its conversion into S-adenosylmethionine (SAM) which is the universal methyl donor, thereby impeding the methylation reactions in the body (Scott 1999).

8.11 Dietary Sources

It is rich in plants, animal sources and microorganisms and copiously found in vegetables like spinach, cauliflower, peas, asparagus, broccoli, peanuts, tomato juice, banana, papaya and citrus fruit. Animal sources are fish, liver, yeast, meat and milk (NIH 2016; USDA 2012; Carmel 2005).

8.12 Folic Acid Metabolism

8.12.1 Intestinal Absorption of Folic Acid

Folic acid is reduced either completely or partially into tetrahydrofolate and dihydrofolate by folate reductase. Folylpolyglutamate synthetase is contained within mitochondria and cytosol. Enzyme has the central role in maintaining the concentration of folylpolyglutamate in cytosol and mitochondria (Cichowicz and Shane 1987; Milman 2012).

Folylpolyglutamate synthetase attaches glutamate residue to the γ-carboxylic residue in THF molecule. It results into formation of polyglutamyl folates which link with methyl, methylene, methenyl or formyl one-carbon residues at N5 and/or

N10 positions. These provide methyl groups during biosynthesis of various compounds (Stokstad 1990).

5-Methyltetrahydrofolate is the main folic acid that circulates in the blood stream (Chanarin and Perry 1969) and is utilized by cells. Transfer of a methyl group from 5-methyl-THF to homocysteine takes place in cells. Methylation of homocysteine is catalysed by methionine synthase in the presence of cobalamin. Tetrahydrofolate is released which again interacts with formic acid in the presence of 10-formyl-THF synthetase, resulting into formation of 10-formyltetrahydrofolate, which furnishes two carbon units in the purine nucleotide metabolism. Purine synthesis reaction is catabolized by β-glycinamide ribonucleotide transformylase enzyme in the first phase; thereafter, the reaction is mediated by 5-amino-4-imidazolecarboxamide ribonucleotide transformylase enzyme. Further, 10-formyltetrahydrofolate dehydrates into 5,10-methenyltetrahydrofolate which is again converted into 5,10-methylenetetrahydrofolate. Additionally, serine hydroxylmethyl transferase enzyme converts serine to glycine resulting into formation of 5,10-methylenetetrahydrofolate which is needed for the de novo synthesis of thymidine monophosphate catalysed by enzyme thymidylate synthase involving a transfer of one-carbon residue to deoxyuridine monophosphate. The enzyme 5,10-methylene-THF reductase converts the 5,10-methylenetetrahydrofolate into 5-methyltetrahydrofolate that has the potential to be oxidized into dihydrofolate, hence burdening the cellular pool of THF. To prevent deficiency of THF in tissues, DHF is rapidly reduced into TFH by enzyme DHF reductase (Seither et al. 1989).

Methotrexate is an antifolate and antimetabolite drug which is used for the treatment of neoplasia. The drug acts as dihydrofolate reductase inhibitor, by which, interconversion of DHF into THF is inhibited; hence, source of THF to the actively proliferating neoplastic cells for the DNA synthesis is curtailed causing death of cells (Stephen et al. 2006).

Folic acid exists in highly oxidized polyglutamate form in the natural foods. Its biological activity is limited to few days. Dietary folates are labile compounds. Food processing and storage and food preparation could lead to loss of biological activity of folate in natural foods. According to Caudill (2010), there is a loss of vitamin activity of folate of up to 40% in the vegetables upon cooking, whereas in cereals and grains, grinding and baking could lead to loss of activity of up to 70% which is much higher in comparison to the former group of food. Folates can be destroyed by cooking, and in grains/cereals, up to 70% of folates can be destroyed by milling and baking.

However, fortified foods contain folate in the monoglutamate form which is a reduced form that can retain its biochemical activity intact even for months and years (Blakley 1969).

Folic acid in foods is conjugated to polyglutamate chain which contains variable number of glutamic acids depending on nature of food. Folate conjugase is an enzyme present in the mucosa of the jejunum. The enzyme has two distinct locations in the body. First location is the surface of brush border epithelium where it has exopeptidase activity by which it cleavages glutamate residues on the terminal portion of polyglutamate chain. Second site is the cytoplasm of enterocytes, where enzyme has endopeptidase activity. It can hydrolyse polyglutamyl folate present in diet into easily absorbable monoglutamyl form. Thereafter, folate monoglutamate is absorbed by the jejunal mucosa (Scott and Weir 1994).

Folate transporters like PCFT and RFC are highly sensitive to the nature of folates in the intestinal lumen. These are highly specific to the folate monoglutamates as their substrates. Folates in the natural foods are in the polyglutamate forms, hence, need hydrolysis into easily absorbable monoglutamate forms.

Brush border surface of the jejunum harbours glutamate carboxypeptidase II which is a zinc containing conjugated enzyme located within the cell membranes of brush border membrane of the jejunum, prostate epithelium, renal proximal convoluted tubules and nervous tissues (Barinka et al. 2004).

Enzyme has an optimum pH of 6.5 in the intestinal lumen. Enzyme splits terminal glutamate residues from the polyglutamate molecule and converts it into monoglutamate folate form (Darcy-Vrillon et al. 1988).

Halsted (1989) perfused the jejunal mucosa with labelled polyglutamate folates. It was found that hydrolysis of polyglutamate folates occurred on the surface of the jejunum. Pteroyl polyglutamate hydrolase was observed on the surface of the jejunal mucosa in human and pig.

Enzyme catalyses the cleavage of pteroyl polyglutamates to monoglutamates. Two different pools of pteroyl polyglutamate hydrolase have been detected in humans. The first category of enzyme is located in the brush border surface of the jejunal mucosa, whereas the second type is present in the cytosol of enterocytes within the lysosomes (Wang et al. 1985).

Negligible amounts of pteroyl polyglutamate hydrolase have been detected in saliva, bile and pancreatic juices whose optimum pH is 4.5 and are associated with very weak enzymatic activity (Klipstein 1967).

Folate absorption is impaired in diseases which affect the structure and function of the jejunal mucosa. Folate malabsorption in this condition is profoundly associated with failure of folate polyglutamate hydrolysis in the jejunum.

Gamma-glutamyl hydrolase is another hydrolytic enzyme with carboxypeptidase activity that is found in the liver and kidney of rats. These organs store large amount of folate polyglutamates. Enzyme exerts its hydrolytic action at a pH in the range between 4.5 and 6.0. Hydrolase activity within these organs is necessary for the cleavage of the polyglutamates so as to release monoglutamates into the systemic circulation (Shafizadeh and Halsted 2007).

8.12.2 *Folate Transporters*

Transportation of folates across the cell membrane has been a topic of interest among academicians and clinicians. Folates are the essential compounds for the growth and development of normal mammalian cells as well as the actively dividing neoplastic cells and pathogenic bacteria and viruses. At physiological pH, folates are the anions whose diffusion across the bilipid layer of cell membrane is slow and cannot account for the rapid intestinal absorption and their import into the body tissues. Further, antimetabolites and antifolates are non-lipophilic drug molecules requiring a facilitative active transport across the cell membranes of neoplastic tissues to exert their effect (Goldman et al. 2010).

Hence, folates are transported by three different mechanisms whose expression in the tissues is determined by different genes. Proton-coupled folate transporter is expressed by SLC46A1 gene on the epithelium and is responsible for intestinal absorption of folates in low-pH environment. Authors (Qiu et al. 2006) proved the malabsorption of folates due to hereditary dysfunction of the gene as a result of mutation in the same gene. Earlier, PCFT was considered a protein for the haem-transportation whose activity was independent of the pH of the medium. They showed its activity in folate uphill transportation.

Later on, Matherly and Hou (2008) demonstrated the presence of a ubiquitously expressed folate transporter called reduced folate transporter being expressed by SLC19A1 gene in the mammalian cells and tissues mainly concerned with the import of folates into body tissues. Any abnormality, either in the expression of RFC or its functioning, heralds a number of untoward consequences in the physiology of organs and/or the growth and development of organs in intrauterine as well as postnatal life terminating into foetal abnormality, cardiovascular disease and neurological manifestations.

Folate receptor-α has high affinity for the folates and reduced folates. It is a protein which is linked to glycosylphosphatidylinositol moiety. According to Kamen and Smith (2004), the receptor is expressed on the epithelial cells of the kidney, choroid plexus and placenta as well as in the epithelial carcinomas. Folates are transgressed across membrane by receptor-mediated endocytosis.

Under physiological pH, folate contains two glutamate deprotonated carboxylic groups rendering it a negatively charged ion with high hydrophilic character that limits the diffusion of folates from the lumen into intestinal mucosa and across the cell membranes in body tissues. Folates' facilitative diffusion requires transporters (Goldman et al. 1968).

Reduced folate carrier is expressed by mammalian tissues, ubiquitously, and is considered as the chief folate transporter in mammalian cells (Matherly and Goldman 2003).

Reduced folate carrier is a facilitative type of mobile carrier having 12 transmembrane domains with amino (N) and carboxy (C) terminals positioned inside the membrane in the cytoplasm (Matherly and Hou 2008).

Gene for the synthesis of human RFC is located on the chromosome 21q22. The carrier is made up of 591 amino acids. RFC is expressed in all tissues, especially, the apical brush border membrane of the small intestine, the basolateral membranes of the proximal convoluted tubules, retinal pigment membrane and the apical membranes of the choroid plexus. RFC is the chief method of folate transportation into cells from the blood circulation (Wang et al. 2001). Reduced folate carrier or protein (SLC19A1) is the member of SLC19 family of solute carriers. RFC behaves as an anion exchanger which functions at optimum pH of 7.4. Further, it has been observed by Ganapathy et al. (2004) that its activity decreases as pH declines.

RFC undertakes specialized functions as folate transports across intestinal mucosal cells (Balamurugan and Said 2006). Absorption of folates across the basolateral membrane of proximal convoluted tubules (Kneuer et al. 2005).

Additionally, RFC also performs folate transportation across the placenta (Sweiry and Yudilevich 1985) and folate transportation across the blood–brain barrier (Spector and Johanson 2006).

RFC-mediated influx of folate is energy dependent. Reduced folate carrier lacks any domain for the attachment of ATP molecules on its surface; hence, the uphill transport of folate by RFC cannot utilize the energy liberated by hydrolysis of ATP molecules. RFC is greatly sensitive to the transmembrane gradient of organic phosphates. These are synthesized during phosphorylation and dephosphorylation reactions of the substrates and remain concentrated within the cells. Concentration of organic phosphates is higher inside the cells in comparison to their outside environment. RFC is the anions gradient sensitive antiporter.

Downhill transportation of driving ions provides energy to the uphill transportation of driven ions. Asymmetric distribution of organic phosphates provides energy for the uphill influx of folates facilitated by reduced folate carrier (Goldman 1971).

Study of Zhao et al. (2002) provided an evidence for the organic phosphate sensitive antiporter ability of reduced folate carrier in mammalian cells. Thiamine is ingressed into cells by SLC19A2 and SLC19A3. Inside the cells, thiamine undergoes phosphorylation into thiamine monophosphate and pyrophosphate. These organic phosphates induce uphill transport of folate by RFC.

Proton-coupled folate transporter is encoded by SLC46A1 gene located on chromosome 17q11.2. The PCFT is a protein consisting of 459 amino acids. It has 12 transmembrane domains along with amino and carboxy terminals located on the internal surface of transmembrane transporter within cytosol. PCFT is a homo-oligomer. Its C229 residue cross-links the oligomeric units of PCFT molecule (Unal et al. 2008).

It is a symporter of proton and folate ions across the cell membrane. It is expressed in the epithelium of the duodenum, jejunum and tumour tissues. The transporter operates at the acidic pH of GIT. It emerged as the main transporter involved in the absorption of folates through the apical brush border membrane of the small intestine. PCFT transporting activity is pH dependent and is inversely linked to hydrogen ion concentration of the intestinal lumen. However, its maximum activity is observed at optimum pH of 5.5. But a pH value below 5.5 results into protonation of carboxylic group of folate molecule, hence causing a decrease in the negative charge of folate. Overall, active transportation of folate below optimum pH by PCFT is decreased, while its passive diffusion independent of PCFT is increased (Visentin et al. 2014).

8.12.3 *Transportation of Folates in the Plasma and Liver*

Within the enterocytes, partial amount of monoglutamates undergoes reduction and methylation. This process is catalysed by dihydrofolate reductase, a rate-limiting enzyme which is pH sensitive and has an optimum pH of 6.0. It converts dihydrofolate into tetrahydrofolate and ultimately 5-methyl-THF is formed inside the enterocytes. It is transported through the basolateral surface of the small intestine and enters the hepatic portal vein. This exit is mediated by multidrug resistance-associated protein 3 (MRP3). MRP1, MRP2 and MRP3 are the proteins that are organic anion transporters helpful in the exit of folates, antifolates like methotrexate and drugs conjugated to ligands like glucuronate or sulphate. These proteins expel folates and antifolates out of cells. Family of MRPs sustains an effective enterohepatic circulation of folates in the body (Borst 2000).

5-Methyl-THF appears in the liver. Major portion of reduced monoglutamates (80%) is released into the systemic circulation. About 20% of reduced monoglutamates are stored in the liver in polyglutamate form. Within blood circulation, folates exist in three forms, as around 30% of the folates are in the unbound form; another 66% of the folates are present in bound form with albumin, whereas the remaining 3% of the folates remain associated with high affinity folate binders.

8.12.4 *Cellular Uptake of Folates*

Proliferating cells of the body rapidly take up folates from the circulation. It is mediated by folate receptor alpha which is located in the cell membrane. There is a folate–folate receptor complex formation which undergoes invagination to form an endosomal vesicle. The pH of vesicle becomes acidic; hence, folate at pH 6.0 within the vesicle is released from the complex and moves into the cytoplasm (Hoffman et al. 2013).

8.13 Congenital Disorders of Folate Metabolism

Congenital folate malabsorption has been reported by Zittoun (1995), Whitehead (2006), Corbeel et al. (1985) and Geller et al. (2002). These patients show signs of megaloblastic anemia, repeated diarrhoeal episodes, malnutrition and neurological manifestations in early months of infancy. Children are mentally retarded. Serum folate and cerebrospinal folate level are declined. Anemia responds to folate treatment. However, neurological signs do not improve significantly on folate administration owing to inability to adequately sustain CSF folate level. Multiple causes have been attributed to this defect of folate metabolism. A defunct folate transporter responsible for defective folate transportation folate in the intestine and brain is responsible for decreased folate level. Defective cellular uptake of folate is also the possible cause for this congenital defect.

Methylenetetrahydrofolate reductase deficiency is the frequent congenital disorder of folate metabolism. Its onset is during early infancy. It is characterized by neurological features like mental retardation and convulsions, microcephaly and delayed psychomotor development. Children in early childhood have marked gait abnormality, psychiatric abnormality like schizophrenic disorder. Vascular abnormality has also been observed in patients. Haematological features are not prominent in this category of defect.

Folate levels are decreased in serum, red blood cells and CSF. It is associated with raised serum homocysteine level and low methionine level. Homocystinuria is a common feature. Response to folate therapy is variable, and the prognosis of disorder is poor (Zittoun 1995; Whitehead 2006; Ramaekers and Blau 2004). Other drugs like cobalamin, methionine, riboflavin, pyridoxine and carnitine have been used to ameliorate the signs and symptoms. However, disease is resistant to most of the drugs asserted above, and overall improvement has been observed by the use of betaine (Rosenblatt and Fenton 2001; Rosenblatt and Whitehead 1999).

Methionine synthase reductase deficiency is another disorder of folate metabolism which is characterized by megaloblastic anemia, hyperhomocysteinaemia and hypomethioninaemia. Its onset is early in infancy.

Occurrence of mutation in MTR gene has been assigned as a cause of methylcobalamin deficiency (Banerjee and Matthews 1990). A rise in serum homocysteine level is a feature of this disorder which is related to manifestations like blindness, neural tube defects and neurological manifestations. These symptoms appear early in childhood, mainly, in the first 2 years of life. A few cases of methionine deficiency exhibit severe encephalopathy (Outteryck et al. 2012).

Methionine synthase reductase deficiency concomitant with the deficiency of another cobalamin-dependent enzyme, namely, methylmalonyl–CoA mutase, affects the patients severely. Hyperhomocysteinaemia, homocysteinuria and methylmalonic aciduria are the biochemical manifestations. Association of two defunct enzymes in the body has deleterious consequences like delayed growth, mental retardation, hypotonia, seizures and megaloblastic anemia (Zittoun 1995; Whitehead 2006; Ramaekers and Blau 2004). Most of the patients die, and those who survive have to withstand neurological abnormalities (Rosenblatt et al. 1997). Parenteral administration of hydroxocobalamin may decrease the severity of this disorder.

Glutamate formiminotransferase-cyclodeaminase deficiency is characterized by heavy excretion of formiminoglutamic acid. It is an autosomal recessive trait which is rare in prevalence and has been reported in less than 20 patients (Whitehead 2006). Patients with severe form of this defect has been associated with increased serum folate level and enhanced excretion of formiminoglutamate (FIGLU) in urine on consumption of L-histidine (Whitehead 2006). Megaloblastic anemia is variable. Mental retardation is usually the outcome of this defect (Zittoun 1995; Whitehead 2006; Ramaekers and Blau 2004).

Dihydrofolate reductase deficiency was observed in three children. It was associated with megaloblastic anemia following birth of infants. Defect was responsive to folinic acid (Zittoun 1995; Whitehead 2006; Ramaekers and Blau 2004). Folate is highly sought in the growth of the central nervous system in infancy. Hence, any abnormality in its metabolism is invariable followed by neurological manifestations of diverse nature.

8.14 Excretion of Folic Acid

Folic acid is excreted in urine in the range between 2 and 5 μg per day. Another important source of excretion is the stools which have about 20% of the unabsorbed folates from food along with 60–90 μg of folates from the bile and a trace amount of unabsorbed folates from the colonic bacteria (Chatterjea and Shinde 2002).

8.15 Deficiency of Folic Acid and Nutritional Anemia in Children

Folate deficiency causes anorexia, loss of body weight, headache and megaloblastic anemia in children. Also irritability and a change in the behaviour of child have been observed (Haslam and Probert 1998). Preschool children face stunted growth. Folate deficiency before conception is fatal for the growth and development of foetus and is associated with serious consequences. According to MRC (1991), Czeizel and Dudas (1992) and Botto et al. (1999, 2005), preconceptual folate deficiency is responsible for more than 50% prevalence of neural tube defects covering spina bifida, anencephaly and meningomyelocele in intrauterine life. Further, authors asserted the growth retardation of the foetus in case of folate deficiency in pregnant women.

Folate deficiency during intrauterine life results into neural tube defects. These are congenital defects of the brain and spine either due to failure of primary neurulation process or secondary neurulation process. Ectodermal layer is converted into neural plate by flattening of ectodermal tissues. It is important for the nervous system formation. Neural plate undergoes primary neurulation in which it converges inwards into a tube and then secondary neurulation wherein its interior is hollowed to form neural tube. Neural tube defects arise due to failure of the sides of the neural tube to seal. Multiple explanations have been given for its explanation (Gordon 1985).

Methionine is activated in the presence of ATP and SAM synthase into high-energy intermediate called S-adenosylmethionine (SAM). It is necessarily necessary for the post-transcriptional methylation of arginine and histidine which are essential for the ectodermal tissues. These are highly sensitive to the level of homocysteine. SAM is also vital for the methylation of DNA, lipids and proteins (MINER et al. 1997.)

After the methylation reaction, SAM is converted into S-adenosyl homocysteine (SAH). There is a fixed ratio of SAM to SAH in the tissues. According to Skovby (1989), rat liver has 13 μmol/g of SAH and 60–90 μmol/g of SAM (Skovby 1989).

SAH is considered a negative feedback regulator of all reactions that synthesize SAM. It is a toxic by-product of the metabolic reactions. To prevent its buildup in tissues, homocysteine is remethylated or condensed into cysteine by transsulphuration pathway. A deficiency of folate is responsible for high level of homocysteine which has lethal effect on the ectodermal tissues resulting in neural tube defect. Neuroepithelial cells of the neural plate undergo rapid morphological changes before closure of neural tube, and cytoskeleton plays an important role in the process. Cytoskeleton includes actin, tubulin and neurofilament, and spatial changes in them are necessary for the changes in shape of neuroepithelial cells. The cytoskeletal proteins have arginine and histidine residues which require methylation in the presence of SAM. Reduced folate thwarts methylation and manifests as impaired organization of the cytoskeletal elements in basal and apical ends of the cells, and ultimately, cellular contraction and migration do not occur resulting into neural tube defect (Moephuli et al. 1997).

References

Allen RH, Stabler SP, Savage DG, Lindenbaum J (1993) Metabolic abnormalities in cobalamin and folate deficiency. FASEB 7:1344

Alpers DH, Russell-Jones G (2013) Gastric intrinsic factor: the gastric and small intestinal stages of cobalamin absorption. A personal journey. Biochimie 95(5):989–994

Anthony AC (2012) Megaloblastic anemias. In: Hoffman R, Benz EJ Jr, Silberstein LE, Heslop HE, Weitz JI (eds) Hematology: basic principles and practice. Elsevier Saunders, Philadelphia. chap 32

Bailey LB, Gregory JF (2006) Folate. In: Bowman B, Russell R (eds) Present knowledge in nutrition. International Life Sciences Institute, Washington, DC, pp 278–301

Balamurugan K, Said HM (2006) Role of reduced folate carrier in intestinal folate uptake. Am J Physiol Cell Physiol 291(1):C189–C193

Baneerjee RV, Matthews RG (1990) Cobalamin-dependent methionine synthase. FASEB J 4(5):1450–1459

Barinka C, Sacha P, Sklenar J, Man P, Bezouska K, Slusher BS, Konvalinka J (2004) Identification of the N-glycosylation sites on glutamate carboxypeptidase II necessary for proteolytic activity. Protein Sci 13(6):1627–1635

Berry RJ, Bailey L, Mulinare J, Bower C (2010) Fortification of flour with folic acid. Food Nutr Bull 31:S22–S35

Birn H (2006) The kidney in vitamin B12 and folate homeostasis: characterization of receptors for tubular uptake of vitamins and carrier proteins. Am J Physiol Renal Physiol 291(1):F22–F36

Bjorke Monsen AL, Ueland PM, Vollset SE, Guttormsen AB, Markestad T, Solheim E, Refsum H (2001) Determinants of cobalamin status in newborns. Pediatrics 108:624–630

Blakley R (1969) In: Newbergen H, Taton EL (eds) The biochemistry of folic acid and related pteridines. North Holland Publishing Company, Amsterdam

Borst P (2000) A family of drug transporters: the multidrug resistance-associated proteins. J Natl Cancer Inst 92(16):1295–1302

Botto LD, Moore CA, Khoury MJ, Erickson JD (1999) Neural tube defects. N Engl J Med 341:1509–1519

Botto LD, Lisi A, Robert-Gnansia E, Erickson JD, Vollset SE, Mastroiacovo P, Botting B, Cocchi G, de Vigan C, de Walle H, Feijoo M, Irgens LM, McDonnell B, Merlob P, Ritvanen A, Scarano G, Siffel C, Metneki J, Stoll C, Smithells R, Goujard J (2005) International retrospective cohort study of neural tube defects in relation to folic acid recommendations: are the recommendations working? Br Med J 330:571–576

Brown KL (2005) Chemistry and enzymology of vitamin B12. Chem Rev 05:2075–2149

Carmel R (2005) Folic acid. In: Shils M, Shike M, Ross A, Caballero B, Cousins R (eds) Modern nutrition in health and disease. Lippincott Williams & Wilkins, Baltimore, pp 470–481

Casterline JE, Allen LH, Ruel MT (1997) Vitamin B-12 deficiency is very prevalent in lactating Guatemalan women and their infants at three months postpartum. J Nutr 127:1966–1972

Caudill MA (2010) Folate bioavailability: implications for establishing dietary recommendations and optimizing status. Am J Clin Nutr 91:1455S–1460S

Centers for Disease Control and Prevention (2004) Spina bifida and anencephaly before and after folic acid mandate – United States, 1995–1996 and 1999–2000. Morb Mortal Wkly Rep 53:362–365

Centers for Disease Control and Prevention (CDC) (2003) Neurologic impairment in children associated with maternal dietary deficiency of cobalamin – Georgia, 2001. Morb Mortal Wkly Rep 52:61–64

Chanarin I, Perry J (1969) Evidence for reduction and methylation of folate in the intestine during normal absorption. Lancet 2:776–778

Chanarin I, Muir M, Hughes A, Hoffbrand AV (1978) Evidence for intestinal origin of transcobalamin II during vitamin B12 absorption. Br Med J i:1453–1455

Chanarin I, Deacon R, Lumb M, Muir M, Perry J (1985) Cobalamin-folate interrelations: a critical review. Blood 66(3):479–489

Chatterjea MN, Shinde R (2002) Vitamins, in: text book of medical biochemistry. Jaypee Brothers Medical Publishers, New Delhi, p 173

Chiara B, Chiara T, Citton V, Manara R, Pompanin S, Binotto G, Adami F (2013) Cobalamin deficiency: clinical picture and radiological findings. Forum Nutr 5(11):4521–4539

Christensen EI, Birn H, Verroust P, Moestrup SK (1998) Membrane receptors for endocytosis in renal proximal tubule. Int Rev Cytol 180:237–284

Cichowicz DJ, Shane B (1987) Mammalian folylpoly-gamma-glutamate synthetase. 1. Purification and general properties of the hog liver enzyme. Biochemistry 26(2):504–512

Clarke R (2008) B-vitamins and prevention of dementia. Proc Nutr Soc 67:75–81

Codazzi D, Sala F, Parini R, Langer M (2005) Coma and respiratory failure in a child with severe vitamin B(12) deficiency. Pediatr Crit Care Med 6:483–485

Combs G (1992) Vitamin B12 in the vitamins. Academic, New York

Cooper BA, Castle WB (1960) Sequential mechanism in the enhanced absorption of vitamin B12 by intrinsic factor in rat. J Clin Invest 39:199–214

Corbeel L, Van den Berghe G, Jaeken J, Van Tornout J, Eeckels R (1985) Congenital folate malabsorption. Eur J Pediatr 143(4):284–290

Cordingley FT, Crawford GP (1986) Giardia infection causes vitamin B12 deficiency. Aust NZ J Med 16(1):78–79

Croft MT, Lawrence AD, Raux-Deery E, Warren MJ, Smith AG (2005) Algae acquire vitamin B12 through a symbiotic relationship with bacteria. Nature 438(7064):90–93

Czeizel AE, Dudas I (1992) Prevention of the first occurrence of neural tube defects by periconceptual vitamin supplementation. N Engl J Med 327:1832–1835

Darcy-Vrillon B, Selhub J, Rosenberg IH (1988) Analysis of sequential events in intestinal absorption of folylpolyglutamate. Am J Physiol Gastrointest Liver Physiol 255:G361–G366

Das KC, Mohanty D, Garewell G (1986) Cytogenetics in nutritional megaloblastic anemia: prolonged persistence of chromosomal abnormalities in lymphocytes after remission. Acta Haematol 76:146

Davis CG, Goldstein JL, Sudhof TC, Anderson RG, Russell DW, Brown MS (1987) Acid-dependent ligand dissociation and recycling of LDL receptor mediated by growth factor homology region. Nature 326:760–765

Donaldson MS (2000) Metabolic vitamin B12 status on a mostly raw vegan diet with follow up using tablets, nutritional yeasts, or probiotics. Ann Nutr Metab 44(5–6):229–234

Downing M, Schweigert BS (1956) Role of vitamin B12 in nucleic acid metabolism. IV. Metabolism of CKlabeled thymidine by Lactobacillus leichmannii. J Biol Chem 220:521

Duthie SJ, McMillan P (1997) Uracil misincorporation in human DNA detected using single cell gel electrophoresis. Carcinogenesis 18:1709

Eitenmiller R, Landen W (1999) Folate. In: Reaw L (ed) Vitamin analysis for the health and food science. CRC Press, Boca Raton, pp 411–465

Emery ES, Homans AC, Colletti RB (1997) Vitamin B12 deficiency: a cause of abnormal movements in infants. Pediatrics 99:255–256

Ganapathy V, Smith SB, Prasad PD (2004) SLC19: the folate/thiamine transporter family. Pflugers Arch 447:641–646

Garcia-Casal MN, Osorio C, Landaeta M, Leets I, Matus P, Fazzino F, Marcos E (2005) High prevalence of folic acid and vitamin B12 deficiencies in infants, children, adolescents and pregnant women in Venezuela. Eur J Clin Nutr 59:1064–1070

Geller J, Kronn D, Jayabose S, Sandoval C (2002) Hereditary folate malabsorption: family report and review of the literature. Medicine (Baltimore) 81:51–68

Goldman ID (1971) The characteristics of the membrane transport of amethopterin and the naturally occurring folates. Ann N Y Acad Sci 186:400–422

Goldman ID, Lichtenstein NS, Oliverio VT (1968) Carrier-mediated transport of the folic acid analogue methotrexate, in the L1210 leukemia cell. J Biol Chem 243:5007–5017

Goldman ID, Chattopadhyay S, Zhao R, Moran R (2010) The antifolates: evolution, new agents in the clinic, and how targeting delivery via specific membrane transporters is driving the development of a next generation of folate analogs. Curr Opin Investig Drugs 11:1409–1423

Goraya JS (1998) Persistence of neurological damage induced by dietary vitamin B-12 deficiency in infancy. Arch Dis Child 78:395

Gordon R (1985) A review of the theories of vertebrate neurulation and their relationship to the mechanics of neural tube birth defects. J Embryol Exp Morpholog 89(Suppl):229–255

Gordon MM, Brada N, Remacha A, Badell I, del Rio E, Baiget M, Santer R, Quadros EV, Rothenberg SP, Alpers DH (2004) A genetic polymorphism in the coding region of the gastric intrinsic factor gene (GIF) is associated with congenital intrinsic factor deficiency. Hum Mutat 23(1):85–91

Gruber K, Kratky C (2006) Cobalamin-dependent methionine synthase: part 4. Cobalt. Wiley, New Jersey

Guerra-Shinohara EM, Morita OE, Peres S, Pagliusi RA, Sampaio Neto LF, D'Almeida V, Irazusta SP, Allen RH, Stabler SP (2004) Low ratio of S-adenosylmethionine to S-adenosylhomocysteine is associated with vitamin deficiency in Brazilian pregnant women and newborns. Am J Clin Nutr 57:1312–1321

Hakami N, Neiman PE, Canellos GP, Lazerson J (1971) Neonatal megaloblastic anemia due to inherited transcobalamin 11 deficiency in two siblings. N Engl J Med 285:1163–1170

Halsted CH (1989) The intestinal absorption of dietary folates in health and disease. J Am Coll Nutr 8:650–658

Harrod VL, Howard TA, Zimmerman SA, Dertinger SD, Ware RE (2007) Quantitative analysis of Howell-Jolly bodies in children with sickle cell disease. Exp Hematol 35(2):179–183

Haslam N, Probert CS (1998) An audit of the investigation and treatment of folate deficiency. J R Soc Med 91(2):72–73

Herbert V, Drivas G, Manusselis C, Mackler B, Eng J, Schwartz E (1984) Are colon bacteria a major source of cobalamin analogues in human tissues? 24-hr human stool contains only about 5 μg of cobalamin but about 100 μg of apparent analogue (and 200 μg of folate). Trans Assoc Am Phys 97:161–171

Hermann L, Nilsson B, Wettre S (2004) Vitamin B12 status of patients treated with metformin: a cross-sectional cohort study. Br J Diabet Vasc Dis 4:401

Hoffman R, Benz Jr. EJ, Silberstein LE, Heslop H, Weitz J, Anastasi J (2013) Hematology: diagnosis and treatment. Elsevier Health Sciences, Amsterdam, pp 478

Institute of Medicine-Food and Nutrition Board (1998) Dietary reference intakes: thiamin, riboflavin, niacin, vitamin B6, folate, vitamin B12, pantothenic acid, biotin, and choline. National Academy Press, Washington, DC

Johanson JE Jr, Reyes FE, Polaski JT, Batey RT (2012) Structure of cobalamin and cobalamin riboswitches. Nature 492:133–137

Kalser MH, Cohen R, Arteaga I, Yawn E, Mayoral L, Hoffert WR, Frazier D (1966) Normal viral and bacterial flora of the human small and large intestine. N Engl J Med 274(9):500–505

Kamen BA, Smith AK (2004) A review of folate receptor alpha cycling and 5-methyltetrahydrofolate accumulation with an emphasis on cell models in vitro. Adv Drug Deliv Rev 56:1085–1097

Kanazawa S, Herbert V, Herzlich V, Drivas G, Manusselis C (1983) Removal of cobalamin analogue in bile by the enterohepatic circulation of vitamin B12. Lancet:707–708

Kittang E, Schjqnsby H (1987) Effect of gastric anacidity on the release of cobalamins from food and their subsequent binding to R-protein. Scand J Gastroenterol 22:1031–1037

Klipstein FA (1967) Intestinal folate conjugase activity in tropical sprue. Amer J clin Nutr 20:1004–1009

Kneuer C, Honscha KU, Honscha W (2005) Rat reduced-folate carrier-1 is localized basolaterally in MDCK kidney epithelial cells and contributes to the secretory transport of methotrexate and fluoresceinated methotrexate. Cell Tissue Res 320(3):517–524

Korenke GC, Hunneman DH, Eber S, Hanefeld F (2004) Severe encephalopathy with epilepsy in an infant caused by subclinical maternal pernicious anemia: case report and review of the literature. Eur J Pediatr 163:196–201

Kozyraki R, Kristiansen M, Silahtaroglu A (1998) The human intrinsic factor–vitamin B12 receptor, cubilin: molecular characterization and chromosomal mapping of the gene to 10p within the autosomal recessive megaloblastic anemia (MGA1) region. Blood 91:3593–3600

Krautler B (2012) Biochemistry of B12-cofactors in human metabolism. Subcell Biochem 56:323–346

Kumar V, Abbas AK (2007) Robbins basic pathology. Elsevier, Amsterdam, p 437

Lam JR, Schneider JL, Zhao W, Corley DA (2013) Proton pump inhibitor and histamine 2 receptor antagonist use and vitamin B12 deficiency. JAMA 310(22):2435–2442

Lichtin AE (2013) Megaloblastic macrocytic anemias. Available at: http://www.msdmanuals.com/professional/hematology-and-oncology/anemias-caused-by-deficient-erythropoiesis/megaloblastic-macrocytic-anemias

Longo MC, Berninger MS, Hartley JL (1990) Use of uracil DNA glycosylase to control carry-over contamination in polymerase chain reactions. Gene 93(1):125–128

Lovblad K, Ramelli G, Remonda L, Nirkko AC, Ozdoba C, Schroth G (1997) Retardation of myelination due to dietary vitamin B12 deficiency: cranial MRI findings. Pediatr Radiol 27:155–158

Lowenstein L, Lalonde M, Deschenes EB, Shapiro L (1960) Vitamin B12 in pregnancy and the puerperium. Am J Clin Nutr 8:265–275

Matherly LH, Goldman DI (2003) Membrane transport of folates. Vitam Horm 66:403–456

Matherly LH, Hou Z (2008) Structure and function of the reduced folate carrier: a paradigm of a major facilitator superfamily mammalian nutrient transporter. Vitam Horm 79C:145–184

Mayo Clinic Vitamin B12 assay serum (n.d.). Available at: http://www.mayomedicallaboratories.com/test-catalog/Clinical+and+Interpretive/9154

McBride Judy (2000) B12 deficiency may be more widespread than thought. Agricultural Research Service. United States Department of Agriculture, Maryland

McKenzie SB, Williams JL (2010) Clinical laboratory hematology. Prentice Hall, Upper Saddle River

McMullin MF, Young PB, Bailie KE, Savage GA, Lappin TR, White R (2001) Homocysteine and methylmalonic acid as indicators of folate and vitamin B12 deficiency in pregnancy. Clin Lab Haematol 23(3):161–165

Medline Plus (2014) RBC indices. Available at: https://www.nlm.nih.gov/medlineplus/ency/article/003648.htm

MedlinePlus (2012) Megaloblastic anemia. Available at: https://www.nlm.nih.gov/medlineplus/ency/article/000569.htm

Michael Whitehead V (2006) Acquired and inherited disorders of cobalamin and folate in children. Br J Haematol 134:125–136

Milman N (2012) Intestinal absorption of folic acid – new physiologic & molecular aspects. Indian J Med Res. 2012 Nov 136(5):725–728

Miner SE, Evrovski J, Cole DE (1997) Clinical chemistry and molecular biology of homocysteine metabolism: an update. Clin Biochem 30:189–201

Mitchell HK, Snell EE, Williams RJ (1941) The concentration of folic acid. J Am Chem Soc 63:2284–2281

Mitsuyama Y, Kogoh H (1988) Serum and cerebrospinal fluid vitamin B12 levels in demented patients with CH3-B12 treatment – preliminary study. Jpn J Psychiatry Neurol 42(1):65–71

Moephuli SR, Klein NW, Baldwin MT, Krider HM (1997) Effects of methionine on the cytoplasmic distribution of actin and tubulin during neural tube closure in rat embryos. Proc Natl Acad Sci 94:543–548

Moestrup SK, Kozyraki R, Kristiansen M, Kaysen JH, Rasmussen HH, Brault D, Pontillon F, Goda FO, Christensen EI, Hammond TG, Verroust PJ (1998) The intrinsic factor-vitamin B_{12} receptor and target of teratogenic antibodies is a megalin-binding peripheral membrane protein with homology to developmental proteins. J Biol Chem 273(9):5235–5242

Morkbak AL, Poulsen SS, Nexo E (2007) Haptocorrin in humans. Clin Chem Lab Med 45(12):1751–1759

MRC Vitamin Study Research Group (1991) Prevention of neural tube defects: results of the Medical Research Council vitamin study. Lancet 338:131–137

Myhre E (1964) Studies on megaloblasts in vitro. Scand J Clin Invest 16:307

National Institutes of Health (2016) Folates: dietary supplements fact sheet. Available at: https://ods.od.nih.gov/factsheets/Folate-HealthProfessional/

O'Broin JD, Temperley IJ, Brown JP, Scott JM (1975) Nutritional stability of various naturally occurring monoglutamate derivatives of folic acid. Am J Clin Nutr 28:438–444

Outteryck O, de Sèze J, Stojkovic T, Cuisset JM, Dobbelaere D, Delalande S, Lacour A, Cabaret M, Lepoutre AC, Deramecourt V, Zéphir H, Fowler B, Vermersch P (2012) Methionine synthase deficiency: a rare cause of adult-onset leukoencephalopathy. Neurology 79(4):386–388

Peters TJ, Hoffbrand AV (1970) Absorption of vitamin B12 by the guinea pig. 1. Subcellular localization of vitamin B12 in the ileal enterocyte during absorption. Br J Haematol 19:369–382

Pruthi RK, Tefferi A (1994) Pernicious anemia revisited. Mayo Clin Proc 69:144

Qiu A, Jansen M, Sakaris A, Min SH, Chattopadhyay S, Tsai E, Sandoval C, Zhao R, Akabas MH, Goldman ID (2006) Identification of an intestinal folate transporter and the molecular basis for hereditary folate malabsorption. Cell 127:917–928

Ramaekers VT, Blau N (2004) Cerebral folate deficiency. Dev Med Child Neurol 46(12):843–851

Randaccio L, Geremia S, Demitri N, Wuerges J (2010) Vitamin B12: unique metalorganic compounds and the most complex vitamins. Molecules 15:3228–3259

Reghu A, Hosdurga S, Sandhu B, Spray C (2005) Vitamin B12 deficiency presenting as oedema in infants of vegetarian mothers. Eur J Pediatr 164:257–258

Rogers LM, Boy E, Miller JW, Green R, Sabel JC, Allen LH (2003) High prevalence of cobalamin deficiency in Guatemalan school children: associations with low plasma holotranscobalamin II and elevated serum methylmalonic acid and plasma homocysteine concentrations. Am J Clin Nutr 77:433–440

Roschitz B, Plecko B, Huemer M, Biebl A, Foerster H, Sperl W (2005) Nutritional infantile vitamin B12 deficiency: pathobiochemical considerations in seven patients. Arch Dis Child Fetal Neonatal Ed 90:F281–F282

Rosenblatt DS, Fenton WA (2001) Inherited disorders of folate and cobalamin transport and metabolism. In: Scriver CR, Beaudet AL, Sly WS, Valle D (eds) The metabolic and molecular bases of inherited metabolic disease. McGraw-Hill, New York, pp 3897–3933

Rosenblatt DS, Whitehead VM (1999) Cobalamin and folate deficiency: acquired and hereditary disorders in children. Semin Hematol 36:19–34

Rosenblatt DS, Aspler AL, Shevell MI, Pletcher BA, Fenton WA, Seashore MR (1997) Clinical heterogeneity and prognosis in combined methylmalonic aciduria and homocystinuria (cblC). J Inherit Metab Dis 20:528–538

Savage DG, Lindenbaum J, Stabler SP, Allen RH (1994) Sensitivity of serum methylmalonic acid and total homocysteine determinations for diagnosing cobalamin and folate deficiency. Am J Med 96(3):239–246

von Schenck U, Bender-Gotze C, Koletzko B (1997) Persistence of neurological damage induced by dietary vitamin B-12 deficiency in infancy. Arch Dis Child 77:137–139

Scott JM (1999) Folate and vitamin. B12 Proc Nutr Soc 58(2):441–448

Scott JM, Weir DG (1994) Folate/vitamin B12 interrelationships. Essays Biochem 8:63–72

Seither RL, Tredent DF, Mikulecky DC, Rape TJ, Goldman ID (1989) Folate-pool interconversions and inhibition of biosynthetic processes after exposure of L1210 leukemia cells to antifolates. J Biol Chem 264:17016–17023

Shafizadeh TB, Halsted CH (2007) γ-Glutamyl hydrolase, not glutamate carboxypeptidase II, hydrolyzes dietary folate in rat small intestine. J Nutr 137:1149–1153

Shaw S, Jayatilleke E, Meyers S, Colman N, Herzlich B, Herbert V (1989) The ileum is the major site of absorption of vitamin B12 analogues. Am J Gastroenterol 84(1):22–26

Simsek OP, Gonc N, Gumruk F, Cetin M (2004) A child with vitamin B12 deficiency presenting with pancytopenia and hyperpigmentation. J Pediatr Hematol Oncol 26:834–836

Skovby F (1989) Inborn errors of metabolism causing homocysteinemia and related vascular involvement. Haemostasis 19(Suppl 1):4–9

Smith R, Cobalt M (1987) In: Mertz W (ed) Trace elements in human and animal Nutrition. Academic, San Diego, pp 143–184

Spector R, Johanson C (2006) Micronutrient and urate transport in choroid plexus and kidney: implications for drug therapy. Pharm Res 23(11):2515–2524

Steinberg SE, Fonda S, Campbell CL, Hillman RS (1983) Cellular abnormalities of folate deficiency. Br J Haematol 54:605. 1983

Stephen H, Sergio L, Khalid I (2006) Dihydrofolate reductase inhibitors as antibacterial agents. Biochem Pharmacol 71(7):941–948

Stokstad ELR (1990) Historical perspective on key advances in the biochemistry and physiology of folates. In: Picciano MF, Stokstad ELR (eds) Folic acid metabolism in health and disease. Wiley-Liss, New York, pp 1–21

Stover PJ (2009) One-carbon metabolism–genome interactions in folate-associated pathologies. J Nutr 139(12):2402–2405

Sweiry JH, Yudilevich DL (1985) Transport of folates at maternal and fetal sides of the placenta: lack of inhibition by methotrexate. Biochim Biophys Acta 19 821(3):497–501

Tanner SM, Li Z, Bisson R, Acar C, Oner C, Oner RC, Cetin M, Abdelaal MA, Ismail EA, Lissens W, Krahe R, Broch H, Grasbeck R, de la Chapelle A (2004) Genetically heterogeneous selective intestinal malabsorption of vitamin B12: founder effects, consanguinity, and high clinical awareness explain aggregations in Scandinavia and the Middle East. Hum Mutat 23:327–333

Tanner SM, Li Z, Perko JD, Oner C, Cetin M, Altay C, Yurtsever Z, David KL, Faivre L, Ismail EA, Grasbeck R, de la Chapelle A (2005) Hereditary juvenile cobalamin deficiency caused by mutations in the intrinsic factor gene. Proc Natl Acad Sci USA 102:4130–4133

Unal ES, Zhao R, Qiu A, Goldman ID (2008) N-linked glycosylation and its impact on the electrophoretic mobility and function of the human proton-coupled folate transporter (*Hs*PCFT). Biochim Biophys Acta 1178:1407–1414

United States Department of Agriculture (2011) USDA National Nutrient Database for Standard Reference.

United States Department of Agriculture (2012) USDA National Nutrient Database for Standard Reference. Available at: https://ndb.nal.usda.gov//

Verroust PJ (2002) Pathophysiology of cubilin: of rats, dogs and men. Nephrol Dial Transplant 17(9):55–56

Viola-Villegas N, Rabideau AE, Bartholomä M, Zubieta J, Doyle RP (2009) Targeting the cubilin receptor through the vitamin B(12) uptake pathway: cytotoxicity and mechanistic insight through fluorescent Re(I) delivery. J Med Chem 52(16):5253–5261

Visentin M, Diop-Bove N, Zhao R, Goldman ID (2014) The intestinal absorption of folates. Annu Rev Physiol 76:251–274

Wang TT, Reisenauer AM, Halsted CH (1985) Comparison of folate conjugase activities in human, pig, rat and monkey intestine. J Nutr 115(6):814–881

Wang Y, Zhao R, Russell RG, Goldman ID (2001) Localization of the murine reduced folate carrier as assessed by immunohistochemical analysis. Biochim Biophys Acta 1513:49–54

Wang YH, Yan F, Zhang WB, Ye G, Zheng YY, Zhang XH, Shao FY (2009) An investigation of vitamin B12 deficiency in elderly inpatients in neurology department. Neurosci Bull 25(4):209–215

Watanabe F, Takenaka S, Kittaka-Katsura H, Ebara S, Miyamoto E (2002) Characterization and bioavailability of vitamin B12-compounds from edible algae. J Nutr Sci Vitaminol 48(5):325–331

Waxman S, Metz J, Herbert V (1969) Defective DNA synthesis in human megaloblastic bone marrow: effects of homocysteine and methionine. J Clin Invest 48:284

Weir DG, Scott JM (1999) Cobalamins physiology, dietary sources and requirements. In: Sadler MJ, Strain JJ, Caballero B (eds). Encyclopedia of human nutrition, vol 1. Academic Press, London, pp 394–401

Weiss R, Fogelman Y, Bennett M (2004) Severe vitamin B12 deficiency in an infant associated with a maternal deficiency and a strict vegetarian diet. J Pediatr Hematol Oncol 26:270–271

Wickremasinghe RG, Hoffbrand AV (1980) Reduced rate of DNA replication fork movement in megaloblastic anemia. J Clin Invest 65:26. 1980

Williams AM, Chosy JJ, Schilling RF (1963) Effect of vitamin B12 in vitro on incorporation of nucleic acid precursors by pernicious anemia bone marrow. J Clin Invest 42:670

Wuerges J, Garau G, Geremia S, Fedosov SN, Petersen TE, Randaccio L (2006) Structural basis fofor mammalian vitamin B12 transport by transcobalamin. PNAS 103(12):4386–4391

Yassin F, Rothenberg SP, Rao S, Gordon MM, Alpers DH, Quadros EV (2004) Identification of a 4-base deletion in the gene in inherited intrinsic factor deficiency. Blood 103:1515–1517

Yeomans ND, Hanson RG, Smallwood RA, Mihaly GW, Louis WJ (1982) Effect of chronic ranitidine treatment on secretion of intrinsic factor. Br Med J 285(6337):264

YYamada K (2013) Cobalt: its role in health and disease. Met Ions Life Sci 13:295–320

Zhang Z, Tian C, Zhou S, Wang W, Guo Y, Xia J, Liu Z, Wang B, Wang X, Golding BT, Griff RJ, Du Y, Liu J (2012) Mechanism-based design, synthesis and biological studies of N5-substituted tetrahydrofolate analogs as inhibitors of cobalamin-dependent methionine synthase and potential anticancer agents. Eur J Med Chem 58:228–236

Zhao R, Gao F, Goldman ID (2002) Reduced folate carrier transports thiamine monophosphate: an alternative route for thiamine delivery into mammalian cells. Am J Phys Cell Phys 282:C1512–C1517

Zittoun J (1995) Congenital errors of folate metabolism. Baillieres Clin Haematol 8(3):603–616

Zydowsky TM (1986) Stereochemical analysis of the methyl transfer catalyzed by cobalamin-dependent methionine synthase from Escherichia coli B. J Am Chem Soc 108(11):3152–3153

Chapter 9
Implication of Vitamin A in Nutritional Anemia

Abstract Vitamin A is a fat-soluble organic compound having nutritional value, and it exists in highly diversified forms like retinal, retinoic acid and retinol. Its widely available provitamin form is the carotenoids. Vitamin A has multiple roles. It is necessary for the normal growth and development, immunity development and acquisition of normal vision. Vitamin A is a fat-soluble vitamin and its digestion, almost, resembles that of lipid digestion. Vitamin A affects and controls the erythropoiesis at different stages in human life. Retinoids regulate the development of haematovascular tissues in yolk sac stage and shape the foetal erythropoiesis. Vitamin A also controls cellular proliferation and differentiation of the haemopoietic stem cells. Vitamin A is engaged in the pathogenesis of anemia in children through varied mechanisms. The conclusive role of vitamin A in the cause of anemia is uncertain.

9.1 Structure of Vitamin A

Vitamin A is a fat-soluble organic compound having nutritional value, and it exists in highly diversified forms like retinal, retinoic acid and retinol, and its widely available provitamin form is the carotenoids in nature (Berdanier 1997; Johnson and Russell 2010).

Around 600 carotenoids exist in nature which can be grouped into oxygenated (xanthophylls) and non-oxygenated (carotenes) carotenoids. Latter group of carotenoids possess beta-ionone ring and conjugated double bonds, hence termed as unsaturated hydrocarbons. Vitamin A and carotenes are highly sensitive to oxidation process in the presence of light and oxygen (Mordi et al. 1993; Ross 2006, 2010).

Generally, carotenoids or the term provitamin A represents the compounds which have biological activity like that of vitamin A. Carotenes are easily convertible into retinol (Oslo 1993).

Beta-carotene is a provitamin A. Carotene is the red–orange pigmented, unsaturated hydrocarbons present in the plants and derives its name from carrot with red colour. It belongs to the class of terpenes which are aromatic hydrocarbons made up of eight isoprene units. They are responsible for colours of the fruits and vegetables. Apart from carrots, these are present in mango, pumpkin and sweet potato.

A. Gupta, *Nutritional Anemia in Preschool Children*,
DOI 10.1007/978-981-10-5178-4_9

Beta-carotene is an important form of carotene and is associated with beta-rings on either end of carotene molecule. It is made up of two retinyl groups. It is converted into retinal by beta-carotene 15,15′-dioxygenase enzyme in the mucosa of the small intestine in humans. Alpha-carotene and gamma-carotene have weak vitamin A activity due to the presence of single retinyl group (Fennema 2008).

Vitamin A possesses multiple roles. It is necessary for the normal growth and development of children, development of the immune system and acquisition of normal vision (Tanumihardjo 2011; Solomon 2006; IOM/FNB 2001).

The retinal form of vitamin A as a prosthetic group associates with the opsin protein to form rhodopsin in the rods of eyes. It is a photosensitive pigment which splits on exposure to light and, hence, generates a nerve impulse (Wolf 2001).

The retinoic form of vitamin A is necessary for the normal epithelization of the skin and mucosa. It acts like a growth factor for cells. Retinyl palmitate is an ester of palmitic acid and retinol and is abundantly found in the animal source of food. It is changed into retinol in the small intestine (Berdanier 1997).

9.2 Dietary Sources of Vitamin A

Animal food contains retinol form of vitamin A. It is a yellow, lipid-soluble organic compound. In its pure alcoholic form, it is unstable; hence, it is associated with fatty acid and exists as retinyl esters. Retinyl palmitate and retinyl acetate are two important commercial forms of vitamin A. Important animal sources are beef, eggs, cod liver oil and milk and its products like cheese, butter and yoghurt (UMMC 2016; Meschino 2012).

Plant sources of food contain beta-carotene, alpha-carotene and gamma-carotene as the provitamin A. These have beta-ionone rings which are a characteristic of vitamin A activity. All herbivores and omnivores possess beta-carotene 15,15′-dioxygenase enzyme in the small intestine. It is an iron-dependent enzyme responsible for cleavage of beta-ionone ring in beta-carotene and converting it into all-*trans*-retinal form. Chief plant sources are oranges, yellow-pigmented vegetables and fruits, broccoli and spinach (DeMan 1999; USNLM 2016).

Vitamin A is available in animal and plant foods. According to USDA (2011), animal foods (per 100 g) like cod liver oil, turkey liver, pork liver, eggs and milk contain 30,000 μg, 8058 μg, 6500 μg, 140 μg and 28 μg of vitamin A, respectively. Plant sources of food (per 100 g) like dandelion green, sweet potato, carrot, broccoli, spinach, pumpkin, papaya and mango contain 5588 μg, 961 μg, 83 μg, 800 μg, 469 μg, 40 μg, 55 μg and 38 μg of vitamin A, accordingly.

9.3 Daily Recommended Allowance

The recommended daily allowance of vitamin A is represented in the form of retinol activity equivalents (RAE) and is expressed in mcg unit. According to Otten et al. (2006), all forms of vitamin A obtained from diet are transformed into retinol. According to authors, 12 mcg of beta-carotene and 24 mcg of alpha-carotene available in foods are comparable with 1 mcg of bioavailable retinol (USDA 2011; NIH 2016).

For the sake of interconversion of RAE into IU, the following standard values have been recommended by NIH (2016) and Otten et al. (2006):

1 IU retinol is equivalent to 0.3 mcg of RAE.
1 IU beta-carotene in diet is equivalent to 0.05 mcg of RAE.

According to Ansstas et al. (2014) and IOM/FNB (2001), the recommended daily allowance is determined to fulfil the requirement of all individuals in a particular age group. RDA of vitamin A for the infants 0–6 months is 400 mcg RAE, for infants 7 months to 12 months is 500 mcg RAE, for children in age group 1–3 years is 300 mcg RAE and for children in age group 4–8 years is 400 mcg RAE of vitamin A for their daily activities. These values are independent of gender of children in their age groups.

9.4 Digestion of Dietary Vitamin A

Vitamin A is a fat-soluble vitamin and its digestion, almost, resembles to that of lipid digestion. Digestion of vitamin A is not observed in the lumen of the stomach (Tyssandier et al. 2003).

Free forms of vitamin A and provitamin A are absorbed into the small intestine. However, esterified forms of vitamin A like retinyl palmitate and retinyl acetate undergo hydrolysis in the GIT into free forms. Gastric lipase does not have lipolytic activity on the retinyl esters (Borel et al. 2001).

These are effectively hydrolysed into the lumen of the small intestine. Cholesterol ester hydrolase and pancreatic lipase are secreted by the pancreas into the duodenum. Pancreatic lipase has higher lipolytic activity over retinyl ester in comparison with cholesterol ester hydrolase (Van Bennekum et al. 1999). Pancreatic lipase-related protein 2 is additionally associated with cleavage of retinyl ester in the duodenum (Reboul et al. 2006). However, it has been found that a few esters of retinol are transported intact across the mucosa of the small intestine. Later on, these are hydrolysed within the enterocytes (Dhuique-Mayer et al. 2007).

Carotenoids associate with lipids and fat-soluble vitamins and form micelles and, hence, pass through the mucosa of the small intestine (Borel 2003). β-Lactoglobulin is a conjugated protein in the cow milk that has the ability to bind with retinol and β-carotene (Perez and Calvo 1995).

Free retinol enters into enterocytes by simple diffusion (During and Harrison 2007). Retinol also undergoes facilitated transport with the help of a carrier protein called stimulated by retinoic acid 6A (STRA6). It is a transmembrane protein which assists transportation of micellar retinol and/or retinol bound to β-lactoglobulin (Kawaguchi et al. 2007).

Carotenoids are absorbed by passive diffusion and facilitated diffusion by scavenger receptor class B type I. This transporter is located in the surface of the brush border membrane of the enterocytes (Lobo et al. 2001). Within the enterocytes, retinol is again esterified into retinyl esters. The reaction is catabolized by lecithin retinol acyl transferase (LRAT). Carotenoids are broken down into retinal form and, thereafter, converted into retinol and subsequently into retinyl esters.

9.5 Role of Vitamin A in Erythropoiesis

According to Evans (2005), vitamin A affects and controls the haematopoiesis at different stages in human life. Retinoids regulate the development of haematovascular tissues in yolk sac stage and shape the foetal erythropoiesis. Vitamin A also controls cellular proliferation and differentiation of the haemopoietic stem cells.

Nuclear receptors regulate the genotype and phenotype of tissues during intrauterine and postnatal life. Retinoids act as ligands which interact with the nuclear receptors to influence their function favourably. Retinoids (Oren et al. 2003) are necessary for the normal erythropoiesis during yolk sac stage and embryonic liver stage. Epidemiological studies strengthen the necessity of vitamin A in normal haematopoiesis.

9.5.1 Effect of Retinoids on Erythropoiesis

Vitamin A affects the growth and differentiation of haemopoietic tissues. Morphological studies by Wolbach and Howe (1925) detected a reduction in the haematopoietic cell mass in the bone marrow of animals deficient of vitamin A. Later on, Douer and Koeffler (1982), in an in vitro study indicating that the retinoids, additionally, are implicated in the growth of haematopoietic tissues, showed that retinoic acid stimulates the growth of normal human granulocyte-macrophage progenitors.

9.6 Vitamin A Deficiency and Anemia

Vitamin A is necessary for the epithelization of the skin and mucosa. Its deficiency affects the growth of erythrocyte progenitors. Vitamin A is engaged in the pathogenesis of Anemia in children through varied mechanisms (Semba and Bloem 2002).

It is involved in the development of immunity, and its inadequacy in the body is negatively implicated in recurrent acute infections of the upper respiratory tract, diarrhoea, malabsorption syndrome of the GIT and anemia of infection, especially in the poor, rural population of the developing countries. Iron-deficiency anemia and vitamin A deficiency are the major public health hazards affecting millions of children in early childhood (Karyadi and Bloem 1996).

An epidemiological study was conducted by Mejia and Chew (1988) to find out the role of vitamin A deficiency in the causation and amelioration of anemia and iron deficiency anemia. In their study, a total of 99 anaemic children in early- and school-age groups were divided into four categories. They were given iron, vitamin A, their combination and placebo as supplements. Vitamin A and iron combination provided a better improvement in the serum Fe and percent transferrin saturation in comparison with individual supplements of Fe and vitamin A. The study clearly indicates the beneficial role of vitamin A in the management of anemia in association with iron.

The conclusive role of vitamin A in the cause of anemia is uncertain. Animal studies have yielded necessary knowledge concerning the impact of vitamin A on the iron status and iron deficiency anemia. Authors (Roodenburg et al. 1994) studied the rats in relation to vitamin A deficiency and anemia. In the group of rats with iron deficiency, it was found that the haemoglobin concentration was reduced. Hence, absorption of iron and TIBC was raised. In the other group of rats with vitamin A deficiency, the haemoglobin level was reduced followed by an increase in iron absorption and decline in TIBC. The rise in iron absorption in vitamin A deficiency is due to the decline in the iron store of the body. However, the total iron binding capacity is not raised due to locking of iron in tissues.

9.7 Pathophysiology of Vitamin A-Induced Anemia

Hypotheses have been proposed by different researchers explaining the possible role of vitamin A in the iron metabolism and in iron deficiency anemia. Vitamin A in the body might affect the iron metabolism at varied stages.

9.7.1 Vitamin A-Stimulated Iron Mobilization

Iron is stored in the body tissues in the form of ferritin and haemosiderin. It is carried by the transferrin in blood circulation and delivered to the bone marrow for erythropoiesis.

Semba et al. (1992) studied the role of vitamin A supplements in the iron mobilization in the body. Researchers (Semba et al. 1992) conducted a randomized, double-blind, placebo-controlled trial on 236 preschool-age children in age group 3–6 years in West Java, Indonesia. Plasma ferritin and haemoglobin concentration were increased by vitamin A supplements.

Authors suggested that vitamin A dose induced mobilization of iron from the hepatocytes which entered into circulation and, thereby, improved the erythropoiesis in the bone marrow.

Beynen et al. (1992) studied the effect of vitamin A-deficient diet on the iron store of the body. Diet with low and high vitamin A was fed to rats for 28 days. Authors found that the iron store in the liver was comparatively higher in low vitamin A-fed diets in rats. However, the iron level in the femur was low.

According to Mejia et al. (1979), supplementation of vitamin A helps improve the haemoglobin concentration along with serum iron level and percentage of transferrin saturation in children who were deficient in vitamin A. Authors speculated that adequate level of vitamin A is necessary for the mobilization of iron from the body stores.

9.7.2 Harmful Effect on Erythropoiesis

Pluripotent stem cells undergo proliferation and differentiation into committed pluripotent stem cells and ultimately differentiation into proerythroblasts and intermediates of the erythrocytes. Apart from other factors that influence erythropoiesis, erythropoietin predominantly affects the growth and differentiation of erythroid progenitors in the bone marrow (Fischer 2003).

Erythropoietin is a polypeptide hormone secreted by juxtaglomerular cells of the kidney that regulates erythropoiesis under normal physiological condition. However, in acute and chronic hypoxic condition, its secretion is enhanced leading to increase in RBC synthesis. It is hypothesized that retinoic acid might influence the secretion of erythropoietin. It was found by Neumcke et al. (1999) that in cell culture, the level of erythropoietin was increased in response to retinoic acid independent of the partial pressure of oxygen.

Human studies have been conducted depicting the role of vitamin A supplementation on haematological parameters, erythropoietin level in individuals suffering from anemia and infection. A double-blind, randomized study was conducted by Zimmermann et al. (2006) on the 81 Moroccan schoolchildren to assess the impact of vitamin A supplements. Researchers (Zimmermann et al. 2006) observed a significant rise in the geometric mean erythropoietin level, haemoglobin level and MCV in the group of children who received vitamin A supplements. A profound decline in (Zimmermann et al. 2006) the prevalence of anemia in children was detected. It is assumed by the researchers that vitamin A and its metabolites in the body influence genes that regulate transcription of proteins involved actively in the iron store in the body and import of iron from tissues (Zimmermann et al. 2006).

In an another study by Cusick et al. (2005), contrasting impact of the vitamin A supplement was observed on the secretion of erythropoietin in anaemic preschool-age group children. Sub-Saharan Africa has high prevalence of anemia whose cause is multifactorial including a very high occurrence of *Plasmodium falciparum* that caused malaria in children (Jaenisch et al. 2012).

Vitamin A deficiency in preschool children is common in Zanzibari area of coastal East Africa. Authors provided vitamin A supplements to anaemic Zanzibari preschool children. Surprisingly, a sharp decline in erythropoietin level was observed by the authors in those children.

It was hypothesized by the researchers that malarial infection suppressed the secretion of erythropoietin in children. According to Phillips and Pasvol (1992), pathogenesis of anemia in individuals suffering from malaria is dependent on a myriad of factors that interplay. A study by Woodruff et al. (1979) showed the suppression of erythropoietin in malaria infection that caused anemia.

9.7.3 Impact on Iron Absorption

Minerals and trace elements are essential for the consistent growth and development of the body. There is an interaction among the dietary minerals in the lumen of the small intestine during absorption. It is hypothesized that vitamin A deficiency may be associated with iron deficiency anemia. According to Garcia-Casal and Layrisse (1998), vitamin A influences favourably the absorption of dietary iron in the small intestine. It might be possible that vitamin A chelates iron and keeps it in the soluble form and, hence, enhances its absorption. Further, authors augment that food fortification with iron and vitamin A stimulates iron absorption in the presence of inhibitors in the lumen of the small intestine.

Authors (Layrisse et al. 2000) studied 170 persons to assess the effect of vitamin A or beta-carotene on iron absorption in the presence of inhibitors. Authors found that vitamin A supplements raised the iron absorption from rice, wheat and corn by 3-, 2.4-, and 1.8-folds, respectively. Beta-carotene enhanced iron absorption up to threefolds from three cereals. It was assumed that vitamin A and carotene formed complexes with iron that stabilized in soluble form in the lumen of the small intestine.

Studies authenticate the complex interaction between dietary iron and vitamin A in the intestinal lumen. Dietary deficiency of vitamin A could retard the absorption of iron and, thus, manifest as anemia. Also, earlier studies indicated the enhanced secretion of erythropoietin in response to vitamin A supplements and, therefore, could help ameliorate iron deficiency anemia in suffering children.

References

Ansstas G, Hakore J, Gopalswamy N (2014) Vitamin A deficiency. Griffing GT (ed) Available at: http://emedicine.medscape.com/article/126004-overview#a5

Berdanier C (1997) Advanced nutrition micronutrients. CRC Press, Boca Raton, pp 22–39

Beynen AC, Sijtsma KW, Van Den Berg GJ, Lemmens AG, West CE (1992) Iron status in rats fed a purified diet without vitamin A. Biol Trace Elem Res 135:81–84

Borel P (2003) Factors affecting intestinal absorption of highly lipophilic food microconstituents (fat-soluble vitamins, carotenoids and phytosterols). Clin Chem Lab Med 41:979–994

Borel P, Pasquier B, Armand M, Tyssandier V, Grolier P, Alexandre-Gouabau MC, Andre M, Senft M, Peyrot J, Jaussan V, Lairon D, Azais-Braesco V (2001) Processing of vitamin A and E in the human gastrointestinal tract. Am J Physiol Gastrointest Liver Physiol 280:G95–G103

Cusick SE, Tielsch JM, Ramsan M, Jape JK, Sazawal S, Black RE, Stoltzfus RJ (2005) Short-term effects of vitamin A and antimalarial treatment on erythropoiesis in severely anemic Zanzibari preschool children. Am J Clin Nutr 82:406–412

DeMan J (1999) Principles of food chemistry. Aspen Publication Inc, Maryland, p 358

Dhuique-Mayer C, Borel P, Reboul E, Caporiccio B, Besancon P, Amiot MJ (2007) Beta-cryptoxanthin from citrus juices: Assessment of bioaccessibility using an in vitro digestion/Caco-2 cell culture model. Br J Nutr 97:883–890

Douer D, Koeffler HP (1982) Retinoic acid enhances colony-stimulating factor induced clonal growth of normal human myeloid progenitor cells in vitro. Exp Cell Res 138(1):193–198

During A, Harrison EH (2007) Mechanisms of provitamin A (carotenoid) and vitamin A (retinol) transport into and out of intestinal Caco-2 cells. J Lipid Res 48:2283–2294

Evans T (2005) Regulation of hematopoiesis by retinoid signalling. Exp Hematol 33(9):1055–1061

Fennema O (2008) Fennema's food chemistry. CRC Press Taylor & Francis, pp 454–455

Fisher JW (2003) Erythropoietin: physiology and pharmacology update. Exp Biol Med 228:1–14

Garcia-Casal MN, Layrisse M (1998) Dietary iron absorption. Role of vitamin A. Arch Lationoam Nutr 48(3):191–196

Institute of Medicine, Food and Nutrition Board (2001) Dietary reference intake for vitamin A, vitamin K, arsenic, boron, chromium, copper, iodine, iron, manganese, molybdenum, nickel, silicon, vanadium and zinc. National Academy Press, Washington, DC

Jaenisch T, Sazawal S, Dutta A, Deb S, Ramson M, Sullivan DJ (2012) Contributions of Polyclonal Malaria, Gametocytemia, and Pneumonia to Infant Severe Anemia Incidence in Malaria Hyperendemic Pemba, Tanzania. Am J Trop Med Hyg 86(6):925–930

Johnson EJ, Russell RM (2010) Beta-carotene. In: Coates PM, Betz JM, Blackman MR (eds) Encyclopedia of dietary supplements. Informa Healthcare, London, pp 115–120

Karyadi D, Bloem MW (1996) The role of vitamin A in iron deficiency anemia and implications for interventions. Biomed Environ Sci 9:316–324

Kawaguchi R, Yu J, Honda J, Hu J, Whitelegge J, Ping P, Wiita P, Bok D, Sun H (2007) A membrane receptor for retinol binding protein mediates cellular uptake of vitamin A. Science 315:820–825

Layrisse M, Garcia-Casal MN, Solano L, Baron MA, Arguello F, Llovera D, Ramirez J, Leets I, Tropper E (2000) New property of vitamin A and beta-carotene on human iron absorption: effect on phytate and polyphenols as inhibitors of iron absorption. Arch Latinoam Nutr 50(3):243–248

Lobo MV, Huerta L, Ruiz-Velasco N, Teixeiro E, de La Cueva P, Celdran A, Martín-Hidalgo A, Vega MA, Bragado R (2001) Localization of the lipid receptors CD36 and CLA-1/SR-BI in

the human gastrointestinal tract: towards the identification of receptors mediating the intestinal absorption of dietary lipids. J Histochem Cytochem 49:1253–1260
Mejia LA, Chew F (1988) Hematological effect of supplementing anemic children with vitamin A alone and in combination with iron. Am J Clin Nutr 48(3):595–600
Mejia LA, Hodges RE, Rucker RB (1979) Role of vitamin A in the absorption, retention and distribution of iron in the rat. J Nutr 109:129–137
Meschino J (2012) Comprehensive guide to vitamin A. Available at: http://www.meschinohealth.com/books/vitamin_a
Mordi RC, Walton JC, Burton GW, Hughes L, Ingold KU, Lindsay DA, Moffatt DJ (1993) Oxidative degradation of 3-carotene and 13b-apo-8′ carotenal. Tetrahedron 49911:928
National Institutes of Health, Office of dietary supplements (2016) Vitamin A. Available at: https://ods.od.nih.gov/factsheets/VitaminA-HealthProfessional
Neumcke I, Schneider B, Fandrey J, Pagel H (1999) Effects of pro- and antioxidative compounds on renal production of erythropoietin. Endocrinology 140:641–645
Olson JA (1993) In: Shils ME, Olson JA, Shike M (eds) Vitamin A, retinoids and carotenoids. Lea & Febiger, Philadelphia, pp 287–307
Oren T, Sher JA, Evan T (2003) Hematopoiesis and retinoids: development and disease. Leuk Lymphoma 44(11):1881–1891
Otten JJ, Hellwig JP, Meyers LD (2006) Dietary reference intakes: the essential guide to nutrient requirements. The National Academies Press, Washington, DC
Perez MD, Calvo M (1995) Interaction of b-lactoglobulin with retinol and fatty acids and its role as a possible biological function for this protein: a review. J Dairy Sci 78:978–988

Absorption of Vitamin A

Phillips RE, Pasvol G (1992) Anemia of Plasmodium falciparum malaria. Baillieres Clin Haematol 5:315–330
Reboul E, Berton A, Moussa M, Kreuzer C, Crenon I, Borel P (2006) Pancreatic lipase and pancreatic lipase-related protein 2, but not pancreatic lipase-related protein 1, hydrolyze retinyl palmitate in physiological conditions. Biochim Biophys Acta 1761:4–10
Roodenburg AJC, West CE, Yu S, Beynen AC (1994) Comparison between time-dependent changes in iron metabolism of rats as induced by marginal deficiency of either vitamin A or iron. Br J Nutr 71:687–699
Ross A (2006) Vitamin A and carotenoids. In: Shils M, Shike M, Ross A, Caballero B, Cousins R (eds) Modern nutrition in health and disease. Lippincott Williams & Wilkins, Baltimore, pp 351–375
Ross CA (2010) Vitamin A. In: Coates PM, Betz JM, Blackman MR (eds) Encyclopedia of dietary supplements. Informa Healthcare, London, pp 778–791
Semba RD, Bloem MW (2002) The anemia of vitamin A deficiency: epidemiology and pathogenesis. Eur J Clin Nutr 56(4):271–281
Semba RD, Muhilal WKP, Winget M, Natadisastra G, Scott A, Sommer A (1992) Impact of vitamin A supplementation on hematological indicators of iron metabolism and protein status in children. Nutr Res 12:469–478
Solomons NW (2006) Vitamin A. In: Bowman B, Russell R (eds) Present knowledge in nutrition. International Life Sciences Institute, Washington, DC, pp 157–183
Tanumihardjo SA (2011) Vitamin A: biomarkers of nutrition for development. Am J Clin Nutr 94(2):658S–665S
Tyssandier V, Reboul E, Dumas JF, Bouteloup-Demange C, Armand M, Marcand J, Sallas M, Borel P (2003) Processing of vegetable-borne carotenoids in the human stomach and duodenum. Am J Physiol Gastrointest Liver Physiol 284:G913–G923

United States Department of Agriculture, Agricultural Research Service (2011) USDA National Nutrient Database for standard reference. Available at: http://www.ndb.nal.usda.gov/

United States National Library of Medicine (2016) Vitamin A. Available at: https://www.nlm.nih.gov/medlineplus/ency/article/002400.htm

University of Maryland Medical Center (2016) Vitamin A, Retinol. Available at: http://umm.edu/health/medical/altmed/supplement/vitamin-a-retinol

Van Bennekum AM, Li L, Piantedosi R, Shamir R, Vogel S, Fisher EA, Blaner WS, Harrison EH (1999) Carboxyl ester lipase overexpression in rat hepatoma cells and CEL deficiency in mice have no impact on hepatic uptake or metabolism of chylomicron-retinyl ester. Biochemistry 38:4150–4156

Wolbach SB, Howe PR (1925) Tissue changes following deprivation of fat soluble A vitamin. J Exp Med 42:753–777

Wolf G (2001) The discovery of the visual function of vitamin A. J Nutr 131(6):1647–1650

Woodruff AW, Ansdell VA, Pettitt LE (1979) Cause of anemia in malaria. Lancet i:1055–1057

Zimmermann MB, Biebinger R, Rohner F, Dib A, Zeder C, Hurrell RF, Chaouki N (2006) Vitamin A supplementation in children with poor vitamin A and iron status increases erythropoietin and hemoglobin concentrations without changing total body iron. Am J Clin Nutr 84:580–586

Chapter 10
Role of Zinc, Copper and Selenium in Nutritional Anemia

Abstract Zinc is an essential trace element necessary for a broad category of enzymatic reactions, enzymes, transcription factors and proteins of the body. It is of prime importance in prenatal and postnatal growth and development. It is profoundly involved in normal endocrine functioning, sexual maturation and development of immunity. Its deficiency in the body either during intrauterine or postnatal life is responsible for recurrent infection of upper respiratory tract, diarrhoea, wasting and stunting in children. Zinc essentially regulates the process of erythropoiesis in the bone marrow. Zinc is the key component of GATA-1, also called as erythroid transcription factor. Zinc is required in different stages of erythropoiesis.

Hypocupremia is responsible for dysregulation of human pluripotent stem cell proliferation and/or an impediment in the cellular differentiation in the bone marrow. Copper deficiency anemia in children is typical of microcytic or normocytic or macrocytic with neutropenia and thrombocytopenia as the main characteristics.

Selenium is an important component of glutathione peroxidase, and its concentration in erythrocytes shows its protective effect on erythrocytes against oxidative damage. This suggests that increased oxidative stress could be another contributing factor for the development of anemia due to selenium deficiency.

10.1 Overview of Zinc as Mineral

Zinc is an essential trace element necessary for a broad category of enzymatic reactions, enzymes, transcription factors and proteins of the body. It is of prime importance in prenatal and postnatal growth and development. It is profoundly involved in normal endocrine functioning, sexual maturation and development of immunity. Its deficiency in the body either during intrauterine or postnatal life is responsible for recurrent infection of upper respiratory tract, diarrhoea, wasting and stunting in children. Zinc deficiency (Prasad 2003) affects about two billion people in the developing world and is associated with many diseases (Prasad et al. 1963).

A. Gupta, *Nutritional Anemia in Preschool Children*,
DOI 10.1007/978-981-10-5178-4_10

Zinc is the second most important trace element in the human body. It is of prime significance for all the living organisms. Its abundance in the body as a trace mineral is next to iron (Hambidge et al. 1986). Zinc is distributed throughout the body tissues. It is the integral part of the zinc metalloproteins in the body and, hence, is perfused in every cell of the body (Plum et al. 2010).

Normal serum zinc concentration is in the range of 0.7–1.3 mg/100 ml, while the concentration of zinc in the interstitial fluid is 2–5 uM (Magneson et al. 1987). Mammalian cells have normal intracellular concentration of zinc in the range of 100–500 uM. Nevertheless, a large amount (90%) of intracellular zinc is present in bound form as zinc metalloprotein and does not participate actively in cellular functions. Residual amount of zinc subsists in free form as Zn++ ion (Renty and Leslie 2009).

Zinc is the structural constituent of transcriptional factors and hormone receptors in the cell membranes and necessary for the integrity of biological membranes of the tissues. Zinc-dependent enzymes have been profoundly implicated in the mitosis, differentiation and growth of cells (Chesters 1978). Zinc is ligated to the amino acids in metalloproteins. The most common amino acids that coordinate with zinc are aspartic acid, cysteine, glutamic acid and histidine, wherein zinc exists as zinc–aspartate, zinc–histidine, zinc–cysteine and zinc–glutamate. These ligands constitute labile highly reactive pool of zinc (Brandt et al. 2009; Ensminger and Konlande 1993; Johnson 2008).

The body of a hypothetical person contains around 2–4 g of zinc (Wapnir 1990). It is prominently distributed in the brain, bones, muscles, liver and kidneys. Tissues of the prostate glands, semen and eyes are highly enriched with zinc concentration. It is necessary for normal physiology of reproductive organs (Berdanier et al. 2007).

Zinc has a conclusive influence in the regulation of apoptosis in mammalian cells. It interacts with the cellular organelles in a highly complex, coordinated and diverse pathway. It can act on nucleus directly or indirectly through mitochondria, thus regulating apoptosis. Zinc can also regulate factors that signal cellular apoptosis and, therefore, can have influence on nucleus and mitochondria. According to Chang et al. (2006), zinc enhances the expression of cytokines and, hence, increases their secretion and induces apoptosis. Zinc acts on receptors on the cells and/or triggers mitochondria-induced apoptogenesis.

Zinc is located in the nuclear assembly. It is a part of the nucleus and nucleolus. Hundreds of enzymes for DNA replication, transcription of RNA and translation of proteins harbour zinc as a prosthetic group (Wu et al. 1992).

Zinc is the major part of the glutaminergic neurons (Frederickson et al. 2000; Danscher 1984; Danscher et al. 1985). Zinc is contained within presynaptic vesicles in neurons, as stated by Finn Mogens. Later on, Haug (1967) researched intensively in this field exploring the zinc-enriched hippocampal region of the brain which is present, mainly, in the telencephalon. These neurons confiscate zinc in the presynaptic vesicles. However, glutaminergic neurons located in the brainstem, cerebellum and thalamus are devoid of zinc-containing vesicles. The hippocampus, amygdala and perirhinal areas are the prominent structures of telencephalon. Zinc influences the normal functioning of these structures and has an effect in the cause of pathology concerning these parts of the brain. According to Frederickson and Mocrieff (1994), zinc in the synaptic cleft determines and modulates the release of the neurotransmitters like glutamate and ultimately influences the level of postsynaptic excitatory potential.

10.2 Dietary Intake

According to Bales and Ritchie (2009), RDA of zinc is 8 mg per day and 11 mg per day for women and men, respectively, in the United States.

In India, RDA of zinc as recommended by ICMR (2010) is in the range of 7–12 mg per day.

According to the Food and Nutrition Board (2000), children in the age group from 1 to 3 years need zinc around 10 mg per day in the food.

Zinc natural sources are wheat, sesame, alfalfa and mustard (Ensminger and Konlande 1993; Ibs and Rink 2003).

Beans, nuts, almonds, whole grains and seeds of pumpkin and sunflower are enriched with zinc. Other food grains like rice, maize, barley and oats are other sources of zinc. However, a large amount of zinc from the food grains is lost during milling and processing (USDA 2007).

Nevertheless, the inhibitors like phytates present in the whole grains and food fibres and some seeds, definitely, hinder the intestinal absorption of zinc (Prasad 2003). Animal sources include liver, milk and its products and fish.

10.3 Zinc Deficiency and Anemia

Zinc deficiency is mostly due to inadequate dietary intake of zinc. However, its deficiency is also associated with certain diseases, like malabsorption syndrome of GIT, chronic liver disease, chronic diabetes mellitus and neoplasia (Prasad 2003). Children in preschool-aged group, elder persons and immunity-compromised persons have higher predisposition towards zinc deficiency. About two billion population in the developing world are suffering from zinc deficiency (Prasad 2003).

Zinc deficiency is manifested as stunted growth of children, wasting, recurrent episodes of diarrhoea, faltered growth and development, impaired cognitive ability, compromised immunity, skin lesions, impaired appetite, altered cognition, impaired host defence properties, defects in carbohydrate utilization and reproductive teratogenesis. Mild zinc deficiency depresses immunity.

The World Health Organization recommends supplements of zinc to be administered to preschool children for the prevention and management of diarrhoea and for interception of different forms of malnutrition in children (WHO 2007).

Micronutrient deficiencies are caused by inadequate dietary intake, increased losses from the body and/or increased requirements (Adelekan 2003).

Poverty, inaccessibility to good food sources, illiteracy, ignorance about the type of foods and their nutritive values, custom and food practices are certain predisposing factors to the consumption of adequate quantity and good-quality nutritious food which renders the children in infancy and early childhood prone to the deficiency of trace minerals like zinc, iron, iodine and selenium (Bwibo and Neumann 2003).

According to Bhan et al. (1996), in developing countries, micronutrient deficiency including zinc is highly prevalent in preschool children. Therefore, iron supplements have profound impact in reducing the burden of disease and mortality in children.

Authors (Ryu et al. 2008; Huber and Cousins 1993; Perkins 1993; Trainor et al. 2000; Nishiyama et al. 1998) posited that zinc essentially regulates the process of erythropoiesis in the bone marrow.

It was hypothesized by Cole et al. (2010) that zinc may influence iron metabolism in the body, and, hence, its deficiency is an additional factor in the cause of nutritional anemia in children. Authors concluded that zinc level can invariably modulate immune response of the body of children. In a study by the authors, it was found that zinc deficiency and anemia were significantly prevalent in children belonging to low-income population, especially, among African Americans. Analysis proved a strong interrelationship between serum zinc and haemoglobin in the children.

The above study indicates the complex interaction between the iron in the body and bone marrow for erythropoiesis and zinc level that has bearing on the iron deficiency anemia. Iron, zinc and other trace elements like nickel, copper and cadmium share common intestinal transporter as DMT. Inadequacy of one trace mineral in diet affects the absorption of another trace mineral by the small intestine either antagonistically or synergistically and, hence, ultimately guides the RBC synthesis.

One group of researchers has found a low serum zinc level in children with iron deficiency anemia (Ece et al. 1997; Gurgoze et al. 2006; Wajeunnesa et al. 2009).

Another group of researchers proclaims no significant correlation between zinc level and iron deficiency anemia in children (Turgut et al. 2007, 2009; Hegazy et al. 2010).

However, zinc level declines with the onset of iron deficiency anemia.

Coexistence of zinc and iron deficiencies in children was observed by Prasad et al. (1963), characterized by enlargement of the spleen and liver with the presence of mental lethargy and retarded growth of children, collectively termed as *Prasad syndrome* (Prasad et al. 1963).

10.4 Pathophysiology of Anemia in Zinc Deficiency

A study done by Huber and Cousins (1993) showed that the demand for zinc is raised in the body in a condition in which the erythropoiesis is induced. The study, indirectly, posits the role of zinc in erythroid differentiation. According to Pevny et al. (1991), GATA-1 is necessary for normal erythropoiesis. Further it was asserted by the authors that deficiency of GATA-1 in the embryonic stem cell stage in chimeric mice leads to failure of maturation of RBCs.

Zinc is the key component of GATA-1, also called as erythroid transcription factor, and is encoded by GATA1 gene (Caiulo et al. 1991). GATA-type zinc finger proteins have an important role in the proliferation, differentiation and growth of cells in the eukaryotes. GATA factors possess more than one zinc finger domain for attaching with DNA and another substrate (Ko and Engel 1993).

Zinc is required in different stages of erythropoiesis and is the key element in varied activities at biomolecular level in the cells. Zinc finger proteins are the short protein motifs containing zinc as the coordinating ion (Laity et al. 2001; Krishna et al. 2003). Zinc finger proteins can bind with DNA, RNA and proteins and are necessary for gene expression (Klug 1999). These are the stable structures which, generally, do not undergo conformational changes after binding with the substrates. Zinc fingers are necessary component of transcription factor in eukaryotes which bind with DNA.

Additionally, authors (Trainor et al. 2000) proved the necessity of zinc for zinc finger transcription factors which are needed for expression of proteins implicated in the last stage of erythroid maturation. Zinc concentration in the body and/or zinc supplementation may have an influence on haemoglobin production (Nishiyama et al. 1998).

GATA-1 is expressed in the erythroid cells, and authors (Pevny et al. 1995) found that the maturation of GATA-1-null erythroid cells was restricted to pro-erythroblast stage indicating the necessity of zinc ion in the form of zinc finger protein, GATA-1, during the course of erythropoiesis.

It is concerned with the regulation of genes, which sequentially control the development, maturation and functions of RBC. In premature erythrocytes, development of cytoskeleton and biosynthesis of globin and haem are the imperative functions which are controlled by the GATA-1-regulated genes (Crispino 2005).

Kruppel-like factor is another zinc finger motif encoded by human genes KLF1–KLF17. The protein is responsible for gene expression by binding with the DNA. Erythroid Kruppel-like factor regulates transcription of gene which is essential for the translation of globin, whose requirement is utmost for the biosynthesis of haemoglobin. Erythroid Kruppel-like factor also undertakes the transcription of gene for the biosynthesis of ferrochelatase enzyme needed for coordination of ferrous ion with tetra-pyrrole ring structure in order to synthesize haem (Hodge et al. 2006).

Based on the past work of Prasad et al. (1963) on the role of zinc deficiency in causing anemia, clinical studies were conducted to find out the impact of independent zinc and iron supplementations and combined supplementations on the amelioration of anemia and on other haematological factors. Profound beneficial effects have been observed by the synergistic dose of zinc and iron on the haemoglobin level and reticulocyte and erythrocyte counts of the anaemic children in comparison to the individual supplements.

Nevertheless, inadequate or overdose of zinc has been negatively implicated in the effects of zinc on the erythropoiesis. Zinc toxicity may result in haematological abnormalities in children. Its prolonged supplementation may be associated with sideroblastic anemia (Forman et al. 1990).

According to Magee and Matrone (1960), zinc level in the range of 0.5–1% in rats has resulted in hypochromic, macrocytic anemia indicating the deleterious effect of zinc on the iron metabolism of rats. Zinc intoxication has an effect on the iron metabolism. It decreases the hepatic store of iron and life span of RBCs.

According to Bloomer et al. (1983), excessive zinc ions in the circulation stimulate the formation of zinc protoporphyrin which is incorporated into immature RBCs during haem synthesis. It replaces haem in the RBCs.

Increased zinc protoporphyrin synthesis inhibits the delta-aminolevulinate synthase-2 enzyme also called as ALAS-2 encoded by gene ALAS2 which guides the biosynthesis of haem under normal physiological condition. Hence, deficiency of ALAS-2 under zinc toxicity impairs haem synthesis resulting in sideroblastic anemia in humans. Conclusively, it is to confer that combined zinc–iron supplements in adequate dose and duration are necessary to overcome iron deficiency anemia in children. However, the concentration of zinc in the circulation must be monitored strictly in order to maintain normal erythropoiesis.

According to Allen et al. (2006), micronutrient deficiency is rampant in developing countries affecting preschool-aged children and pregnant women. Deficiency predisposes these individuals to high prevalence of anemia, diarrhoea, compromised immunity and increased occurrence of respiratory tract infections. Vitamin and micronutrient inadequacy in growing children is responsible for high disease burden which deters the physical, intellectual, emotional and economic development of children, hence faltering human capital development. Iron deficiency predisposes to perinatal mortality (Stoltzfus 2003).

Zinc interacts with other trace elements in the body, hence affecting their absorption and metabolism. Excessive zinc in the body inhibits intestinal absorption of copper and decreases its bioavailability. Manifestation of copper deficiency appears in the body (Bremner et al. 1976).

Surplus zinc induces copper deficiency in the liver, heart and other tissues of the body. It reduces activities of copper-dependent metalloenzymes (Duncan et al. 1953).

10.5 Contradictory Role of Zinc in Iron Deficiency Anemia

Synergism and antagonism among the trace elements like iron, zinc, copper, cadmium and chromium determine the absorption and metabolism of these elements in the body and finally influence the iron status in the body and erythropoiesis.

In a study by Gurgoze et al. (2006) on the iron-deficit anaemic children, it was found by the authors that serum zinc level in the children was significantly low.

Authors suggested that the coexisting nutritional deficiencies of iron and zinc in the diet owing to their common food sources become the basis for the low level of zinc in iron deficiency anemia in children. Another possible mechanism might be the increased synthesis of zinc protoporphyrin in the RBCs instead of haem; hence, serum zinc level declined.

Increasing zinc protoporphyrin production (ZPP) in IDA leading to lower serum zinc values has been proposed as an alternative mechanism.

10.6 Copper

10.6.1 Overview of Copper as Mineral

Copper is another trace mineral that is essential to all living organisms. Copper is found in the liver, muscle and bone in humans. The adult human body contains around 100–140 mg of copper depending on the body weight which is distributed in different body parts. Skeletal muscles, bones and liver contain approximately 65 mg, 23 mg and 18 mg of copper, respectively. RDA for copper in adults is 3.0 mg/day (NRC/FNB 1980).

Copper concentration in the foetal liver is much higher in comparison to the liver in postnatal life in humans. It is focused in the mitochondria and remains associated with protein and is called as neonatal hepatic mitochondrocuprein by Porter (1966). It is a storage form of copper instead of an enzyme in the foetal liver. After birth, mitochondrocuprein dissociates, thereby releasing copper (Woddowso 1974).

Copper occurs in bound state with protein in erythrocytes and brain and liver cells in the form of erythrocuprein, cerebrocuprein and hepatocuprein. Important dietary sources of copper include kidney, liver, eggs, cereals, legumes, nuts and vegetables. Dietary copper is absorbed from the duodenum. Metallothionein in the intestinal mucosal cells helps in the absorption of copper. Zinc, cadmium, molybdenum and phytates retard the absorption of copper. It is transported to the liver in bound form with albumin. In blood circulation, copper exists in two forms. Around 95% of copper remains tightly bound to alpha globulin called as ceruloplasmin whose normal serum level is around 25–50 mg/100 ml. Its one molecule possesses six to eight atoms of copper, equally distributed as cuprous and cupric ions. Unlike transferrin, copper in ceruloplasmin is not easily available to other tissues. Another 4% of copper is loosely attached to albumin (Linder et al. 1998).

A routine diet provides around 5 mg of copper, and the daily requirement for an adult person is around 2.5 mg of copper. Under normal condition, the liver excretes around 85% of the dietary copper along with bile. It enters the duodenum. Reabsorption of copper from the duodenum does not take place, and probably it is the mechanism of copper homeostasis in the body, whereby the copper from GIT is excreted in faeces. Urinary excretion of copper is negligible (Percival and Harris 1990). Copper is one of the structural constituents of the cytochrome c oxidase enzyme in the mitochondrial inner membrane. The enzyme is required in the terminal stage of electron transport chain for oxidative phosphorylation. Enzyme is involved in the transfer of electrons to oxygen molecule, hence reducing it into water molecule (Vest et al. 2013).

Superoxide dismutase is dependent on the presence of copper for its catalytic activity. The enzyme is characterized for its dismutation potential of superoxide free radicals produced in the body as a result of metabolic reactions. These radicals are converted into either molecular oxygen or hydrogen peroxide by addition or removal of electrons from the radicals.

Ceruloplasmin and hephaestin are the highly essential enzymes with ferroxidase activity. These enzymes oxidize their substrates with the simultaneous reduction of dioxygen into water. The oxidizing activity is due to the presence of copper and iron ions in the enzymes. Ceruloplasmin is importantly involved in the oxidation of ferrous form into ferric form, which, in turn, forms the basis of iron binding to apotransferrin in circulation, as well as the iron storage in the form of ferritin in tissues. Hephaestin in the form of ferroxidase is indispensable for conversion of ferrous into ferric iron within intestinal mucosa cells, a process needed for its transport from inside of enterocytes into enterohepatic circulation (Wierzbicka and Gromadzka 2014).

Copper has a role in the iron absorption, necessary for activity of enzymes concerning biosynthesis of haem and globin. Copper deficiency results in a number of symptoms like neutropenia, low RBC count, low haemoglobin, leucopenia and low neutrophil count. Neutrophils are responsible for first-line defence, and hence, recurrent infections are the common consequence in low copper state of the body (Halfdanarson et al. 2008).

10.6.2 Copper and Iron Deficiency Anemia

Copper deficiency anemia in children is typical of microcytic or normocytic or macrocytic with neutropenia and thrombocytopenia as the main characteristics (Kumar 2006). The prominent pathogenesis of copper deficiency anemia is the impaired transportation of copper in the body as a result of weak activity of copper-dependent enzymes like ceruloplasmin and hephaestin in the body (Kodama and Fujisawa 2009). Cytochrome c oxidase enzyme is the key and terminal enzyme responsible for electron transport to oxygen and is copper-dependent enzyme. Deficiency of the enzyme in animal models as studied by Klevay (2006) showed the decreased haem synthesis in the body. Also, iron ions aggregated to form unusual patterns around the erythroblasts during erythropoiesis and are called sideroblastic anemia.

Hypocupremia, according to Tokuda et al. (1986), is responsible for irregular human pluripotent stem cell proliferation and/or an impediment in the cellular differentiation in the bone marrow. The study of Graham and Cordano (1976) proved that hypocupremia in children affects the iron metabolism. Initially, copper deficiency impairs the absorption of iron and is responsible for restricted transportation of iron in circulation and limited bioavailability of iron. Later on, hypocupremia aggravates the supply of iron to the bone marrow and impairs the normal process of erythropoiesis.

Copper deficiency has been considered the likely cause of anemia, but it was not until the completion of a series of case–control studies of copper deficiency in infants recovering from malnutrition (Cordano 1998).

Copper deficiency in humans is relatively low and according to Williams (1983) is manifested as hypochromic, normocytic or macrocytic anemia, neutropenia, thrombocytopenia and hypoferremia and raised susceptibility to bone fracture.

Conclusively, it can be inferred that hypocupremia-induced anemia is due to impaired iron absorption and decreased iron transportation from the tissues as a result of diminished copper-dependent ferroxidase activity in the body, which ultimately is expressed as reduced erythropoiesis in the bone marrow.

10.7 Selenium

10.7.1 Overview

Selenium is a trace element essential for all living organisms including humans and is an important component of conjugated protein named as selenoproteins involved in multiple functions as DNA synthesis, reproduction, antioxidative property and immunity (Sunde 2012; Roman et al. 2014; Hatfield et al. 2014; Hoffmann 2007; Papp et al. 2007, 2010).

Authors Schubert and Lahey (1959) asserted that cytological examination of the bone marrow showed the presence of vacuoles in erythroid and myeloid progenitors and their consequent delayed maturation. There are as many as 15 selenoproteins that have been discovered in various tissues like GIT, thyroid glands, liver, kidneys, CNS, plasma, muscle, sperm tail and adipose tissues (Levander 1986).

Selenium for humans is abundantly present in plant-based foods. The concentration of selenium in plants is dependent on its soil content. Selenium subsists in inorganic and organic forms in nature. Selenate and selenite are the main inorganic forms of selenium which are concentrated in soil, and hence, plants can extract and accumulate them. Selenomethionine and selenocysteine are the organic states of selenium (Sunde 2006; EFSA 2009).

Cereals contain around 1 μg of selenium by weight. Contrarily, animal-based foods are the poorer source of selenium (Pennington and Schoen 1995).

Plants convert inorganic forms of selenium into organic forms like selenocysteine and selenomethionine, while the latter form of selenium is widely present in the human body, prominently in the muscles, liver, nails, kidneys, bones and blood. According to Terry and Diamond (2012), voluntary muscles represent the main tissue of selenium storage in the body, accounting for approximately 28–46% of the body selenium.

Recommended dietary allowance for selenium during the first 6 months is 15 μg and thereafter, from 7 months to 3 years is 20 μg and in the age group between 4 and 8 years is 30 μg, according to NIH (2016).

Another important dietary source is the Brazil nuts which provide approximately 544 μg of selenium in one serving (1 oz). Wholewheat one slice of bread provides 13 μg of selenium as per NIH (2016).

In addition to dietary sources of selenium, it is available in multivitamin preparations. Further, selenium-rich yeast is a source of selenium. Mostly, the supplements contain selenomethionine which is easily absorbed by the human body (IOM/FNB 2000).

Dietary selenium is absorbed from the duodenum through brush border surfaces of enterocytes. Selenomethionine is more easily absorbed in comparison to other forms of selenium. Afterwards, selenium is circulated in plasma in bound form with beta lipoprotein. Selenomethionine is a biologically active form and, hence, can be taken up by myoglobin, nucleoproteins, myosin, cytochrome c and other body tissues. Around two thirds (80%) of the selenite in the dietary sources is absorbed from the small intestine (WHO 1987).

Selenomethionine has comparatively higher rate of absorption (90%) in the small intestine than the selenite (Patterson et al. 1993). The bioavailability of selenium from the wheat is around 80%, while its absorption and bioavailability from Brazil nuts (Mutanen 1986) are a little higher (90%).

Selenium has multiple functions in the human body. It is the structural constituent of glutathione peroxidase enzyme which is fundamentally involved in the reduction of hydrogen peroxide into water and prevents the peroxidation of lipids, proteins and DNA molecules. It is a selenium-dependent antioxidant (Steinbrenner and Sies 2013).

10.7.2 Selenium and Iron Deficiency Anemia

It was hypothesized by Allen et al. (2000) that deficiency of iron alone and/or other trace minerals is responsible for anemia in children. Further it was asserted by the authors that poor compliance by the patients, deficiency of multiple micronutrients and improper regimen of the iron supplements are responsible for persistence of anemia in children.

It has been observed by Hamedani et al. (1987) in the animal studies that ideal concentration of glutathione peroxidase in erythrocytes shows its protective effect to erythrocyte against oxidative damage. This suggests that increased oxidative stress could be another contributing factor to the development of anemia due to selenium deficiency.

Selenium is contained in glutathione peroxidase, an enzyme that plays a major role in the protection against free radicals and oxidative stress. A possible biological mechanism by which selenium deficiency could cause anemia is through (Tron et al. 2005) upregulation of haem oxygenase-1, which catalyses the initial step of haem metabolism and reduces haem to biliverdin, carbon monoxide and free divalent iron. Haem oxygenase-1 has anti-inflammatory properties and may cause a chronic inflammatory syndrome. In addition, haem oxygenase-1 releases Fe2+ from the core of the haem molecule, leading to the rapid expression of the iron-sequestering protein ferritin, as well as an ATPase pump that actively removes intracellular iron from the cell (Mostert et al. 2003).

A study by Nhien et al. (2008) was conducted to investigate the interrelationship among micronutrient deficiency and anemia involving children in age group of 1–3 years. Serum concentrations of copper, zinc, selenium and magnesium and Hb concentration were estimated. Serum selenium deficiency was correlated with anemia in children.

A descriptive study was conducted by Yetgin et al. (1992) on children with iron deficiency anemia. Serum selenium concentration was detected by spectrofluorometric method. Low serum selenium level was found in anaemic children.

Gurgoze et al. (2004) evaluated the serum selenium level in iron deficiency anaemic children. Iron deficiency anemia is very common in children. Low serum selenium levels were found by the authors in children with IDA in comparison to control group of children. Rashid and Ghaznavi (2015) found low serum selenium levels in their study on primary school children. This was associated with anemia in children. Selenium deficiency was linked to stunting and anemia in children. It was observed by Gashu et al. (2015).

References

Adelekan DA (2003) Multiple micronutrient deficiencies in developing countries. Nutrition 19:473–474

Allen LH, Rosado JL, Casterline JE, Lopez P, Munoz E, García OP (2000) Lack of hemoglobin response to iron supplementation in anemic Mexican preschoolers with multiple micronutrient deficiencies. Am J Clin Nutr 71(6):1485–1494

Allen L, Benoist B, Dary O, Hurrel R (2006) Guidelines on food fortification with micronutrients. WHO, Geneva

Bales CW, Ritchie CS (2009) Handbook of clinical nutrition and aging. Springer, New York, p 151

Berdanier CD, Dwyer JT, Feldman EB (2007) Handbook of nutrition and food. CRC Press, Boca Raton

Bhan MK, Bhandari N, Bhatnagar S, Bahl R (1996) Epidemiology & management of persistent diarrhoea in children of developing countries. Indian J Med Res 104:103–114

Bloomer JR, Reuter RJ, Morton KO, Wehner JM (1983) Enzymatic formation of zinc-protoporphyrin by rat liver and its potential effect on hepatic heme metabolism. Gastroenterology 85:663–668
Brandt EG, Hellgren M, Brinck T, Bergman T, Edholm O (2009) Molecular dynamics study of zinc binding to cysteines in a peptide mimic of the alcohol dehydrogenase structural zinc site. Phys Chem 11(6):975–983
Bremner I, Young BW, Mills CF (1976) Protective effect of zinc supplementation against copper toxicosis in sheep. Br J Nutr 36:551–561
Bwibo NO, Neumann CG (2003) The need for animal source foods by Kenyan children. J Nutr 133:3936S–3940S
Caiulo A, Nicolis S, Bianchi P, Zuffardi O, Bardoni B, Maraschio P, Ottolenghi S, Camerino G, Giglioni B (1991) Mapping the gene encoding the human erythroid transcriptional factor NFE1-GF1 to Xp11.23. Hum Genet 86(4):388–390
Chang KL, Hung TC, Hsieh BS, Chen YH, Chen TF, Cheng HL (2006) Zinc at pharmacologic concentrations affects cytokine expression and induces apoptosis of human peripheral blood mononuclear cells. Nutrition 22:465–474
Chesters JK (1978) Biochemical functions of zinc in animals. World Rev Nutr Diet 32:135–164
Cole CR, Grant FK, Swaby-Ellis ED, Smith JL, Jacques A, Northrop-Clewes CA, Caldwell KL, Pfeiffer CM, Zeigler TR (2010) Zinc and iron deficiency and their interrelations in low-income African American and Hispanic children in Atlanta. Am J Clin Nutr 91(4):1027–1034
Cordano A (1998) Clinical manifestations of nutritional copper deficiency in infants and children. Am J Clin Nutr 67(5):10125–10165
Crispino JD (2005) GATA1 in normal and malignant hematopoiesis. Semin Cell Dev Biol 16(1):137–147
Danscher G (1984) Autometallography: a new technique for light and electron microscopic visualization of metals in biological tissues (gold, silver, metal sulphides and metal selenides). Histochemistry 81:331–335
Danscher G, Howell G, Perez-Clausell J, Hertel N (1985) The dithizone, Timm's sulphide silver and the selenium methods demonstrate a chelatable pool of zinc in CNS: a proton activation (PIXE) analysis of carbon tetrachloride extracts from rat brains and spinal cords intravitally treated with dithizone. Histochemistry 83:419–422
Duncan GD, Gray LF, Daniel LJ (1953) Effect of zinc on cytochrome oxidase activity. Proc Soc Exp Biol Med 83:625–627
Ece A, Uyanik BS, Iscan A, Ertan P, Yigitoglu MR (1997) Copper and decreased serum Zinc levels in children with Iron deficiency anemia. Biol Trace Elem Res 59(1–3):31–39
Ensminger AH, Konlande JE (1993) Foods and nutrition encyclopedia. CRC Press, Boca Raton, pp 2368–2369
European Food Safety Authority (2009) Scientific opinion of the panel on food additives and nutrient sources added to food on L-selenomethionine as a source of selenium added for nutritional purposes to food supplements, following a request from the European Commission. EFSA J 082:1–38
Food and Nutrition Board (2000) Dietary reference intakes of vitamin A, vitamin K, As, B, Cr, Cu, I, Fe, Zn. National Academy Press, Washington, DC
Forman WB, Sheehan D, Cappelli S, Coffman B (1990) Zinc abuse: an unsuspected cause of sideroblastic anemia. West J Med 152:190–192
Frederickson CJ, Mocrieff DW (1994) Zinc-containing neurons. Biol Signals 3(3):127–139
Frederickson CJ, Suh SW, Silva D, Frederickson CJ, Thomson RB (2000) Importance of zinc in the central nervous system: the zinc-containing neurons. J Nutr 130(5):1471S–1483S
Gashu D, Stoecker BJ, Bougma K, Adish A, Haki GD, Marquis GS (2015) Stunting, selenium deficiency and anemia are associated with poor cognitive performance in preschool children from rural Ethiopia. Nutr J 15:38
Graham GG, Cordano A (1976) Copper deficiency in human subjects. In: Prasad AS, Oberleas D (eds) Trace elements in human health and disease, Zinc and copper, vol I. Academic, New York, pp 363–372

Gurgoze MK, Ayguna AD, Olcucu A, Dogana Y, Yilmaza E (2004) Plasma selenium status in children with iron deficiency anemia. J Trace Elem Med Biol 18(2):193–196
Gurgoze MK, Olcucu A, Aygun AD, Taskin E, Kilic M (2006) Serum and hair levels of zinc, selenium, iron, and copper in children with iron-deficiency anemia. Biol Trace Elem Res 111(1–3):23–29
Halfdanarson TR, Kumar N, Li CY, Phyliky RL, Hogan WJ (2008) Hematological manifestations of copper deficiency: a retrospective review. Eur J Haematol 80(6):523–531
Hambidge KM, Casey CE, Krebs NF (1986) Zinc. In: Mertz W (ed) Trace elements in human and animal nutrition. Academic, New York, pp 1–37
Hamedani P, Hashmi K, Manji M (1987) Iron depletion and anemia: prevalence, consequences, diagnostic and therapeutic implications in a developing Pakistani population. Curr Med Res Opin 10(7):480–485
Hatfield DL, Tsuji PA, Carlson BA, Gladyshev VN (2014) Selenium and selenocysteine: roles in cancer, health, and development. Trends Biochem Sci 39(3):112–120
Haug FM (1967) Electron microscopical localization of the zinc in hippocampal, mossy fiber synapses by a modified sulfide silver procedure. For Hist 8:355–368
Hegazy AA, Zaher MM, Abd El-Hafez M, Morsy AA, Saleh RA (2010) Relation between anemia and blood levels of lead, copper, zinc and iron among children. BMC Res Notes 3: 133
Hodge D, Coghill E, Keys J, Maguire T, Hartmann B, McDowall A, Weiss M, Grimmond S, Perkins A (2006) A global role for EKLF in definitive and primitive erythropoiesis. Blood 107:3359–3370
Hoffmann PR (2007) Mechanisms by which selenium influences immune responses. Arch Immunol Ther Exp 55(5):289–297
Huber KL, Cousins RJ (1993) Zinc metabolism and metallothionein expression in bone marrow during erythropoiesis. Am J Phys 264:E770–E775
Ibs KH, Rink L (2003) Zinc-altered immune function. J Nutr 133(5 Suppl 1):1452S–1456S
Indian Council Of Medical Research (2010) Nutrient requirement and recommended dietary allowances for Indians. Available at: http://www.icmr.nic.in/final/rda-2010.pdf
Institute of Medicine, Food and Nutrition Board (2000) Dietary reference intakes: vitamin c, vitamin e, selenium, and carotenoids. National Academy Press, Washington, DC
Johnson LE (2008) Copper. Merck manual home health handbook. Merck Sharp & Dohme Corp
Klevay LM (2006) Myelodysplasia, myeloneuropathy, and copper deficiency. Mayo Clin Proc 81(1):132–132
Klug A (1999) Zinc finger peptides for the regulation of gene expression. J Mol Biol 293(2):215–218
Ko LJ, Engel JD (1993) DNA binding specificities of the GATA transcription factor family. Mol Cell Biol 13(7):4011–4022
Kodama H, Fujisawa C (2009) Copper metabolism and inherited copper transport disorders: molecular mechanisms, screening, and treatment. Metallomics 1(1):42–52
Krishna SS, Majumdar I, Grishin NV (2003) Structural classification of zinc fingers survey and summary. Nucleic Acids Res 31(2):532–550
Kumar N (2006) Copper deficiency myelopathy (human swayback). Mayo Clin Proc 81(10):1371–1138
Laity JH, Lee BM, Wright PE (2001) Zinc finger proteins: new insights into structural and functional diversity. Curr Opin Struct Biol 11(1):39–46
Levander OA (1986) Selenium. In: Mertz W (ed) Trace elements in human and animal nutrition. Academic, Orlando, pp 209–279
Linder MC, Wooten L, Cerveza P, Cotton S, Shulze R, Lomeli N (1998) Copper transport. Am J Clin Nutr 67(5):965S–971S
Magee AC, Matrone G (1960) Studies on growth, copper metabolism and iron metabolism on rats fed high levels of zinc. J Nutr 72:233–242
Magneson GR, Puvathingal JM, Ray WJ (1987) The concentrations of free Mg2+ and free Zn2+ in equine blood plasma. J Biol Chem 262:11140–11148

Mostert V, Hill KE, Burk RF (2003) Loss of activity of the selenoenzyme thioredoxin reductase causes induction of hepatic heme oxygenase-1. FEBS Lett 541:85–88

Mutanen M (1986) Bio-availability of selenium. Ann Clin Res 18:48–54

National Institutes of Health (2016) Selenium: dietary supplement fact sheet. Available at: https://ods.od.nih.gov/factsheets/Selenium-HealthProfessional/#h2

National Research Council, Food Nutrition Board (1980) Copper. Recommended dietary allowances. NRC/NAS, Washington, DC, pp 151–154

Nhien NV, Khan NC, Ninh NX, Huan PV, Hop LT, Lam NT, Ota F, Yabutani T, Hoa VQ, Motonaka J, Nishikawa T, Nakaya Y (2008) Micronutrient deficiencies and anemia among preschool children in rural Vietnam. Asia Pac J Clin Nutr 17(1):48–55

Nishiyama S, Irisa K, Matsubasa T, Higashi A, Matsuda I (1998) Zinc status relates to hematological deficits in middle-aged women. J Am Coll Nutr 17:291–295

Papp LV, Lu J, Holmgren A, Khanna KK (2007) From selenium to selenoproteins: synthesis, identity, and their role in human health. Antioxid Redox Signal 9(7):775–806

Papp LV, Holmgren A, Khanna KK (2010) Selenium and selenoproteins in health and disease. Antioxid Redox Signal 12(7):793–795

Patterson BH, Zech LA, Swanson CA, Levander OA (1993) Kinetic modelling of selenium in Humans using stable isotope tracers. J Trace Elem Electrolytes Health Dis 7:117–120

Pennington J, Schoen S (1995) Contributions of food groups to estimated intakes of nutritional elements: results from the FDA total diet studies, 1982–1991. Int J Vitam Nutr Res 66(4):342–349

Percival SS, Harris ED (1990) Characterization of the cellular uptake mechanism. Am J Phys Cell Phys 258(1):C140–C146

Perkins A (1993) Erythroid Kruppel like factor: from fishing expedition to gourmet meal. Int J Biochem Cell Biol 31:1175–1192

Pevny LM, Simon C, Robertson E, Klein WH, Tsai SF, Agati VD, Orkin SH, Costantini F (1991) Erythroid differentiation in chimaeric mice blocked by a targeted mutation in the gene for transcription factor GATA-1. Nature 349:257–260

Pevny LC, Lin S, Agati VD, Simon MC, Orkin SH, Costantini F (1995) Development of hematopoietic cells lacking transcription factor GATA-1. Development 121:163–172

Plum LA, Rink L, Hajo H (2010) The essential toxin: impact of zinc on human health. Int J Environ Res Public Health 7(4):1342–1365

Porter H (1966) In: Peisach J, Aisen P, Blumberg WE (eds) In the biochemistry of copper. Academic Press, New York, p 159

Prasad AS (2003) Zinc deficiency: has been known of for 40 years but ignored by global health organization. Br Med J 326(7386):409–410

Prasad AS, Miale A Jr, Farid Z, Sanstead HH, Schulert A (1963) Zinc metabolism in patients with the syndrome of iron deficiency anemia, hepatosplenomegaly, dwarfism, and hypogonadism. J Lab Clin Med 61:537–549

Rashid N, Ghaznavi S (2015) Association of levels of serum selenium with anemia in primary school children. Biomedica 31(2)

Renty BF, Leslie CC (2009) The important role of the apoptotic effects of zinc in the development of cancers. J Cell Biochem 106(5):750–757

Roman M, Jitaru P, Barbante C (2014) Selenium biochemistry and its role for human health. Metallomics 6(1):25–54

Ryu MS, Lichten LA, Liuzzi JP, Cousins RJ (2008) Zinc transporters ZnT1 (Slc30a1), Zip8 (Slc39a8), and Zip10 (Slc39a10) in mouse red blood cells are differentially regulated during erythroid development and by dietary zinc deficiency. J Nutr 138(11):2076–2083

Schubert WK, Lahey ME (1959) Copper and protein depletion complicating hypoferic anemia of infancy. Pediatrics 24:710–733

Steinbrenner H, Sies H (2013) Selenium homeostasis and antioxidant selenoproteins in brain: implications for disorders in the central nervous system. Arch Biochem Biophys 536(2):152–157

Stoltzfus RJ (2003) Iron deficiency: global prevalence and consequences. Food Nutr Bull 24:S99–103

Sunde RA (2006) Selenium. In: Bowman B, Russell R (eds) Present knowledge in nutrition. ILSI, Washington, DC, pp 480–497
Sunde RA (2012) Selenium. In: Ross AC, Caballero B, Cousins RJ, Tucker KL, Ziegler TR (eds) Modern nutrition in health and disease. Lippincott Williams & Wilkins, Philadelphia, pp 225–237
Terry EN, Diamond AM (2012) Selenium. In: Erdman JW, Macdonald IA, Zeisel SH (eds) Present knowledge in nutrition. Wiley-Blackwell, Washington, DC, pp 568–587
Tokuda Y, Yokoyama S, Tsuji M, Sugita T, Tajima T, Mitomi T (1986) Cu deficiency in an infant on prolonged total parenteral nutrition. JPEN J Parenter Enteral Nutr 10:242–244
Trainor CD, Ghirlando R, Simpson MA (2000) GATA zinc finger interactions modulate DNA binding and transactivation. J Biol Chem 275:28157–28166
Tron K, Novosyadlyy R, Dudas J, Samoylenko A, Kietzmann T, Ramadori G (2005) Upregulation of heme oxygenase-1 gene by turpentine oil-induced localized inflammation: involvement of interleukin-6. Lab Investig 85:376–387
Turgut S, Polat A, Inan M, Turget M, Emmungil G, Bican M, Karakus TY, Genec O (2007) Interaction between anemia and blood levels of iron, zinc, copper, cadmium and lead in children. Indian J Pediatr 74(9):827–830
Turgut S, Hacioglu S, Emmungil G, Turgut G, Keskin A (2009) Relations between iron deficiency anemia and serum levels of copper, zinc, cadmium and lead. Pol J Environ Stud 18(2):273–277
USDA National Nutrient Database for Standard Reference (2007) Zinc content of selected foods per common measure. USDA, Washington, DC
Vest KE, Hashemi HF, Cobine PA (2013) Chapter 13 The copper metallome in eukaryotic cells. In: Banci, L (ed) Metallomics and the cell. Metal ions in life sciences. Springer, Dordrecht
Wajeunnesa M, Begum N, Ferdousi S, Akhter S, Quarishi SB (2009) Serum zinc and copper in iron deficient adolescents. J Bangladesh Soc Physiologist 4(2):77–80
Wapnir RA (1990) Protein nutrition and mineral absorption. CRC Press, Boca Raton
Wierzbicka D, Gromadzka G (2014) Ceruloplasmin, hephaestin and zyklopen: the three multicopper oxidases important for human iron metabolism. Postepy Hig Med Dosw 68:912–924
Williams DM (1983) Copper deficiency in humans. Semin Hematol 20:118–128
Woddowso EM (1974) Trace elements in foetal and early postnatal development. Proc Nutr Soc 33:275
World Health Organization (1987) Selenium. WHO, Geneva, Environmental Health Criteria, 58
World Health Organization (2007) The impact of zinc supplementation on childhood mortality and severe morbidity. WHO
Wu FY, Huang WJ, Sinclair RB, Powers L (1992) The structure of the zinc sites of Escherichia coli DNA-dependent RNA polymerase. J Biol Chem 267:25560–25567
Yetgin S, Hincal F, Basaran N, Ciliv G (1992) Serum selenium status in children with iron deficiency anemia. Acta Haematol 88(4):185–188

Chapter 11
Effect of Diarrhea on Anemia in Children

Abstract Diarrhea is an infective and inflammatory condition of gastrointestinal tract characterized by passage of malformed watery stools frequently. Unlikely, the habit of passing normal consistency and shaped stools more than one time a day is not considered as diarrhea. Chronic diarrhea is rampant in developing countries. Preschool children residing in slum areas, villages, and remote areas where the medical facility and awareness are lacking are highly susceptible toward infections of respiratory tract and GIT. It is a common occurrence that they suffer from diarrheal episodes repeatedly in a year, especially in developing countries. Diarrhea is the prime cause of malnutrition in children under 5. Vicious cycle between malnutrition and diarrhea drains and debilitates preschool children. Emaciated children are further predisposed to bacterial and viral infections. Poor sanitization, poor quality of drinking water, defecation in open in villages, leaching and drainage of human excreta in water bodies, poor medical facilities, and inaccessibility and ignorance to hygiene are a few prominent and profound factors for prevalence and incidence of diarrhea.

11.1 Overview

Diarrhea is an infective and inflammatory condition of GIT (Gupta 2014a, b). According to WHO (2014), diarrhea is the second leading cause of mortality in preschool children. Generally, a diarrheal episode terminates within a week (NDDI 2013). Infection spreads by use of contaminated food or drinking water (Jill et al. 2010). Diarrhea is a main cause of childhood mortality in the developing countries. It impairs the immunity of affected preschool children (Bern et al. 1992; UNICEF 2005).

According to Gibbons and Fuchs (2007), chronic diarrhea is rampant in developing countries. Its attack may persist for more than 2 weeks and is prevalent in children in 2–3 year age group. Chronic diarrhea induces inflammation in the mucosa of duodenum and, therefore, impairs the absorptive capability of small intestine.

However, its persistence is accompanied by severe effects on the health of preschool children. It dehydrates and drains minerals from the body. Children become malnourished with compromised immunity, and these children are highly vulnerable to chronic diarrhea (WHO 2014).

A. Gupta, *Nutritional Anemia in Preschool Children*,
DOI 10.1007/978-981-10-5178-4_11

11.2 Descriptive Research Study

A descriptive study was conducted by the author (Gupta 2014) on children between 2 years and below 5 years in Fazilka city in India. The study was focused to deliver prevalence of diarrhea in children and to ascertain the correlation between diarrhea and wasting among children. The author observed overall prevalence of (5.5%) diarrhea in children. The author found that children (89% and 85%) from schools, anganwadi, and slum areas had normal nutrition status as shown in Table 11.1. Further, the author observed that prevalence of wasting was 11% and 15% in children, as in Table 11.1. The author observed prevalence of diarrhea (37%, 30%, and 72%) in children as shown in the table. Further, it showed a higher prevalence (46%) of diarrhea in children who suffered from wasting.

The author observed that 86% of children in the study were normal. Another 46% of children suffered from diarrhea and wasting. Children (13%) had only wasting, as shown in Table 11.2.

Diarrhea physically weakens the body, impairs immune status, and deteriorates the ability of gut mucosa to absorb food. These manifestations repeat periodically; thereby, affected children become malnourished. These children develop predisposition to infections.

A study by Gupta (2014) depicted the prevalence of 46% diarrhea in children who were suffering from wasting below the age of 5 years. In another work by authors (Joshi et al. 2011) in the district of Bahraich, Uttar Pradesh, India, almost similar prevalence (55.6%) of diarrhea was observed in malnourished children. Both observational studies on malnourished children under 5 exemplified the presence of high prevalence of diarrhea in malnourished children, although the prevalence between two studies varies which can be attributed to variation in availability of food, poverty, literacy, socioeconomic status, sample size, and medical facility.

Table 11.1 Prevalence of wasting and diarrhea among children

	Normal children		Wasting in children		Diarrhea in children	
Strata	(n/N)	(n)%	(n/N)	(n)%	(n/N)	(n)%
School children	213/240	89	27/240	11	10/27	37
Anganwadi children	107/127	85	20/127	15	6/20	30
Slum children	62/73	85	11/73	15	8/11	72

Reproduced with permission from Gupta (2014a)

Table 11.2 Demonstrating prevalence of normalcy, wasting, and diarrhea among children

Category	Overall prevalence
Normal children	86%
Children with wasting	13%
Children with diarrhea and wasting	46%

Reproduced with permission from Gupta (2014a)

Findings by UNICEF (2006) that 39% of malnourished children suffered from diarrheal episodes frequently and managed with oral rehydration therapy, additionally, supplement the findings in the present study.

Study by Gupta (2014a) showed highly significant correlation between diarrhea and wasting ($p < 0.001$) among children. Similar high association between diarrhea and wasting in children had been observed by Siddique et al. (1991), EI Samani et al. (1998), and Sachdev et al. (1991).

11.3 Diarrhea and Anemia

Preschool children residing in slum areas, villages, and remote areas where the medical facility and awareness are lacking are highly susceptible toward infections of respiratory tract and GIT. It is a common occurrence that they suffer from diarrheal episodes repeatedly in a year, especially in developing countries (Joshi et al. 2011; Banerjee et al. 2004; Park 2000; NICED 2005). Diarrheal attacks worsen the physical and nutritional status of affected children. These children are predisposed to wasting and underweight, and the chronicity of the disease is responsible for stunting of children. Diarrhea weakens the absorptive capability of small intestine, and the children are deprived of the dietary micronutrients and, hence, endure deficiency of iron and other minerals in their body. Children have impaired immunity and higher tendency to nutritional anemia. Diarrhea is a predisposing factor for malnutrition among preschool children (WHO 2013).

Vicious cycle between malnutrition and diarrhea drains and debilitates preschool children. Emaciated children are further predisposed to bacterial and viral infections. Diarrhea in preschool children is more frequently caused by the infection of *E. coli*, *Salmonella*, Paratyphi, and *Shigella* through contaminated foods and drinking water. Poor environmental factors, namely, open defecation, improper waste disposal, and poor hygiene, are profoundly implicated in the spread of diarrhea especially in poor settings (Jill et al. 2010).

Diarrhea weakens the immunity of children and is the leading reason behind the high morbidity and mortality of preschool children (Bern et al. 1992; UNICEF 2005; WHO 2007; UNICEF and WHO 2010; Woldie and Kbede 2015).

According to Gupta (2014b), poor sanitization, poor quality drinking water, defecation in open in villages, leaching and drainage of human excreta in water bodies, poor medical facilities, and inaccessibility and ignorance to hygiene are a few prominent and profound factors for prevalence and incidence of diarrhea in developing countries. It is a easily controllable but life-threating disease responsible for high disease burden in children under the age of 3 years (Lakshminarayanan and Jayalakshmi 2015; Jayalakshmy et al. 2011).

Authors (Howard et al. 2007) investigated the occurrence of diarrhea, fever, and anemia in children. Cross-sectional study was conducted in rural areas of Indonesia on children in age group from 6 to 59 months. Prevalence of anemia was observed in more than 50% of the population, and these anemic children were suffering from wasting and stunting. Diarrheal episodes were significantly associated with anemia in children. Diarrhea is the cause for micronutrient deficiency anemia in children under 5. Disease impairs the absorption of dietary nutrients, and it exhausts the body store of iron and other micronutrients.

Another study by Haile et al. (2015) found high probability of anemia in children who were suffering from diarrhea as determined by preceding 2 weeks history of diarrhea before the study. Study was conducted on children below 2 years of age in Northeast Ethiopia. Anorexia and malabsorption result into anemia.

11.4 Conclusion

Diarrhea is profoundly correlated with wasting in children. It derails the absorptive ability of small intestinal mucosa. It exaggerates the loss of minerals from body. Diarrhea must be managed promptly at the base level through proper education and motivation of population by doctors, paramedicals, and government sector.

References

Banerjee B, Hazara S, Bandopadhyaya D (2004) Diarrhea management among under fives. Indian Pediatr 41:255–260

Bern C, Martines J, De Zoyas I, Glass RL (1992) The magnitude of the global burden of diarrhoeal disease. PLoS One 8(5):e64713

El Samani EZ, Willet WC, Ware JH (1998) Association of malnutrition and diarrhea in children aged under five years: a prospective follow-up study in a rural Sudanese community. Am J Epidemiol 128:93–105

Gibbons T, Fuchs GJ (2007) Chronic enteropathy: clinical aspects. Nestle Nutr Workshop Ser Pediatr Program 59:89–101

Gupta A (2014a) Study of the prevalence of diarrhea in children under age of five years: its association with wasting. Indian J Sci Res 7(1):1315–1318

Gupta A (2014b) Prevalence of diarrhoea and its association with wasting, age, and gender in children below five years of age. IJMRR 2(4):291–295

Howard CT, de Pee S, Sari M, Bloem MW, Semba RD (2007) Association of diarrhea with anemia among children under age five living in rural areas of Indonesia. J Trop Pediatr 53(4):238–244

Jayalakshmy R, Roy G, Premarajan KC (2011) Incidence and risk factors of acute diarrheal disease among under-five children in urban slums in Pondicherry z – one year follow-up study. Indian J Matern Child Health 13:1–11

Jill WA, Wenjing T, Lofgren J, Forsberg B (2010) Diarrhoeal diseases in low and middle income countries, incidence, prevention and management. Infectious Diseases Journal 4(133):113–124

Joshi PC, Kaishal S, Aribam BS, Khattri P, D' Aoust O, Singh MM, Marx M, Guha-Sapir D (2011) Recurrent floods and prevalence of diarrhea among under five children: observations from Bahraich district, Uttar Pradesh, India. Glob Health Action 4:10

Lakshminarayanan S, Jayalakshmi R (2015) Diarrheal diseases among children in India: current scenario and future perspectives. J Nat Sci Biol Med 6(1):24–28

National Digestive Diseases Information Clearing house (2013) Diarrhoea. Available at http://digestive.niddk.nih.gov/ddiseases/pubs/diarrhea

NICED (2005) Estimation of the burden of diarrhoeal diseases in India. MHFW, New Delhi, Government of India, pp 182–188. Available at: http://www.whoindia.org/LinkFiles/Commision on Macroeconomic and Health Bg P2 Estimation of the burde of diarrheal diseases in India.pdf

Park K (2000) Park's textbook of preventive and social medicine. Banarsidas Bhanot Publishers, Jabalpur

Sachdev HPS, Kumar S, Singh KK, Satyanarayana L, Puri RK (1991) Risk factors for fatal diarrhea in hospitalized children in India. J Pediatr Gastr Nutr 12:76–81

Siddique AK, Baqui AH, Eusof A, Zaman K (1991) 1988 floods in Bangladesh: pattern of illness and causes of death. J Diarrhoeal Dis Res 9:310–314

UNICEF (2005) The state of the world children: childhood under threat. Available at: http:// www.unicef.org/publications/files/SOWC_2005

UNICEF (2006) Progress for children report -a statistical review. Available at: http://www.unicef.org/india/media_3766.htm

UNICEF, WHO (2010) Global, regional, and national causes of child mortality: an updated systematic analysis for 2010 with time trends since 2000. Lancet 380(9850):1308

Woldie H, Kebede Y, Tariku A (2015) Factors associated with anemia among children aged 6-23 months attending growth monitoring at Tsitsika Health Center, Wag-Himra zone, Northeast Ethiopia. J Nutr Metab 2015:9

World Health Organisation (2007) The WHO child growth standards. Available at: http://www.who.int/childgrowth/en/

World Health Organisation (2013) Diarrhoeal diseases. Available at: http://www.who.int/mediacentre/factsheets/fs3 30/en/

World Health Organisation (2014) Available at: http://www.who.int/mediacentre/factsheets/fs330/en/

Chapter 12
Effect of Malnutrition on Nutritional Anemia

Abstract Child nutrition refers to dietary needs of children and encompasses that children must consume quality foods and in the appropriate quantity which are essential for their normal growth and development and are vital for the prevention of malnutrition like stunting, wasting, underweight, and obesity. Poor dietary intake and repeated infections are responsible for stunted growth of children. Irrespective of government policies for the benefit of population in terms of good nutrition and good physical and mental health of children, childhood malnutrition has been the major public health hazard throughout the world. Protein-energy malnutrition is a specific type of malnutrition, wherein inadequate intake of calories and/or protein is responsible for ill effects on the body of children.

12.1 Overview

Child nutrition refers to dietary needs of children and encompasses that children must consume quality foods and in the appropriate quantity which are essential for their normal growth and development and are vital for the prevention of malnutrition like stunting, wasting, underweight, and obesity.

Good nutrition signifies adequate quantity and quality of food consumed to assure physical, mental, and social health of children. According to United Nations Development Program (UNDP 1990), Human Development Index monitors the health, income, poverty, and education of nations over certain period of time. Authors (Alkire 2002; McGillivray 2007) have included empowerment, physical safety, political and civil freedom, and gender equality.

The United Nations World Summit for Children Goals appreciated child growth as a public health indicator which is meant to monitor the nutritional status and health of children in population (UNICEF 1993). Poor dietary intake and repeated infections are responsible for stunted growth of children. These children have higher tendency to diarrheal episodes and secondary infections of upper respiratory tract, malaria, and pneumonia (De Onis and Blossner 2003).

Irrespective of government policies for the benefit of population in terms of good nutrition and good physical and mental health of children, childhood malnutrition has been the major public health hazard throughout the world (Black et al. 2013). Malnutrition, according to World Food Program (2008), is defined as a state in

A. Gupta, *Nutritional Anemia in Preschool Children*,
DOI 10.1007/978-981-10-5178-4_12

which the physical function of an individual is impaired to the point where he or she can no longer maintain natural bodily capacities such as growth, pregnancy, lactation, learning abilities, physical work, as well as resisting and recovering from diseases.

Malnutrition comprises undernutrition and overnutrition. However, it is largely perceived as undernutrition for the sake of field study. Malnutrition and infection are the chief impending causes of child mortality in developing countries especially in poor, illiterate population in remote settings. Malnutrition is responsible for more than 50% of child mortality per year, according to Neumann et al. (2004).

According to World Health Organization, malnutrition refers to the cellular imbalance between the supply of nutrients and energy and the body's demand for them to ensure growth, maintenance, and specific functions.

Protein-energy malnutrition is a specific type of malnutrition, wherein inadequate intake of calories and/or protein is responsible for ill effects on the body of children. Broadly, all the organs are affected in protein-energy malnutrition. Its type may manifest as predominantly protein malnutrition, which is called Kwashiorkor. The term implies sickness of the weaning (Scheinfeld et al. 2016) and is strictly due to inadequacy in protein intake. Edema is the characteristic feature of Kwashiorkor. Another type of PEM is the Marasmus, which is primarily caused by inadequate calories intake by children. This Greek word signifies wasting or withering (Scheinfeld et al. 2016). Marasmus children are acutely emaciated. Still another type of PEM is the intermediate of earlier two forms, diagnosed as Marasmus-Kwashiorkor. Possibly, the type is associated with poor prognosis. It is characterized by inadequate intake of protein as well as calories. PEM is worldwide in its distribution and is observed in children and adults (Franco et al. 1999).

Stunting is the chronic malnutrition, wherein the height for age is low in reference to the standard. It is the sequel of prolonged inadequacy of food consumption which arises due to varied reasons (World Food Program 2008). In contrast to stunting, acute malnutrition refers to inadequate intake of food over a short period of time and is manifested as low weight for height in comparison to standard. It is called wasting and is the after effect of famine, food shortage, floods, and/or acute infection affecting the children (World Food Program 2008). Underweight is still another type of malnutrition which is characterized by low weight for age in reference to standard. It is the consequence of either long-standing stunting and/or wasting or two which together affect the metabolism of body negatively and result into underweight (World Food Program 2008). These children are unusually thin with respect to age. Malnutrition may be the ill effect of insufficient consumption of macronutrients and micronutrients. Vitamins and minerals are essential for proper growth and development of children. Nutritional anemia is the nutritional disorder arising due to insufficient intake of diet rich in iron. According to WHO, anemia refers to hemoglobin concentration of less than fifth percentile in accordance to age (Janus and Moerschel 2010). Preschool children are severely affected due to iron deficiency anemia which curtails the growth and cognitive development (Villalpando et al. 2003). Academic performance, work capacity, and social health of children are hampered (Villalpando et al. 2003). Iron deficiency and deficiency of folate and vitamin B_{12} are the commonest causes of nutritional anemia in under age 5 (Villalpando et al. 2003; Cornet et al. 1998; Fleming and Werblinska 1982).

As per data provided by You et al. (2012), in 2012, about 6.6 million children below the age of 5 years died. High child mortality is observed in developing countries. These deaths are attributed to lack of medical facility for timely management of acute infections and inaccessibility to affordable diagnostic tests and interventions (You et al. 2012). About 3.7 million (Ezzati et al. 2004) children under 5 die due to underweight. Millennium Development Goal emphasizes on the approaches to reduce the burden of underweight in children under 5 years in the world (Ezzati et al. 2004). A report by UNICEF (2012) asserted that in 2000, about 197 million children under 5 were stunted. The figure has declined to about 162 million in 2012.

A report (UNICEF 2012) stated that around 99 million children under 5 were underweight and around 51 million children suffered from wasting. Globally, about 17 million children under 5 were severely wasted as per report (UNICEF 2012). South Asia had a prevalence of 16% wasting in children under 5 (UNICEF 2012). According to UNICEF (2012), the prevalence of wasting has declined by around 12% in comparison to the data of 1999 (58 million). Further, as per UNICEF (2012), the prevalence of underweight in children has declined by 15%. This downward trend of malnutrition in children is slow (UNICEF 2013). Prevalence of stunting has declined by 40% in Asia and Latin America (UNICEF 2012), whereas stunting prevalence is around 10–15% in Africa and Oceania (UNICEF 2012). According to de Benoist et al. (2008), worldwide anemia prevalence in children is estimated to be 47.4% involving around 239 million children under 5. Africa witnessed the highest prevalence of anemia as 67.6%, and Southeast Asia recorded 65.5% prevalence of anemia in children (Benoist et al. 2008).

In a study by Christofides et al. (2005) among Canadian children, anemia was estimated in the range between 3.5 and 10.5%. Authors (Cusick et al. 2008) reported a prevalence of 3.6% of anemia in the United States. In European countries like Sweden and Germany, the prevalence of anemia, according to Benoist et al. (2008), was 8.6% and 7.8%, respectively.

Authors (George et al. 2000) estimated a correlation between anemia and nutritional health among preschool children in Kerala state in India. Authors observed that around 10% of the participants suffered from anemia and another 505 of the participants were malnourished. Authors posited a confirmatory association between malnutrition and anemia in children. Poverty and illiteracy were widespread among the families who participated in the study.

Another study was conducted by Jain et al. (2000) among children in early childhood period who resided in urban slum areas at Meerut in India. Authors described that 25% of the study participants suffered from severe anemia (Hb <7 g/100 ml), whereas 50% of the study participants had Hb concentration in between 7 and 9.9 g/100 ml. The rest of the participants were suffering from mild anemia with Hb level less than 11 g/100 ml.

Malnutrition is a deranged state of nutrition. It is attributed to either inadequacy of nutrients or superfluous nutrients resulting into undernutrition or overnutrition, respectively. It impairs the physical, mental, and intellectual growth and development of children. As per UNICEF (1993, 2012, 2013), malnutrition is the foremost hazard to the public health. Its occurrence is very high in developing countries and is the key factor for rising mortality and morbidity in children under 5.

Low calorie intake owing to poverty and poor access to nutritious foods and/or loss of micronutrients from the body due to a disease manifest as acute malnutrition called wasting (Bloss et al. 2004). Additionally, inadequate food supply either due to natural calamity or policy of state government and long-standing diseases like tuberculosis, diarrhea, or AIDS are the factors for wasting in children (Gupta 2011).

Pallor is a physical sign of anemia which can be easily observed by a paramedical health worker. Anemia due to deficiency of nutritional components is prevalent in children below the age of 5 years and is associated with malnutrition in children. A cohort study was undertaken by Gupta (2015) to investigate the occurrence of anemia and its effect on the nutritional status of under 5-aged children. The author found that malnourished children exhibited high occurrence of pallor in nail beds in comparison to healthy children.

The first 2 years of postnatal life is highly critical in terms of growth and development of children. Normal physiological process during this phase of life requires adequate supply of dietary nutrients. However, it is deplorable that in villages, rural areas, and slum inhabitants where the poverty is widespread, there is inadequacy in quality and quantity of diet that is provided to the growing children. Further, the lack of knowledge toward the value and beneficial effects of breastfeeding thwarts the normal growth and development of children.

Retarded children have weak immunity, and as a consequence, these children are highly inclined to infections (De Onis and Blossner 2003). A study was undertaken an author in Egyptian rural inhabitation among the children between age groups 6 months to 15 years. Authors found that more than 50% of the children were suffering from diarrhea. Complete weaning below 6 months of age was observed in around one fifth of the children. High prevalence of anemia was observed in children who had frequent diarrheal episodes and were under the age of 2 years. Adequate iron store of body is necessary during preschool age group. Contrarily, anemia in this age group is highly detrimental to the physical and mental growth and development of children (Villalpando et al. 2003).

12.2 Cross-Sectional Research Study

Anemia is described as a decline in concentration of hemoglobin and/or a decrease in count of erythrocytes below a reference value. According to authors (Walter et al. 1989), improper weaning and a supplementary diet which is poor in quality and quantity, definitely, lead to anemia in children.

Table 12.1 Demonstrating pallor prevalence among children

Age	Pallor (n %)	Normal (n %)
2–3 years	16.5% (28/169)	83.4% (141/169)
>3 years–<5 years	32% (87/271)	68% (184/271)

Reproduced with permission from Gupta (2015)

Pallor is examined in conjunctiva, nail beds, crease of palm, and labial mucosa. Additionally, authors (Nardone et al. 1990) posited that pallor of conjunctiva has high diagnostic value to anemia in comparison to the pallor of nail beds or palm. According to authors (Strobach et al. 1988), paleness of conjunctiva is an easily discernable clinical sign that depicts moderate to severe anemia.

A report by WHO and UNICEF (1995) has asserted that sign of pallor is a simple method to detect anemia. Although it is not a perfect technique, however, it is a simple, easy, and noninvasive clinical method to detect anemia. Anemic children have reduced appetite. It deprives essential minerals and enough calories for their growth and development. Children have compromised immunity and become vulnerable to infections. The vicious cycle promulgates and terminates in severe anemia.

The author (Gupta 2015) conducted a cross-sectional study on children between 2 years and below 5 years age group in the city of Fazilka in India to detect effect of pallor on nutritional level of children. The presence of pallor over nail beds is an important clinical sign of anemia. The author selected a total of 440 children. The author posited overall prevalence (26%) of pallor among children.

The author (Gupta 2015) observed that the prevalence of pallor was higher in age group (>3–<5 years) in comparison to age group (2–3 years), as shown in Table 12.1. According to the author (Christoffel 1981), children, after 3 years of age, develop deciduous dentition. They have the tendency to consume junk foods instead of nutritious food items. These children are deprived of necessary minerals and energy for their growth as the junk food is devoid of minerals, primarily iron.

The author (Gupta 2015) observed odds ratio (OR = 43) between group of children with pallor and wasting and another group of children with wasting without pallor. The study demonstrated a much higher odds of developing wasting in children who suffered from pallor. Further analysis in the study by Gupta (2015) showed odds ratio (OR = 9.8) between children with pallor and underweight and children without pallor and underweight.

The findings in the study by Gupta (2015) were supported by the work of Singh (2014), which observed that anemia was higher in children who were underweight.

Correlation between pallor and malnutrition was authenticated by the work of Bhoite and Iyer (2011) which depicted that the proportion of anemic children increased as the severity of underweight increased. In their study, 25% prevalence of anemia was noticed in mild underweight children, which increased to a prevalence 42% of anemia in severely underweight category.

Anemia has a multifactorial etiology. Iron deficiency anemia is more prevalent in preschool children. Anemic children become susceptible to infections. According to Kotecha (2011), immunity of anemic children weakens, and their appetite declines. This vicious cycle perpetuates to malnutrition.

12.3 Anemia, Diarrhea, and Malnutrition

Basic knowledge about the factors that endanger the physical and mental growth of children helps to carve the remedial and interceptive programs. Malnutrition during preschool age deters the growth of children accompanied by poor cognitive behavior and predisposition to infections like diarrhea and respiratory tract infection. Physical, mental, intellectual, and economical growth of children is delayed. Micronutrient deficiency is the main cause of anemia in children, and its prevalence is higher in malnourished children. Low dietary iron, decreased absorption of dietary iron, and malabsorption syndrome are the causes of iron deficiency anemia especially in the poor section of society. Further hygiene, sanitation, water quality, dietary beliefs, and customs have high impact on the health of children in rural inhabitation.

12.4 Conclusion

Anemia and underweight in children are correlated. These are major health problems involving children below the age of 5 years.

References

Alkire S (2002) Dimensions of human development. World Dev 30(2):181–205

Bhoite R, Iyer M (2011) Magnitude of malnutrition and iron deficiency anemia among rural school children: an appraisal. Asian J Exp Biol Sci 2(2):354–361

Black R E, Victora C G, Walker S P, Bhutta Z A, Christian P, de Onis M, Ezzati M (2013) Maternal and child nutrition 1 maternal and child undernutrition and overweight in low-income and middle-income countries. Lancet 382(9890):427–451

Bloss E, Wainaina F, Bailey RC (2004) Prevalence and predictors of underweight, stunting, and wasting among children aged 5 and under in western Kenya. J Trop Pediatr 50(5):260–270

Christoffel KA (1981) Pediatric perspective on vegetarian nutrition. Clin Pediatr 20:632–643

Christofides A, Schauer C, Zlotkin SH (2005) Iron deficiency anemia among children: Addressing a global public health problem within a Canadian context. Paediatr Child Health 10(10):597–601

Cornet M, Le Hesran J-Y, Fievet N, Cot M, Personne P, Gounoue R, Beyeme M, Deloron P (1998) Prevalence of and risk factors for anemia in young children in southern cameroon. Am J Trop Med Hyg 58(5):606–611

Cusick SE, Mei Z, Freedman DS, Looker AC, Ogden CL, Gunter E, Cogswell ME (2008) Unexplained decline in the prevalence of anemia among us children and women between 1988–1994 and 1999–2002. Am J Clin Nutr 88(6):1611–1617

de Benoist B, Mclean E, Egil I, Cogswell M (eds) (2008) Worldwide prevalence of anaemia 1993–2005: WHO Global Database on Anaemia. WHO Press, Geneva. Available at: http://apps.who.int/iris/bitstream/10665/43894/1/9789241596657_eng.pdf

De Onis M, Blossner M (2003) The world health organization global database on child growth and malnutrition: methodology and applications. Int J Epidemiol 32(4):518–526

De Onis M, Monteiro C, Clugston G (1993) The worldwide magnitude of protein-energy malnutrition: an overview from the WHO Global Database on Child Growth. Bull World Health Organ 71(6):703–712

Ezzati M, Lopez AD, Rodgers A, Murray CJ (2004) Comparative quantification of health risks. Global and regional burden of disease attributable to selected major risk factors. World Health Organization, Geneva

Fleming A, Werblinska B (1982) Anemia in childhood in the guinea savanna of nigeria. Ann Trop Paediatr 2(4):161–173

Franco V, Hotta JK, Jorge SM, Dos Santos JE (1999) Plasma fatty acids in children with grade III protein–energy malnutrition in its different clinical forms: Marasmus, marasmic kwashiorkor, and kwashiorkor. J Trop Pediatr 45(2):71–75

George KA, Kumar NS, Lal JJ, Sreedevi R (2000) Anemia and nutritional status of pre-school children in Kerala. Indian J Pediatr 67(8):575–578

Gupta A (2011) Prevalence of wasting in children under the age of five years in Fazilka in Punjab. J Pure Appl Sci Technol 4(2):31–36

Gupta A (2015) To analyse effects of pallor (anemia) on nutritional status of children under age of five years. Golden Res Thoughts 4(7):1–9

Jain S, Chopra H, Bhatnagar M, Singh JV (2000) Anemia in children: early iron supplementation. Indian J Pediatr 67(1):19–21

Janus J, Moerschel SK (2010) Evaluation of anemia in children. Am Fam Physician 81(12):1462

Kotecha PV (2011) Nutritional anemia in young children with focus on Asia and India. Indian J Community Med 36(1):8–16

McGillivray M (2007) Human well-being: concept and measurement. Palgrave Macmillan, Basingstoke

Nardone DA, Roth KM, Mazur DJ, McAfee JM (1990) Usefulness of physical examination in detecting the presence or absence of anemia. Arch Intern Med 150:201–204

Neumann CG, Gewa C, Bwibo NO (2004) Child nutrition in developing countries. Pediatr Ann 33(10):658

Scheinfeld NS, Lin A, Mokashi A (2016) Protein-Energy Malnutrition. Available at: http://emedicine.medscape.com/article/1104623-overview#showall

Singh SPC (2014) Prevalence of nutritional anemia in primary school children in urban slum areas of Hyderabad, Andhra Pradesh, India. BBB 2(1):147–154

Strobach RS, Anderson SK, Doll DC, Ringenberg S (1988) The value of the physical examination in the diagnosis of anemia. Arch Intern Med 148:831–832

UNICEF (1993) Child malnutrition: progress toward the world summit for children goal. UNICEF, New York

UNICEF (2012) Levels and trends in child malnutrition. UNICEF-WHO-The World Bank Joint Child Malnutrition Estimates, New York

UNICEF (2013) Improving child nutrition: the achievable imperative for global progress. UNICEF, New York, p 18

United Nations Development Programme (1990) Human development report 1990. Oxford University Press, New York/Oxford. Available at: http://hdr.undp.org/sites/default/files/reports/219/hdr_190_en_complete_nostats.pdf

Villalpando S, Shamah-Levy T, Ramırez-Silva CI, Mejıa-Rodrıguez F, Rivera JA (2003) Prevalence of anemia in children 1 to 12 years of age: results from a nationwide probabilistic survey in mexico. Salud Publica Mex 45(Suppl 4):490–498
Walter T, de Andraca I, Chadud P, Perales CG (1989) Iron deficiency anemia: adverse effects on infant psychomotor development. Pediatrics 84(1):7–17
World Food Program (2008) World Food Program, Hunger Glossary. Available at: http://www.wfp.org/hunger/glossary
World Health organization (2008) Worldwide prevalence of anemia 1993–2005: WHO global database on anemia. In: Benoist B, McLean E, Egll I, Cogswell M (eds). WHO
World Health Organization, UNICEF (1995) Management of Childhood Illness. WHO, Geneva
You D, New J, Wardlaw T (2012) Levels and trends in child mortality report 2013. estimates developed by the un inter-agency group for child mortality estimation

Chapter 13
Infants' Consumption of Cow Milk and Anemia

Abstract Milk proteins are nutritionally significant for the children under 5 years of age. Proteins are the necessary building blocks of body tissues and provide amino acids essential for the growth and development of body of preschool children and their higher cognitive functions. Casein is important milk protein which is phosphoprotein in nature. It is precipitated in stomach and is the cause of delayed absorption of casein protein in children and infants. Conversely, whey protein's digestibility and absorption are comparatively rapid from mucosa of GIT. Human breast milk is an important source of taurine and cystine amino acids. It is advisory for infants that who have allergy to cow milk should be provided human breast milk. It has been proved that feeding of infants with cow milk early in life might provoke immunological response in a few infants which in turn become a breeding ground for pathogenesis of iron deficiency and zinc depletion in infant's body. Otherwise, exclusive breast-feeding of infants during initial 6 months and relative restriction from the start of cow milk are worthwhile practices to prevent cow milk protein allergy and associated clinical manifestations.

13.1 Overview of Cow Milk and Human Breast Milk

Milk proteins have enormous nutritional value for infants and children. These are essential components of food for humans and furnish essential amino acids for growth and development of preschool children. Milk protein has higher proportion of amino acids in comparison to other proteins. These are easily digested and rapidly absorbed by infants (Hambraeus 1982, 1992).

Casein is principal protein in milk. Other milk proteins are called whey proteins. Important whey proteins of milk are α-lactalbumin and β-lactoglobulin. In cow milk, casein protein represents around 80% of the total proteins in milk, whereas whey proteins outweigh in human milk, and according to Kunz and Lonnerdal (1990), these constitute around 60–70% of total milk proteins. Caseins and whey proteins are the two main groups of milk protein. Caseins constitute (Sindayikengera and Xia 2006) about three fourth of total protein level in cow milk. At isoelectric pH of 4.6, casein precipitates, while the whey proteins stay soluble in milk. Casein protein is heat stable (Sindayikengera and Xia 2006).

A. Gupta, *Nutritional Anemia in Preschool Children*,
DOI 10.1007/978-981-10-5178-4_13

According to the author (Pipes 1997), proportion of whey protein is almost half the proportion of casein protein in human milk, whereas proportion of whey protein is almost one fourth of the proportion of casein protein in cow milk. Therefore, it can be concluded that concentration of whey protein in human breast milk is twice its concentration in cow milk.

Whey proteins constitute around one fifth of total milk proteins. Whey proteins are also called milk serum proteins. It exists in soluble state in the stomach even at low pH value (Walstra and Jenness 1984) and/or its interaction with enzyme rennin in the stomach (Barth and Behnke 1997). Nature of former whey proteins is called acid whey, and the latter group of whey proteins is called rennet whey (Morr 1989). According to Mulvihill (1992), whey proteins have globular shape and remain soluble in milk over a wide range of pH values.

Casein undergoes curding in the stomach of infants to form calcium paracaseinate. Its proteolysis in infants is difficult. It forms a thick layer in the stomach which is not easily digestible (Anderson et al. 1985). According to Mahe et al. (1966), delayed digestion of precipitated casein, in turn, is responsible for delayed absorption of amino acids by intestinal mucosa (Boirie et al. 1997; Fruhbeck 1998).

According to authors (Mahe et al. 1966), digestibility of whey proteins is higher than the casein protein. Digestion of whey proteins in GIT is rapid in comparison to casein. Since casein concentration in cow milk is higher than human milk, it is converted into a large amount of insoluble coagulum in lumen of small intestine. Its degradation by enzyme, pepsin takes comparatively longer time. Overall, absorption of cow milk is delayed (Anderson et al. 1985; Boirie et al. 1997; Fruhbeck 1998; Walstra and Jenness 1984; Barth and Behnke 1997).

Considering the nutritional value, whey proteins have higher nutritional value in comparison to casein protein. Whey proteins have better amino acid profile than casein (Sindayikengera and Xia 2006). Quality of protein, digestibility and absorption are determined by the amino acid composition of proteins. Moreover, absorption decides the bioavailability of amino acids in the body (Sindayikengera and Xia 2006). According to authors (Boirie et al. 1997), absorptive capability of small intestine is dependent on the nature of protein ingested. Absorption of amino acids affects the biosynthesis and biodegradation of proteins in postprandial sate. Authors (Boirie et al. 1997) conducted a study by the use of whey protein and casein protein and labelled them with leucine. These proteins were given to the adult participants under investigation. Authors observed that protein synthesis in postprandial state was 68% in case of whey protein, whereas it was 31% in casein protein.

Taurine is sulphur-containing nonessential amino acid in human's body. It is present in adequate amount in breast milk and almost negligible amount in cow milk (Atkinson and Lonnerdal 1989). Renal tubules are not fully developed in infants. Therefore, reabsorption of taurine by renal tubules is inadequate. It is advisable to provide preterm infants a diet rich in taurine. Adequate serum taurine level is necessary for biosynthesis of bile salts in the liver (Chesney et al. 1998a, b; Whatkin et al. 1983).

Inadequate concentration of taurine in plasma might be a reason for decreased secretion of bile salts by the liver. Bile salts are immensely helpful in the emulsification and absorption of dietary lipids in the small intestine (Howard and Thompson 1992; Spencer et al. 2005).

Lipids are important source of calories and fulfil almost half of the total energy needs per day. Also, lipids are the essential constituent of cow milk, whereas human breast milk has linoleic acid and other polyunsaturated fatty acids in abundance in comparison to cow milk. Linoleic acid is necessary for daily calorie needs of infants (AAP 1992).

Iron quantity in human milk and cow milk is low and is in the range between 0.3 and 1 mg/L, as per authors (Leung and Chan 2001). However, availability of iron in breast milk is comparatively higher than cow milk, and iron is easily absorbed from human breast milk. High phosphate content in cow milk is responsible for hypocalcaemic tetany in newborn infants. So it is reasonable to start cow milk after infant attains 1 year of age (Pipes 1997).

Infants who are provided cow milk early in life are vulnerable to high renal solute load. These infants are given proteins and minerals in amount which is higher than the physiological demand of infants. Excess of mineral are excreted in urine. So the urine of infants who are fed on whole cow milk contains higher amount of solutes in comparison to infants who are fed on breast milk or formulas (Ziegler 1990). High renal solute load raises the urinary osmolality which is almost double than is observed in infants on breast-feeding (Fuchs et al. 1992).

However, evidence is lacking to prove harmful effect of increased renal solute load in infants. So it is safer to feed the infants with human milk in comparison to cow milk. Further with the early and constant use of cow milk, infants are under threat from dehydration due to raised urinary osmolarity. This condition aggravates in case of vomiting and diarrhoea (Ziegler 1990).

13.2 Intake of Cow Milk and Anemia

Infants who have allergy to proteins in cow milk should not be provided cow milk. These preparations contain intact cow milk proteins (AAP 1992). Infants who are exposed to cow milk develop allergy to milk proteins. According to studies by Jakobsson et al. (1985) and Panganelli et al. (1986), antigens of cow milk have been found in human milk.

Trace amount of cow milk protein may provoke allergic response in susceptible infants. Atopic dermatitis is a common manifestation due to cow milk protein allergy. Authors investigated this hypothesis. It was found that prevalence of atopic dermatitis was lower in infants whose mothers consumed diet free from cow milk protein (Paganelli et al. 1986). Contrarily, authors (Host et al. 1988; Hattevig et al. 1989; Lija et al. 1989; Sigurs et al. 1992) asserted that dietary restriction from cow milk protein yielded no significant effect as related to atopic dermatitis. However, it is uncertain to the use of cow milk in infancy with history of atopic dermatitis.

Infants feeding on cow milk early in life are predisposed to iron deficiency and depletion of zinc in infant's body (Ziegler 2007). Infants have high demand for iron due to rapid growth. Infants who are fed on cow milk have natural tendency to iron deficiency owing to low iron amount in cow milk (Ziegler 2011). Hence, rapid growth, raised demand for iron and concomitant administration of cow milk would, definitely, lead to iron deficiency anemia in infants. It is true in age group of 6–24 months (Hermoso et al. 2011). These infants are vulnerable to impeded growth and development (Szajewska et al. 2010).

Several authors have proved that children in age group between 6 months and 24 months have higher predisposition towards anemia (Osorio et al. 2001; WHO 1998). Premature weaning, cow milk and early feeding on cereals are a few factors for anemia in infants (Sadowitz and Oski 1983; Tunnessen and Oski 1987; Pizarro et al. 1991; Kazal 2002).

Human breast milk constitutes significant nutrition to infants and children up to 24 months of life. Exclusive breast-feeding (WHO 1998) during 6 months of life can adequately fulfil iron demands of term infants. Afterwards, breast-feeding is supplemented (WHO 1998) with foods that are rich in micronutrients. It has been proved that consumption of cow milk, consistently, presents as a risk factor for anemia in children (Hadler et al. 2004; Male et al. 2001).

It was assessed by authors (Hadler et al. 2004) among children in age group between 6 months and 12 months that feeding on cow milk resulted into anemia in the city of Goiânia. Consumption of milk fortified with iron helps to reduce the possibility of iron deficiency anemia among infants and children.

Authors (Male et al. 2001) demonstrated the premature start of cow milk as a reason for anemia in children. Cow milk is also responsible for its interference in absorption of dietary iron lumen of the small intestine (Woodruff et al. 1972; Hurrel et al. 1989; Hallberg et al. 1992a, b).

Whole cow milk is low in iron content. Its use in early infancy predisposes to iron deficiency anemia (Dallman et al. 1980). Bioavailability of iron in human breast milk is very high. It is due to the presence of lactoferrin in milk, and it increases the transport of iron from milk to enterocytes. Also, low calcium and phosphorus in milk enhance iron bioavailability. Contrarily, milk proteins in cow milk act as strong inhibitors of iron absorption. Therefore, iron absorption from human breast milk is high (50%) in comparison to very low (10%) of iron absorption from whole cow milk (Oski 1993).

In infants, start of cow milk during 12 months of life is a potential risk factor for iron deficiency and iron deficiency anemia (Sadowitz and Oski 1983; Tunnessen and Oski 1987; Pizarro et al. 1991). According to a study by Penrod et al. (1990), it has been proved that cow milk as a food to infants resulted in higher incidence of iron deficiency by the age of 1 year. Comparatively, infant formula-feeding at 6 months of age had lesser chances of iron deficiency in 1 year of age (Dallman et al. 1980).

It is advisable to keep away from cow milk in early 12 months of life to prevent iron deficiency anemia (Louis and Kazal 2002). Further, infant diets as breast milk, cow milk, vegetables and fruits are all low in iron. It is advisable to fortify staple foods that are used for infant feeding. Cow milk is implicated for iron deficiency anemia. It might be due to different mechanisms which might act and operate synergistically. Low content of iron in milk of cow, high content of mineral calcium and phosphoprotein, casein in cow milk which hinder absorption of iron by infants and low level of ascorbic acid in cow milk which favours iron absorption collectively operate to lower the absorption of iron and reduce the bioavailability of iron in the body of infants who are fed on cow milk (Ziegler 2007, 2011). It has been advocated that bioavailability of iron in cow milk is low. According to Ziegler (2007, 2011), iron level in human milk is also low; however, its bioavailability in breast milk is excellent. It is posited by authors that bioavailability of iron in human versus cow milk is 50% versus 10%. Additionally, intestinal blood loss has been observed in 40% of infants who feed on cow milk. The author (Ziegler 2007) asserted that casein in comparison to whey protein is not easily absorbed by the small intestinal mucosa in infants. It might result into haemorrhage in mucosa. This condition, additionally, is responsible for depletion of iron store in infants. Iron store during first 2 years of age is related to iron status at 6 years of age (Gunnarsson et al. 2005).

It can be well inferred that the period of higher probability of iron deficiency during childhood is the same as the period of active growth development and brain development and it is the age between 6 months and 2 years of age. Therefore, iron deficiency in this period is critical to the normal growth and development of children (Hermoso et al. 2011).

A new study was conducted related to the fortification of cow milk. Authors (Sazawal et al.2010; Rolland-Cachera et al. 1999) conducted a double-blinded, randomized clinical trial over children in age between 1 year and 4 years to assess the effect of fortified cow milk. The trial continued for 1 year. It was observed by authors that children who received iron- and zinc-fortified cow milk had around 88% of lower probability to develop iron deficiency.

It is hypothecated that early start of whole cow milk in infancy might be the reason for occult blood loss from the GIT. Authors (Ziegler et al. 1990) conducted a case–control study to find out the effect of whole cow milk on the health of small intestinal mucosa. Authors randomly allocated whole cow milk and infant formulas to infants at 6 months of age. It was observed by Ziegler et al. (1990) that occult blood loss was higher in infants who were fed on whole cow milk in comparison to infants who were fed on formula. Authors (Ziegler et al. 1990) concluded that although infants lose significant amount of blood in stools, however, consumption of whole cow milk aggravates the occult blood loss.

Bovine serum albumin is present in cow milk. It is responsible for immunogenicity in genetically predisposed children and enteric blood loss. It is a heat labile protein. Authors (Wilson et al. 1974) proved that by the use of heat-processed cow milk formula, enteric blood loss is limited. Heat processing of milk denatures the bovine albumin. Whole cow milk is tolerated by the infants later in life. Age predilection is common in the use of whole cow milk and appearance of occult blood. This feature of whole cow milk when started early in infancy reduces the iron store of the body and becomes another cause for iron deficiency anemia.

Several studies confirmed that addition of whole cow milk to the diet of infants who were previously fed on breast milk would lead to occult blood loss. It has been estimated that occult bleeding constitutes around 3 ml of blood loss per day and 0.9 mg of iron loss per day (De Angelis and Ctenas 1993). This drawback of whole cow milk appears early in infancy and definitely disappear during later period of infancy (Ziegler et al. 1990, 1999; Fomon et al. 1990; Jiang et al. 2000; Oliveira and Osoro 2005).

The amino acids taurine and cystine are present in much higher concentrations in human milk than in whole cow milk. These amino acids may be essential for premature infants (Sturman and Chesney 1995; Rassin et al. 1978).

References

American Academy of Pediatrics (1992) The use of whole cow's milk in infancy. Pediatrics 89:1105–1109

Anderson GH, Morson-Pasut LA, Bryan H, Clegorn G, Tanaka P, Yeung D, Zimmerman B (1985) Age of introduction of cow's milk to infants. J Pediatr Gastroenterol Nutr 4:692–698

Atkinson S, Lonnerdal B (1989) Protein and non protein nitrogen in human milk. CRC Press, Boca Raton, Florida, pp 139–140

Barth CA, Behnke U (1997) Nutritional significance of whey and whey components. Nahrung 41:2–12

Boirie Y, Dangin M, Gachon P, Vasson MP, Maubois JL (1997) Slow and fast dietary proteins differently modulate post prandial protein accretion. Proc Natl Acad Sci 94(26):14930–14935

Chesney RW, Helms RA, Christensen M, Budreau AM, Han X, Sturman JA (1998a) An updated view of the value of taurine in infant nutrition. Adv Pediatr 40:179–200

Chesney RW, Helms RA, Christensen M, Budreau AM, Han X, Sturman JA (1998b) The role of taurine in infant nutrition. Adv Exp Med Biol 442:463–476

Dallman PR, Siimes MA, Stekel A (1980) Iron deficiency in infancy and childhood. Am J Clin Nutr 33:86–118

De Angelis RC, Ctenas ML. Biodisponibilidade de ferro na alimentação infantil. Temas de Pediatria, Nestlé, 52; 1993

Fomon SJ, Sanders KD, Ziegler EE (1990) Formulas for older infants. J Pediatr 116:690–696

Fruhbeck G (1998) Slow and fast dietary protein. Nature 391(6670):843–945

Fuchs GJ, Gastanaduy AS, Suskind RM (1992) Comparative metabolic study of older infants fed infant formula, transition formula, or whole cow's milk. Nutr Res 12:1467–1478

Gunnarsson BS, Thorsdottir I, Palsson G (2005) Iron status in 6-yold children: association with growth and earlier iron status. Eur J Clin Nutr 59:761–767

Hadler MC, Colugnati FA, Sigulem DM (2004) Risks of anemia in infants according to dietary iron density and weight gain rate. Prev Med 39:713–721

Hallberg L, Rossander-Hulthén L, Brune M, Gleerup A (1992a) Bioavailability in man of iron in human milk and cow's milk in relation to their calcium contents. Pediatr Res 31:524

Hallberg L, Rossander-Hultén L, Brune M, Gleerup A (1992b) Calcium and iron absorption: mechanism of action and nutritional importance. Eur J Clin Nutr 46:317–327

Hambraeus L (1982) Nutritional aspects of milk proteins. In: Fox PF (ed) Developments in dairy chemistry–1: proteins. Applied Science Publishers, London, pp 289–313

Hambraeus L (1992) Nutritional aspects of milk proteins. In: Fox PF (ed) Advanced dairy chemistry–1: proteins. Elsevier Applied Science, London, pp 457–490

Hattevig G, Kjellman B, Sigurs N, Bjorksten B, Kjellman N (1989) Effect of maternal avoidance of eggs, cow's milk and fish during lactation upon allergic manifestations in infants. Clin Exp Allergy 19(1):27–32

Hermoso M, Vucic V, Vollhardt C, Arsic A, Roman-Vinas B, Iglesia-Altaba I, Gurinovic M, Koletzko B (2011) The effect of iron on cognitive development and function in infants, children and adolescents: a systematic review. Ann Nutr Metab 59:154–165

Host A, Husby S, Osterballe O (1988) A prospective study of cow's milk allergy in exclusively breast-fed infants. Acta Paediatr Scand 77:663–670

Howard D, Thompson DF (1992) Taurine: an essential amino acid to prevent cholestasis in neonates. Ann Pharmacother 26:1390–1392

Hurrel RF, Lynch SR, Trinidad TP, Dassenko SA, Cook JD (1989) Iron absorption in humans as influenced by bovine milk proteins. Am J Clin Nutr 49:546–552

Jakobsson I, Lindberg T, Benediktsson B, Hansson BG (1985) Dietary bovine beta-lactoglobulin is transferred to human milk. Acta Paediatr Scand 74(3):342–345

Jiang T, Jeter JM, Nelson SE, Ziegler EE (2000) Intestinal blood loss during cow milk feeding in older infants. Arch Pediatr Adolesc Med 154:673–678

Kazal OA Jr (2002) Prevention of iron deficiency in infants and toddlers. Am Fam Physician 66(7):1217–1225

Kunz C, Lonnerdal B (1990) Human milk protein: analysis of casein and casein subunits by anion exchange chromatography, gel electrophoresis and specific staining methods. Am J Clin Nutr 51(1):37–46

Leung AK, Chan KW (2001) Iron deficiency anemia. Adv Pediatr Infect Dis 48:385–408

Lija G, Dannaeus A, Foucard T, Graff-Lonnevig V, Johansson SG, Oman H (1989) Effects of maternal diet during late pregnancy and lactation on the development of atopic diseases in infants up to 18 months of age – in-vivo results. Clin Exp Allergy 19(4):473–479

Louis A, Kazal JR (2002) Prevention of iron deficiency in infants and toddlers. Am Fam Physician 66(7):1217–1225

Mahe S, Roos N, Benamouzig R, Davin L, Luengo C, Gabnon L, Gausserges N, Rautureaux JL, Tome D (1966) Gastrojejunal kinetics and the digestion of [15N]beta-lactoglobulin and casein in humans: the influence of the nature and quantity of the protein. Am J Clin Nutr 63:546–552

Male C, Persson LA, Freeman V, Guerra A, van't Hof MA, Haschke F (2001) Prevalence of iron deficiency in 12-mo-old infants from 11 European areas and influence of dietary factors on iron status (Euro-Growth study). Acta Paediatr 90:492–498

Morr CV (1989) Whey proteins: manufacture. In: Fox PF (ed) Developments in dairy chemistry–4: proteins. Elsevier Science Publishers, London, pp 245–248

Mulvihill DM (1992) Production, functional properties and utilization of milk protein products. In: Fox PF (ed) Advanced dairy chemistry–1: proteins. Elsevier Science Publishers, London, pp 369–404

Oliveira MA, Osorio MM (2005) Cow's milk consumption and iron deficiency anemia in children. J Pediatr 81(5):361–367

Oski FA (1993) Iron deficiency in infancy and childhood. N Engl J Med 329:190–193

Osorio MM, Lira PI, Ashworth A (2001) Prevalence of anemia in children 6–59 months old in the state of Pernambuco, Brazil. Pan Am J Public Health 10:101–107

Paganelli R, Cavangni G, Pallone F (1986) The role of antigenic absorption and circulating immune complexes in food allergy. Ann Allergy 57:330–336

Penrod JC, Anderson K, Acosta PB (1990) Impact of iron status of introducing cow's milk in the second six months of life. J Pediatr Gastroenterol Nutr 10:462–467

Pipes PL (1997) Infant feeding and nutrition. In: Trahms CM, Pipes PL (eds) Nutrition in infancy and childhood. McGraw Hill, New York, pp 98–129

Pizarro F, Yip R, Dallman PR, Olivares M, Her-trampf E, Walter T (1991) Iron status with different infant feeding regimens: relevance to screening and prevention of iron deficiency. J Pediatr 118:687–692

Rassin DK, Sturman JA, Guall GE (1978) Taurine and other free amino acids in milk of man and other mammals. Early Hum Dev 2:1–13

Rolland-Cachera MF, Deheeger M, Bellisle F (1999) Increasing prevalence of obesity among 18-year-old males in Sweden: evidence for early determinants. Acta Paediatr 88:365–367

Sadowitz PD, Oski FA (1983) Iron status and infant feeding practices in an urban ambulatory center. Pediatrics 72:33–36

Sazawal S, Dhingra U, Dhingra P (2010) Micronutrient fortified milk improves iron status, anemia and growth among children 1–4 years: a double masked, randomized, controlled trial. PLoS One 5:e12167

Sigurs N, Hattevig G, Kjellman B (1992) Maternal avoidance of eggs, cow's milk and fish during lactation: effect on allergic manifestations, skin-prick tests, and specific IgE antibodies in children at age 4 years. Pediatrics 89:735–739

Sindayikengera S, Xia WS (2006) Nutritional evaluation of caseins and whey proteins and their hydrolysates from Protamex. J Zhejiang Univ Sci B 7(2):90–98

Spencer AU, Yu S, Tracy TF (2005) Parenteral nutrition-associated cholestasis in neonates: multivariate analysis of the potential protective effect of taurine. J Parenter Enter Nutr 29:337–343

Sturman JA, Chesney RW (1995) Taurine in pediatric nutrition. Pediatr Clin N Am 42:879–897

Szajewska H, Ruszczynski M, Chmielewska A (2010) Effects of iron supplementation in nonanemic pregnant women, infants, and young children on the mental performance and psychomotor development of children: a systematic review of randomized controlled trials. Am J Clin Nutr 91:1684–1690

Tunnessen WW Jr, Oski FA (1987) Consequences of starting whole cow milk at 6 months of age. J Pediatr 111(6 pt 1):813–816

Walstra P, Jenness R (1984) Dairy chemistry and physics. Willey, New York, pp 114–122

Watkins JB, Jarvenpaa AL, Szczepanik-Van Leeuwen P (1983) Feeding the low-birth weight infant: effects of taurine, cholesterol, and human milk on bile acid kinetics. Gastroenterology 85:793–800

Wilson JF, Lahey ME, Heiner DC (1974) Studies on iron metabolism. V. Further observations on cow's milk-induced gastrointestinal bleeding in infants with iron-deficiency anemia. J Pediatr 84:335–344

Woodruff CW, Wright SW, Wrigth RP (1972) The role of fresh cow's milk in iron deficiency: II. Comparison of fresh cow's milk with a prepared formula. Am J Dis Child 124:26

World Health Organization (1998) Complementary feeding of young children in developing countries. A review of current scientific knowledge. WHO, Geneva

Ziegler EE (1990) Milk and formulas for older infants. J Pediatr 117:S76–S79

Ziegler EE (2007) Adverse effects of cow's milk in infants. Nestle Nutr Workshop Ser Pediatr Program 60:185–196

Ziegler EE (2011) Consumption of cow's milk as a cause of iron deficiency in infants and toddlers. Nutr Rev 69(Suppl 1):S37–S42

Ziegler EE, Fomson SJ, Nelson SE, Rebouche CJ, Edward BB, Rogers RR, Lehman LJ (1990) Cow milk feeding in infancy: further observations on blood loss from the gastrointestinal tract. J Pediatr 116(1):11–18

Ziegler EE, Jiang T, Romero E, Vinco A, Frantz JA, Nelson SE (1999) Cow's milk and intestinal blood loss in late infancy. J Pediatr 135:720–726

Chapter 14
Interventional Strategies for Prevention of Nutritional Anemia

Abstract Iron supplementations are the iron formulations in the form of tablets, syrups, capsules or injections recommended by physicians to prevent and/or treat iron deficiency anemia. It is the effective technique to combat iron deficiency anemia among vulnerable population in developing countries. Iron and folic acid (IFA) supplementation has been advocated as a measure to resolve the nutritional deficiency in the target population at a national level. Food fortification is the food enrichment method with the micronutrients. It might help improve the micronutrient status of the target population provided the fortified foods should be easily accessible and affordable and must be available at the local distribution network. Dietary pattern and selection of foods from the food groups should be so as to balance the presence of iron absorption inhibitors and enhancers. Albendazole, an anti-helminthic drug, is suggested for deworming of children. Infants up to 6 months of age should be fed on breast milk. It provides immunity to the child and reduces the prevalence of diarrhoeal episodes. Hand washing, defecation in latrines and safe and proper disposal of human excreta are other good practices that help prevent and/or minimize diarrhoeal episodes in children.

14.1 Iron Supplementation

Iron supplementations are the iron formulations in the form of tablets, syrups, capsules or injections recommended by physicians to prevent and/or treat iron deficiency anemia. It is the effective technique to combat iron deficiency anemia among vulnerable population in developing countries. It is recommended to compensate the dietary inadequacy among poor population and/or the physiological conditions like infancy, preschool children and pregnancy, where the requirement for iron is high due to raised metabolic activities in the body. Failure to provide iron supplements during these physiological conditions would result into iron deficiency in the body (Stoltzfus and Dreyfuss 1998).

Iron supplementation is focused over to prevent as well as manage iron deficiency anemia in children and women. Target groups for iron supplementation should be infants, preschool children, school children, pregnant women and adolescent girls. Iron supplementation can be categorized into prophylactic as well as therapeutic iron supplementation (Viteri 1995).

A. Gupta, *Nutritional Anemia in Preschool Children*,
DOI 10.1007/978-981-10-5178-4_14

The former method helps to complement iron in the target groups and corrects iron deficiency in the body of the individuals. This method prevents the occurrence of iron deficiency anemia in children and women. The latter method of iron supplementation is beneficial to treat iron deficiency anemia in children and women. Prophylactic iron supplementation should be practised at a large scale through the involvement of NGOs, medicals and paramedicals, school teachers and health workers so as to reach the potential target groups in the society (WHO 2001).

According to (MoHFW, GoI 2013), iron and folic acid (IFA) supplementation has been advocated as a measure to resolve the nutritional deficiency in the target population at a national level. For children in age group between 0 and 5 years, 20 mg elemental iron and 100 microgram (mcg) folic acid per ml of liquid formulation and age-appropriate deworming for 100 days were recommended by MoHF.

However, reports of NFHS-2 and 3 indicated a rise in the nutritional anemia in children in India. Probably, compliance to IFA supplementation was not appropriate. There might be hindrances in the execution of IFA supplementation at various levels. Therefore, the population having high potential to nutritional anemia lacked the benefits.

Later on, the Ministry of Health and Family Welfare developed the National Iron + Initiative. The programme is focused over IFA supplementation for children between age group of 6 and 60 months along with pregnant and lactating women. For children, MoHFW, GoI (2013) recommends iron supplementation twice a week in age group between 6 months and 5 years.

Children will be administered 1 ml of IFA syrup which contains 20 mg of elemental iron and 100 µg of folic acid (MoHFW, GoI 2013) by the accredited social health activists (ASHA).

Further, for children in age between 5 and 10 years, MoHFW, GoI (2013) recommends 45 mg elemental iron along with 400 µg of folic acid in tablet formulation to be administered by teachers and anganwadi workers to school children and out-of-school children once a week.

According to Stoltzfus and Dreyfuss (1998), IFA supplementations are based on the prevalence of anemia. Authors suggested 12.5 mg iron along with 50 µg of folic acid should be administered once a day to normal-weight children in age group between 6 and 24 months in area where the prevalence of anemia is less than 40%. IFA supplementation should be started at the age of 6 months and be continued till the age of 12 months. However, for the low birth weight children, IFA supplementation should be started at an earlier age (2 months) in comparison to normal-weight children (6 months). Furthermore, in population with anemia prevalence more than 40%, as per Stoltzfus and Dreyfuss (1998), IFA ought to be started at the age of 6 months and be continued to 24 months for normal children, and it should be started at an age of 2 months in case of low birth weight children. Iron supplementation for children between 2 and 5 years, as suggested by Stoltzfus and Dreyfuss (1998) is 2 mg per kg of body weight per day.

14.2 Food Fortification

Food fortification is the food enrichment method with the micronutrients. It might help improve the micronutrient status of the target population provided the fortified foods should be easily accessible and affordable and must be available at the local distribution network that may cater to the needs of people in the remote areas in any country. Food enrichment, hence, is a long-term procedure to prevent the deficiency of micronutrients in children. A variety of fortified foods are available like rice, wheat flour and table salt which have been enriched with iron compounds. In India, fortified table salt with iodine and iron is available.

14.3 Dietary Diversification

It is aimed at the use of foods rich in iron and/or the iron absorption enhancers which were underutilized earlier by the micronutrient deficiency-prone population. Availability of iron in the plant-based foods is low; therefore, foods from animal sources like, eggs, fish, etc., must be added in the daily dietary regimen to improve the absorption of dietary iron. Further enhancers like ascorbic acid, present in citrus fruits, should be consumed along with the staple diet. Dietary pattern and selection of foods from the food groups should be so as to balance the presence of iron absorption inhibitors and enhancers (MoHFW, GoI 2013). Addition of pulses, jaggery, sprouting seeds, meats, fish and citrus fruits, definitely, improves the dietary pattern in daily life to ward off the deficiency of micronutrients in children. However, cost, awareness and motivation are the guiding principles as far as the poor population is concerned in the initiate in food diversification. This segment of the society is highly vulnerable to dietary deficiency of micronutrients and deficiency of three principles; consequently, the path towards food diversification is obstructed.

14.4 Traditional Foods Rich in Iron

1. Chickpea (chana saag)
2. Spinach (palak)
3. Amaranth (kantewali chaulai ka saag)
4. Onion stalks (pyaz ki kali)
5. Mustard leaves (sarson ka saag)
6. Fenugreek leaves
7. Bengal Gram (whole kala chana)
8. Jaggery

14.5 Prevention and Management of Diarrhoea and Helminthic Infection

14.5.1 Prevention of Diarrhoea in Children

Diarrhoea in children is manageable and preventable with the adoption of healthy and hygienic habits in daily life. According to WHO (2005), infants up to 6 months of age should be fed on breast milk. It provides immunity to the child and reduces the prevalence of diarrhoeal episodes. Further, after the age of 6 months, WHO (2005) recommends the use of complementary foods, and the choice of the foods is based on the dietary habits customary to the place of living. Complementary foods supplement the breast milk to fulfil the energy demand of growing children. Water is decidedly essential for the human consumption, cooking and other purposes and is the nectar of human life. Safe and quality drinking water protects the children from faecal–oral infections. Water should be collected from safe source and should be stored in the container with lid (WHO 2005). Hand washing, defecation in latrines and safe and proper disposal of human excreta are other good practices that help prevent and/or minimize diarrhoeal episodes in children.

Although diarrhoea is a fatal disease, it is preventable. It is a global health problem with comparatively high mortality and morbidity in developing countries than developed countries. Diarrhoea kills millions of children in developing countries each year. According to CDC (1992), Bern et al. (1992) and Claeson and Merson (1990), diarrhoea had been the major cause of child mortality in the United States around a century back, where a child would have suffered around 7–15 episodes (CDC 1992; Bern et al. 1992; Claeson and Merson 1990) of diarrhoea annually till the age of 5 years. Further, asserted by CDC (1992), Bern et al. (1992) and Claeson and Merson (1990), 9% prevalence of diarrhoea is noticed in hospitalized children, and around 300–500 children (CDC 1992; Bern et al. 1992; Claeson and Merson 1990) die due to diarrhoea in the United States.

Diarrhoea is highly fatal in population in developing countries and is responsible for the death of children under 5. According to an estimate by Rohde and Northrup (2008), child mortality was five million due to diarrhoea. In another estimate by Synder and Merson (1982), children below 5 suffered around 2.2 episodes of diarrhoea annually amounting to 4.6 million deaths. Later on, Bern et al. (1992) supported earlier estimate by documenting 2.6 episodes of diarrhoea per child annually with mortality of 3.3 million per year. Kosek et al. (2003) upheld that diarrhoea is responsible for 21% of the total deaths of children under 5 and is the leading cause of 2.5 million deaths per year. Further estimate by Murray et al. (2001) in 2000 asserted a decline in diarrhoeal mortality to 1.4 million per year. Further study by Boschi-Pinto et al. (2008) estimated 17.5–21% of all deaths caused by diarrhoea in children below the age of 5 amounting to 1.5 million mortality per annum. African countries and Southeast Asian regions have higher predisposition to diarrhoeal mortality in children.

According to Claeson and Merson (1990), Glass et al. (1991), Ho et al. (1988) and Gangarosa et al. (1985), in the United States, around 325–425 deaths occur due to diarrhoea, and about 85% of total diarrhoea-related mortality is observed during infancy.

Today, causes of diarrhoea in children are well established. *Rotavirus* is the commonest pathogenic agent for acute diarrhoea in infancy (Cohen 1991). Norwalk-like viruses, astrovirus, adenovirus and calicivirus are equally responsible for childhood diarrhoea. Bacteria like *Salmonella*, *Shigella*, *Campylobacter*, *Yersinia* and *Escherichia coli* along with protozoans such as *Giardia*, *Cryptosporidium* and *Entamoeba histolytica* have pathogenic potential.

Diarrhoea is associated with dehydration, and proper management of acute diarrhoea is necessary to save the life of children. Oral rehydration therapy for rehydration is the universally accepted diarrhoea management programme (Pierce and Hirschhorn 1977). Adequate nutrition is supplementary to ORT for diarrhoea treatment. However, the use of antidiarrhoeal drugs and antibiotics depends on the cause and severity of diarrhoea. Additionally, intravenous therapy is the choice for cases having severe dehydration (CDC 1992).

Oral rehydrate solution can be used for diarrhoea management irrespective of cause, age and gender of patients. According to Hirschhorn (1980), the World Health Organization and United Nations International Children's Emergency Fund decided to introduce single ORS containing sodium 90 mmol/L, potassium 20 mmol/L, chloride 80 mmol/L, base 30 mmol/L and glucose 111 mmol/L. This solution was recommended with an idea that a single solution among populations in different countries would be adopted easily instead of multiple solutions. A study by Hirschhorn (1980) actually proved the efficacy of single solution ORS in managing acute diarrhoea of different aetiology. Later on, *Rotavirus* as another cause of acute diarrhoea was highlighted. It is associated with glucose malabsorption. So, the use of ORS was disputable (CDC 1992). However, a study by Molla et al. (1981), Sack et al. (1982) and Pizarro et al. (1983) proved the usefulness of ORS in acute diarrhoea caused by *Rotavirus*. Thereafter, a WHO Global Diarrhoeal Diseases Control Programme was formulated with emphasis on the immediate use of ORS in acute diarrhoea. It has been asserted that episodes of diarrhoea and pneumonia are higher in infants and children either suffering from HIV infection or exposed to it. According to Thea and Qazi (2008), Amadi (2001) and Chokephaibulkit (2001), diarrhoea is rampant in HIV-infected children than uninfected children.

According to WHO (2010), diarrhoea in HIV-infected and/or HIV-exposed children can be managed as:

Prevention of Diarrhoea

Recommendation 1

Supplements of vitamin A are advised for all HIV-infected and HIV-exposed infants and children in age between 6 months and 5 years in dose of 100,000 IU for children between 6 and 12 months and 200,000 IU for children aged >12 months to be given every 6 months (WHO 2010).

Strong recommendation: low-quality evidence
Treatment of diarrhoea (WHO 2010)

Recommendation 2 (WHO 2010)

Low-osmolarity oral rehydration solution is advised for the treatment of dehydration or IV fluid for severe dehydration in HIV-infected and HIV-exposed infants and children with diarrhoea (WHO 2010).
Strong recommendation: high-quality evidence (WHO 2010)

Recommendation 3 (WHO 2010)

Elemental zinc supplements in a dose of 10 mg per day for infants <6 months and 20 mg per day for children >6 months of age for 10–14 days are advised to all HIV-infected and HIV-exposed children with diarrhoea (WHO 2010).
Strong recommendation: high-quality evidence (WHO 2010).

Recommendation 4 (WHO 2010)

Ciprofloxacin at a dose of 15 mg/kg is advised for 3 days treating bloody diarrhoea in all HIV-infected and HIV-exposed infants and children (WHO 2010).
Strong recommendation: moderate-quality evidence (WHO 2010).

Recommendation 5

Multiple micronutrient supplements per day for 2 weeks are advised to all HIV-infected and HIV-exposed infants and children with persistent diarrhoea (WHO 2010).

Further, the above recommendations by WHO (2010) group for management of pneumonia and diarrhoea in HIV-infected infants and children almost resemble as those for management in uninfected children.

However, use of ORS for acute diarrhoea management is limited to 30–40% of children worldwide (Forsberg et al. 2007; Ram 2008).

14.5.2 Prevention of Helminthic Infection

Deworming of children is essential. Helminthic infections impair the normal functioning of GIT and are responsible for malnutrition, malabsorption syndrome, anemia of infection and other ailments in children. According to MoHFW, GoI (2013), Albonico et al. (2006) and WHO (2004), albendazole, an antihelminthic drug, is suggested for deworming of children in a dose of 200 mg for children in age group of 1–2 years, while for children above 2 years of age, it is advocated in a dose of 400 mg. It is also available in syrup form, wherein a dose of 10 ml contains 400 mg albendazole. Otherwise, the tablet should be crushed and dissolved in water and should be administered to children. Deworming should be practised (MoHFW, GoI 2013) twice a year. According to Albonica et al. (2002), pyrantel pamoate in a single dose of 250 mg guided by 10 mg/kg criteria is an effective antihelminthic drug. The author asserted that a combination of pyrantel-oxantel is better in its efficacy in treating trichiura infection in comparison to single dose of pyrantel. According to Albonico et al. (2003), antihelminthic drug like levamisole in a single dose of 80 mg by weight as per criteria 2.5 mg/kg is effective in primary school children.

Deworming in a large population is an economically feasible antihelminthic regimen (Warren et al. 1993). Another population-based approach is the deworming at the level of schools. This programme covers even those children who have not been treated earlier. These drug regimens, together, help to eradicate helminthic infection from the community (Bundy et al. 1990; Chan et al. 1992; Kabatereine et al. 2006; Narain et al. 2004; Stephenson et al. 2000).

Large-scale environmental sanitation programmes are complex, making interventions directly aimed at the transmission of STHs challenging to implement (Barreto et al. 2007; Barreto et al. 2010; Mascarini-Serra ct al. 2010; Moraes et al. 2003). These interventions directly affect the transmission of several diseases in both the public and private domains (Cairncross et al. 1996).

Poor masses living in isolated areas and seclusion invariably need health education and periodic motivation to create awareness regarding hygiene, sanitation and medical facilities. Health education may improve health-oriented behaviour in the population (Bull et al. 2001; Walsh et al. 1993).

Contrarily, the author (O'Cathain et al. 2002) suggested that attempts to provide health education to poor people through various methods might not help improve the behaviour of the affected population.

However, suggested solutions to prevent soil-transmitted infection, diarrhoea and malnutrition might not be adopted by poor population surviving in utmost paucity which makes it difficult to satisfy the basic needs. Therefore, construction of latrines, management quality food and drinking water and purchase of drugs become highly unapproachable for poor people. So health education and motivation alone are no guarantee for prevention of diarrhoea and helminthic infection in children in poor settings. Printed materials depicting health education steps are disseminated to awaken the population in villages, remote settings and slum areas. These methods help to change harmful rituals, change mind-set and adopt healthy lifestyle measures (Mascarini-Serra 2011; Bull et al. 2001; Walsh et al. 1993; Kreuter et al. 2000, Kreuter and Wray 2003).

Helminthic infection is primarily soil-transmitted infection. Human excreta contaminate food and drinking water. Transmission occurs either in hand-to-mouth mode or by ingestion of contaminated food and drinking water (indirectly). Sanitation and hygiene are the exclusive methods to prevent from soil-transmitted infection (Albonico et al. 2006; Steimann et al. 2015). Proper disposal of human excreta and domestic waste and adoption of personal hygiene practice have soothing effect on the faecal-born infections (Hong et al. 2006; Hotez et al. 2008). Sanitation reduces the prevalence of soil-transmitted infection and episodes of diarrhoea in children. The hypothesis holds true in studies conducted in Sri Lanka, urban slums of Bangladesh and Salvador, Brazil (Henry 1988; Sorensen et al. 1994; Henry et al. 1993; Morases et al. 2004). Proper sewerage and drainage in a city definitely ameliorates the incidence of helminth infections in children by reducing the rate of transmission (Ziegelbauer et al. 2012; Fernando and Balasuriya 1976; Soeripto 1991; Moraes and Cairncross 2004; Moraes et al. 2003; Korkes et al. 2009).

In 1997, in the city of Salvador in Northeast Brazil, a large sanitation programme was started. The intervention work was aimed at extending the sewerage facility to households from an earlier 26% to present 80%. Authors (Mascarini-Serra 2010) studied the effect of the programme on the incidence and prevalence of helminthic infection in school children. Authors confirmed the beneficial effect of this city-wide sanitation programme. Significant reduction in incidence of *Ascaris*, *Trichuris* and hookworm infection was observed by authors.

An extensive meta-analysis was performed by authors (Fewtrell et al. 2005). The study was aimed at finding out the effect of improved water quality, proper sanitation and adequate hygiene in combating the diarrhoeal frequency and helminth infection. Authors observed that all of the above stated methods were beneficial in reducing the prevalence of diarrhoeal episodes. However, quality of water had the largest impact on prevalence of diarrhoea.

References

Albonico M, Bickle Q, Haji HJ, Ramson M, Khatib KJ, Montresor A, Savioli L, Taylor M (2002) Evaluation of the efficacy of pyrantel-oxantel for the treatment of soil-transmitted nematode infections. Trans R Soc Trop Med Hyg 96:685–690

Albonico M, Bickle Q, Ramsan M, Montresor A, Savioli L, Taylor M (2003) Efficacy of mebendazole and levamisole alone or in combination against intestinal nematode infections after repeated targeted mebendazole treatment in Zanzibar. Bull World Heath Organ 81:343–352

Albonico M, Montresor A, Crompton DW, Savioli L (2006) Intervention for the control of soil-transmitted helminthiasis in the community. Adv Parasitol 61:312–348

Amadi B (2001) Intestinal and systemic infection, HIV, and mortality in Zambian children with persistent diarrhoea and malnutrition. J Pediatr Gastroenterol Nutr 32:550–554

Barreto M, Genser B, Strina A, Teixeira MG, Assis AMO, Rego RF (2007) Effect of city-wide sanitation programme on reduction in rate of childhood diarrhoea in northeast Brazil: assessment by two cohort studies. Lancet 370:1622–1628

Barreto ML, Genser B, Strina A, Teixeira MG, Assis AM, Rego RF, Teles CA, Prado MS, Matos S, Alcântara-Neves NM (2010) Impact of a citywide sanitation program in Northeast Brazil on intestinal parasites infection in young children. Environ Health Perspect 118(11):1637–1642

Bern C, Martines J, de Zoyza I, Glass RI (1992) The Magnitude of global problem of diarrhoeal disease: a ten year update. Bull World Health Organ 70(6)

Boschi-Pinto C, Velebit L, Shibuya K (2008) Estimating child mortality due to diarrhoea in developing countries. Bull World Health Organ 86:710–717

Bull FC, Holt CL, Kreuter MW, Clark EM, Scharff D (2001) Understanding the effects of printed health education materials: which features lead to which outcomes? J Health Commun 6:265–279

Bundy DA, Wong MS, Lewis LL, Horton J (1990) Control of geohelminths by delivery of targeted chemotherapy through schools. Trans R Soc Trop Med Hyg 81:115–120

Cairncross S, Blumenthal U, Kolsky P, Moraes L, Tayeh A (1996) The public and domestic domains in the transmission of disease. Tropical Med Int Health 1:27–34

Centers for Disease Control and Prevention (1992) The management of acute diarrhea in children: oral rehydration, maintenance, and nutritional therapy. MMWR Recomm Rep 41(RR-16):001

Chan L, Kan SP, Bundy DA (1992) The effect of repeated chemotherapy on the prevalence and intensity of Ascaris lumbricoides and Trichuris trichiura infection. Southeast Asian J Trop Med Public Health 23(2):228–234

Chokephaibulkit K (2001) Intestinal parasitic infections among human immunodeficiency virus infected and -uninfected children hospitalized with diarrhoea in Bangkok, Thailand. Southeast Asian J Trop Med Public Health 32:770–775

Claeson M, Merson MH (1990) Global progress in the control of diarrheal diseases. Pediatr Infect Dis J 9:345–345

Cohen MB (1991) Etiology and mechanisms of acute infectious diarrhea in infants in the United States. J Pediatr 118:S34–S39

Fernando MA, Balasuriya S (1976) Prevalence of soil-transmitted helminths in the Hindagala community health project area. Ceylon Med J 21(3):177–183

Fewtrell L, Kaufmann RB, Kay D, Enanoria W, Haller L, Colford JM Jr (2005) Water, sanitation, and hygiene interventions to reduce diarrhoea in less developed countries: a systematic review and meta-analysis. Lancet Infect Dis 5:42–52

Fosberg BC, Petzold MG, Tomson G, Allebeck P (2007) Diarrhoea case management in low- and middle-income countries—an unfinished agenda. Bull World Health Organ 85:42–48

Gangarosa RE, Glass RI, Lew JF, Boring JR (1985) Hospitalization involving gastroenteritis in United States,1985: the special burden of disease among the elderly. Am J Epidemiol 135:281–290

Glass RI, Lew JF, Gangarosa RE, LeBaron CW, Ho MS (1991) Estimates of morbidity and mortality rates for diarrheal diseases in American children. J Pediatr 118:S27–S33

Henry FJ (1988) Re-infection with Ascaris lumbricoides after chemotherapy: a comparative study in three villages with varying sanitation. Trans R Soc Trop Med Hyg 82:460–464

Henry FJ, Huttly SR, Ahmed MU, Alam A (1993) Effect of chemotherapy in slums and villages in Bangladesh. Southeast Asian J Trop Med Public Health 24:307–312

Hirschhorn N (1980) The treatment of acute diarrhea in children: an historical and physiological perspective. Am J Clin Nutr 33:637–663

Ho MS, Glass RI, Pinsky PF (1988) Diarrheal deaths in American children – are they preventable? JAMA 260:3281–3285

Hong SH, Misek DE, Wang H, Puravs E, Hinderer R, Giordano TJ (2006) Identification of a specific vimentin isoform that induces an antibody response in pancreatic cancer. Biomark Insights 1:175–183

Hotez PJ, Bottazzi ME, Franco-Paredes C, Ault SK, Periago MR (2008) The neglected tropical diseases of Latin America and the Caribbean: a review of disease burden and distribution and a roadmap for control and elimination. PLoS Negl Trop Dis 24:2–300

Kabatereine NB, Tukahebwa E, Kazibwe F, Namwangye H, Zaramba S, Brooker S, Stothard JR, Kamenka C, Whawell S, Webster JP (2006) Progress towards countrywide control of schistosomiasis and soil-transmitted helminthiasis in Uganda. Trans R Soc Trop Med Hyg 100(3):208–215

Korkes F, Kumagai FU, Belfort RN, Szejnfeld D, Abud TG, Kleinman A, Florez GM, Szejnfeld T, Chieffi PP (2009) Relationship between intestinal parasitic infection in children and soil contamination in an urban slum. J Trop Pediatr 55(1):42–45

Kosek M, Bern C, Guerrant R (2003) The global burden of diarrhoeal disease as estimated from studies published between 1992 and 2000. Bull World Health Organ 81:197–204

Kreuter MW, Wray RJ (2003) Tailored and targeted health communication: strategies for enhancing information relevance. Am J Health Behav Suppl 3:S227–S232

Kreuter MW, Oswald DL, Bull FC, Clark EM (2000) Are tailored health education materials always more effective than non-tailored materials? Health Educ Res 15(3):305–315

Mascarini-Serra L (2011) Prevention of soil-transmitted helminth infection. J Glob Infect Dis 3(2):175–182

Mascarini-Serra LM, Telles CA, Prado MS, Mattos SA, Strina A, Alcantara-Neves NM, Barreto ML (2010) Reductions in the prevalence and incidence of geohelminth infections following a city-wide sanitation program in a Brazilian Urban Centre. PLoS Negl Trop Dis 4(2):e588
Ministry of Health and Family Welfare, Government of India (2013) Guidelines for control of iron deficiency anemia: national iron + initiative. MoHFW, New Delhi
Molla AM, Rahaman M, Sarker SA (1981) Stool electrolyte content and purging rates in diarrhea caused by rotavirus, enterotoxigenic E. coli and V. cholerae in children. J Pediatr 98:835–838
Moraes LR, Cairncross S (2004) Environmental interventions and the pattern of geohelminth infections in Salvador, Brazil. Parasitology 129(Pt 2):223–232
Moraes LR, Cancio JA, Cairncross S, Huttly S (2003) Impact of drainage and sewerage on diarrhoea in poor urban areas in Salvador, Brazil. Trans R Soc Trop Med Hyg 97(2):153–158
Moraes LR, Cancio JA, Cairncross S (2004) Impact of drainage and sewerage on intestinal nematode infections in poor urban areas in Salvador, Brazil. Trans R Soc Trop Med Hyg 98:197–204
Murray CJL, Lopez AD, Mathers CD, Stein C (2001) The global burden of disease 2000 project: aims, methods and data sources. Geneva WHO
Narain K, Medhi GK, Rajguru SK, Mahanta J (2004) Cure and reinfection patterns of geohelminthic infections after treatment in communities inhabiting the tropical rainforest of Assam, India. Southeast Asian J Trop Med Public Health 35(3):512–517
O'Cathain A, Walters SJ, Nicholl JP, Thomas KJ, Kirkham M (2002) Use of evidence based leaflets to promote informed choice in maternity care: randomised controlled trial in everyday practice. Br Med J 324:643
Pierce NF, Hirschhorn N (1977) Oral fluid is a simple weapon against dehydration: how it works and how to use it. WHO Chron 31:87–93
Pizarro D, Posada G, Villavicencio N (1983) Oral rehydration in hypernatremic and hyponatremic diarrheal dehydration: treatment with oral glucose/electrolyte solution. Am J Dis Child 137:730–734
Ram PK (2008) Declines in case management of diarrhoea among children less than five years old. Bull World Health Organ 86:161–240
Rohde JE, Northrup RS (2008) Taking science where disease is. CIBA Found Symp 42:339–358
Sack DA, Rhoads M, Molla A (1982) Carbohydrate malabsorption in infants with rotavirus diarrhea. Am J Clin Nutr 36:1112
Soeripto N (1991) Reinfection and infection rates of soil-transmitted-helminths in Kemiri Sewu, Yogyakarta, Indonesia. Southeast Asian J Trop Med Public Health 22(2):216–221
Sorensen E, Ismail M, Amarasinghe DK, Hettiarachchi I, Dassenaieke CT (1994) The effect of the availability of latrines on soil-transmitted nematode infections in the plantation sector in SriLanka. Am J Trop Med Hyg 51:36–39
Steinmann P, Yap P, Utzinger J, Du ZW, Jiang JY, Chen R, Wu FW, Chen JX, Zhou H, Zhou XN (2015) Control of soil-transmitted helminthiasis in Yunnan province, People's Republic of China: experiences and lessons from a 5-year multi-intervention trial. Acta Trop 141(Pt B):271–280
Stephenson LS, Holland CV, Cooper ES (2000) The public health significance of Trichuris trichiura. Parasitology 121(Suppl):S73–S95
Stoltzfus RJ, Dreyfuss ML (1998) Guidelines for the use of iron supplements to prevent and treat iron deficiency anemia. International Nutritional Anemia Consultative Group (INACG), ILSI PRESS, Washington, DC
Synder JD, Merson MH (1982) The magnitude of global problem of acute diarrhoeal disease: A review of acute surveillance data. Bull World Health Organ 10:605–613
Thea D, Qazi S (2008) Neonatal mortality—4 million reasons for progress. Lancet 371:1893–1895
Viteri F (1995) Iron deficiency in children: new possibilities for its control. Int Child Health 6:49–62
Walsh DC, Rudd RE, Moeykens BA, Moloney TW (1993) Social marketing for public Health. Health Aff 12:104–119

Warren KS, Bundy DA, Anderson RM, Davis AR, Henderson DA, Jamison DT (1993) Helminth infections. In: Jamison DT, Mosley WH, Measham AR, Bobadilla JL (eds) Disease control priorities in developing countries 2001. Oxford University Press, Oxyford, pp 131–160

World Health Organization (2005) The treatment of diarrhea: a manual for physicians and other senior health workers. WHO Press, Geneva

World Health Organization (2010) WHO recommendations on the management of diarrhoea and pneumonia in HIV-infected infants and children: integrated management of childhood illness (IMCI). WHO, Geneva. Available at: http://www.unicef.org/aids/files/hiv_diarrhoea_and_pneumonia.pdf

World Health Organization, Dept. of Nutrition for Health and Development (2001) Iron deficiency anemia assessment, prevention and control, a guide for programme managers. WHO, Geneva

World Health Organization Model Formulary (2004) Based on the 13th model list of essential medicines. WHO, Geneva, pp 82–85

Ziegelbauer K, Speich B, Mäusezahl D, Bos R, Keiser J, Utzinger J (2012) Effect of sanitation on soil-transmitted helminth infection: systematic review and meta-analysis. PLoS Med 9(1):e1001

MIX
Papier aus verantwortungsvollen Quellen
Paper from responsible sources
FSC® C105338

If you have any concerns about our products,
you can contact us on
ProductSafety@springernature.com

In case Publisher is established outside the EU,
the EU authorized representative is:
Springer Nature Customer Service Center GmbH
Europaplatz 3, 69115 Heidelberg, Germany

Printed by Libri Plureos GmbH
in Hamburg, Germany